T0324712

Geometric Realizations of Curvature

ICP Advanced Texts in Mathematics ISSN 1753-657X

Series Editor: Dennis Barden *(Univ. of Cambridge, UK)*

Published

Vol. 1 Recent Progress in Conformal Geometry
by Abbas Bahri & Yong Zhong Xu

Vol. 2 The Geometry of Curvature Homogeneous Pseudo-Riemannian Manifolds
by Peter B. Gilkey

Vol. 3 Dynamics and Symmetry
by Michael J. Field

Vol. 4 The Solution of the $k(GV)$ Problem
by Peter Schmid

Vol. 5 Selected Topics on Continuous-Time Controlled Markov Chains and Markov Games
by Tomás Prieto-Rumeau & Onésimo Hernández-Lerma

Vol. 6 Geometric Realizations of Curvature
by Miguel Brozos Vázquez, Peter B. Gilkey & Stana Nikcevic

ICP Advanced Texts in Mathematics – Vol. 6

Geometric Realizations of Curvature

Miguel Brozos Vázquez
Universidade da Coruña, Spain

Peter B Gilkey
University of Oregon, USA

Stana Nikcevic
University of Belgrade, Serbia

Imperial College Press

ICP

Published by

Imperial College Press
57 Shelton Street
Covent Garden
London WC2H 9HE

Distributed by

World Scientific Publishing Co. Pte. Ltd.
5 Toh Tuck Link, Singapore 596224
USA office: 27 Warren Street, Suite 401-402, Hackensack, NJ 07601
UK office: 57 Shelton Street, Covent Garden, London WC2H 9HE

British Library Cataloguing-in-Publication Data
A catalogue record for this book is available from the British Library.

ICP Advanced Texts in Mathematics — Vol. 6
GEOMETRIC REALIZATIONS OF CURVATURE
Copyright © 2012 by Imperial College Press

All rights reserved. This book, or parts thereof, may not be reproduced in any form or by any means, electronic or mechanical, including photocopying, recording or any information storage and retrieval system now known or to be invented, without written permission from the Publisher.

For photocopying of material in this volume, please pay a copying fee through the Copyright Clearance Center, Inc., 222 Rosewood Drive, Danvers, MA 01923, USA. In this case permission to photocopy is not required from the publisher.

ISBN-13 978-1-84816-741-4
ISBN-10 1-84816-741-5

Printed in Singapore.

Preface

Many questions in modern differential geometry can be phrased as questions of geometric realizability; one studies whether or not certain algebraic objects have corresponding geometric analogues. One must examine the relationship between the algebraic category and the geometric setting to investigate the geometric consequences resulting from the imposition of algebraic conditions on the curvature. The decomposition of certain spaces of curvature tensors under the appropriate structure groups is crucial and motivates many investigations. Although we will primarily focus on the curvature tensor, there are other tensors which arise naturally and which also play an important role in our study. As we will often work in the indefinite setting, the structure groups are in general non-compact. This imposes some minor technical difficulties. In this book, we have attempted to organize some of the results in the literature which fall into this genre; as the field is a vast one, we have not attempted an exhaustive account but have rather focused on only some of the results in order to be able to give a coherent account.

We begin in Chapter 1 by introducing some notation and stating the main results of the book. We also outline in some detail the main results of the book and relate various results to the whole. The remainder of the book consists for the most part in establishing the results given here. In Chapter 2, we turn our attention to representation theory and derive the main results we shall need. Chapter 2 is self-contained with the exception of the results of H. Weyl and others concerning invariance theory for the orthogonal and unitary groups in the positive definite setting; the corresponding results in the higher signature setting and in the para-complex setting are then derived from these results. In Chapter 3, we present some classic results from differential geometry.

In Chapter 4, we work in the real affine setting and in Chapter 5, we work in the (para)-complex affine setting. In Chapter 6 we perform a similar analysis for real Riemannian geometry and in Chapter 7 we study (para)-complex Riemannian geometry. To the greatest extent possible, we present results in the para-complex and in the complex settings in parallel. We present following Chapter 7 a list of the main notational conventions used throughout the book. Following this list, we have included a lengthy bibliography. The book concludes with an index.

Each chapter is divided into sections; the first section of a chapter provides an outline to the subsequent material in the chapter. Theorems, lemmas, corollaries, and so forth are labeled by section. Equations which are cited are labeled by section; equations which are not cited are not labeled.

To comply with stylistic requirements for this series, a few non-standard usages have been employed for which we are not responsible. To begin with, the bibliographic style will be unfamiliar to almost all mathematical readers. For example, [Brozos-Vázquez et al. (2009)] refers to work by Brozos-Vázquez, Gilkey, Kang, Nikčević, and Weingart. On the other hand, [Brozos-Vázquez et al. (2009a)] refers to work by Brozos-Vázquez, Gilkey, Nikčević, and Vázquez-Lorenzo. The words *"para-Hermitian (+) or pseudo-Hermitian (−)"* have been used rather than the customary *"para/pseudo-Hermitian"*. There are a few other similar instances which we hope will not disturb the reader unduely. Es lo que hay.

Much of this book reports on previous joint work with various authors. It is an honor and a privilege to acknowledge the contribution made by these colleagues: N. Blažić, N. Bokan, E. Calviño-Louzao, J. C. Díaz-Ramos, C. Dunn, B. Fiedler, E. García-Río, R. Ivanova, H. Kang, E. Merino, J.H. Park, E. Puffini, K. Sekigawa, U. Simon, G. Stanilov, I. Stavrov, Y. Tsankov, M. E. Vázquez-Abal, R. Vázquez-Lorenzo, V. Videv, G. Weingart, D. Westerman, T. Zhang, and R. Zivaljevic. In addition to pleasant professional collaborations, they have enriched the personal lives of the authors.

Projects MTM2009-07756 and INCITE09 207 151 PR (Spain) have supported the research of M. Brozos-Vázquez. Project MTM2009-07756 (Spain) and DFG PI 158/4-6 (Germany) have supported the research of P. Gilkey. Project MTM2009-07756 (Spain) and project 144032 (Serbia) have supported the research of S. Nikčević.

This book is dedicated to Ana, to Ekaterina, to George, and to Susana.

P. Gilkey, S. Nikčević, and M. Brozos-Vázquez February 2012

Contents

Chapter 1

Introduction and Statement of Results

A central area of study in differential geometry is the examination of the relationship between purely algebraic properties of the Riemann curvature tensor and the underlying geometric properties of the manifold. Many authors have worked in this area in recent years. Nevertheless, many fundamental questions remain unanswered. When dealing with a geometric problem, it is frequently convenient to work first purely algebraically and pass later to the geometric setting. For this reason, many questions in differential geometry are often phrased as problems involving the geometric realization of curvature.

The decomposition of the appropriate space of tensors into irreducible modules under the action of the appropriate structure group is central to our investigation and we review the appropriate results in each section. Many of the results in the book, although they involve non-linear analysis, are closely tied to the representation theory of the appropriate group and the corresponding linear subspaces. In contrast, other results are non-linear in their very formulation since one is studying orbit spaces under the structure group; these need not be linear subspaces.

In the remainder of Chapter 1, we summarize briefly the main results of this book to put them into context for the reader. We shall discuss the basic curvature decomposition results leading to various geometric realization results in a number of geometric contexts. This ensures that the various relations between these theorems are clearly and concisely presented; further details are presented subsequently.

We now outline briefly the contents of Chapter 1. In Section 1.1, we present some basic notational conventions. In Section 1.2, we sketch some of the representation theory we shall need; Chapter 2 will be devoted to the proof of these results.

The results of Section 1.3 and of Section 1.4 will be established in Chapter 3 and in Chapter 4. In Section 1.3 we treat affine geometry. Theorem 1.3.1 gives the decomposition [Strichartz (1988)] of the space of generalized curvature operators as a general linear module. The dimension of these modules is given in Theorem 1.3.2. The decomposition of Theorem 1.3.1 motivates the associated geometric realization results discussed in Theorem 1.3.3. In Theorem 1.3.4, we establish a basic geometric realization result for the curvature and the covariant derivative of the curvature of an affine connection or, equivalently, a connection with vanishing torsion tensor. In Section 1.4, we study mixed structures; this is the geometry of an affine connection in the presence of an auxiliary non-degenerate inner product. The curvature decomposition [Bokan (1990)] is stated in Theorem 1.4.1, the dimensions of the relevant modules are given in Theorem 1.4.2, and the associated geometric realization result presented in Theorem 1.4.3.

The results of Section 1.5 will be proved in Chapter 5. We return to affine geometry to treat (para)-Kähler affine curvature tensors. To emphasize the similarities, we shall for the most part present the complex and the para-complex settings in parallel. We present the curvature decomposition as (para)-complex general linear modules in Theorem 1.5.1 and as unitary modules in Theorem 1.5.2 [Matzeu and Nikčević (1991)] and [Nikčević (1992)]. This leads to the geometric realization result given in Theorem 1.5.3. The dimensions of these modules are stated in Theorem 1.5.4.

The results of Section 1.6 and of Section 1.7 will be established in Chapter 6. Section 1.6 treats Riemannian geometry. The Fiedler generators [Fiedler (2003)] for the space of Riemannian algebraic curvature tensors are given in Theorem 1.6.1. The fundamental curvature decomposition [Singer and Thorpe (1969)] is given in Theorem 1.6.2, and an associated geometric realization theorem by metrics of constant scalar curvature is presented in Theorem 1.6.3. In Section 1.7, we study Weyl geometry; this is midway in a certain sense between affine and Riemannian geometry. The extra curvature symmetry of Weyl geometry is given in Theorem 1.7.1, the curvature decomposition as an orthogonal module is given in Theorem 1.7.2, and the basic geometric realization result is given in Theorem 1.7.3. Theorem 1.7.4 gives various characterizations of trivial Weyl structures.

The results of Section 1.8, of Section 1.9, of Section 1.10, of Section 1.11, and of Section 1.12 will be established in Chapter 7. In Section 1.8, we turn our attention to (para)-complex geometry. The curvature decomposition

[Tricerri and Vanhecke (1981)] of the space of Riemann curvature tensors in the pseudo-Hermitian and in the para-Hermitian settings is given in Theorem 1.8.1. A geometric realization theorem is then presented in this context in Theorem 1.8.3. The dimensions of the associated modules are given in Theorem 1.8.2. If the almost (para)-complex structures $J_\pm$ are integrable, then there is an extra curvature condition [Gray (1976)]; we shall discuss this further in Theorem 1.9.1 in Section 1.9. The relevant geometric realizability results are outlined in Theorem 1.9.2 and rely on the curvature decompositions given previously. Theorem 1.9.3 is an algebraic fact related to these conditions.

(Para)-Kähler geometry is treated in Section 1.10. The (para)-Kähler curvature condition is given in Theorem 1.10.1 and the associated geometric realizability results are presented in Theorem 1.10.2. Additional curvature decomposition results are given in Theorem 1.10.3. In Section 1.11, we discuss Weyl geometry in the Kähler setting either for a complex or for a para-complex structure. We also discuss an analogous algebraic condition giving rise to curvature Kähler–Weyl geometry. We shall restrict our attention to dimensions $m \geq 6$ as the situation in dimension $m = 4$ is quite different. In Theorem 1.11.1, we show any Weyl structure which is (para)-Kähler is trivial and in Theorem 1.11.2, we give a similar characterization solely in terms of curvature. Theorem 1.11.4 is a similar result at the purely algebraic level. Theorem 1.11.3 generalizes Theorem 1.7.2 and Theorem 1.8.1 to Weyl geometry in the (para)-complex setting.

In Section 1.12, we change focus. Let $\nabla\Omega$ be the covariant derivative of the (para)-Kähler form. This has certain universal symmetries. In Theorem 1.12.1 we show that if H is any 3-tensor with these symmetries, then H is geometrically realizable as the covariant derivative of the (para)-Kähler form of some almost para-Hermitian manifold or of some almost pseudo-Hermitian manifold. This is based on an appropriate decomposition result (see Theorem 1.12.3); the relevant dimensions of the irreducible modules involved are given in Theorem 1.12.4.

Finally, in Section 1.13, we give a brief summary of results contained in [De Smedt (1994)] concerning hyper-Hermitian geometry for the sake of completeness.

It is worth giving a bit of an explanation about what we mean by geometric realizability since this is the focus of the book. Let $\{T_1, ..., T_k\}$ be a family of tensors on a vector space V. The structure $(V, T_1, ..., T_k)$ is said to be *geometrically realizable* if there exists a manifold M, if there exists a point P of M, and if there exists an isomorphism $\phi : V \to T_P M$ such that $\phi^* L_i(P) = T_i$ where $\{L_1, ..., L_k\}$ is a corresponding geometric family of tensor fields on M. Thus, for example, if $k = 1$ and if $T_1 = \langle\cdot,\cdot\rangle$ is a non-degenerate inner product on V, then a geometric realization of $(V, \langle\cdot,\cdot\rangle)$ is a pseudo-Riemannian manifold (M, g), a point P of M, and an isomorphism $\phi : V \to T_P M$ so that $\phi^* g_P = \langle\cdot,\cdot\rangle$.

1.1 Notational Conventions

In addition to the notation introduced here, more notation will be introduced subsequently as needed; a summary of the common notational conventions used in this book is to be found at the end just before the bibliography. Let M be a smooth manifold of dimension $m \geq 4$; there are often similar results in dimensions $m = 2$ and $m = 3$ that we will sketch in passing. Let V be a real vector space of dimension m. Let V^* be the associated *dual vector space*. We shall let $\{e_i\}$ be a basis for V and we shall let $\{e^i\}$ be the associated dual basis for V^*; when we wish to consider orthogonal bases, we will make this explicit. Setting $x^i := e^i(\cdot)$ defines coordinates $(x^1, \ldots, x^m)$ on V. Let $\partial_{x_i} := \frac{\partial}{\partial x_i}$. Adopt the *Einstein convention* and sum over repeated indices. We say that $x = c^i \partial_{x_i}$ is a *coordinate vector field* if the coefficients c^i are constant; this notion is independent of the particular basis chosen for V. If $\theta^2 \in \otimes^2 V^*$ and if $\theta^4 \in \otimes^4 V^*$, we expand

$$\theta^2 = \theta^2_{ij} e^i \otimes e^j \quad \text{and} \quad \theta^4 = \theta^4_{ijkl} e^i \otimes e^j \otimes e^k \otimes e^l$$

to define the components of these tensors. In defining tensors, if there are obvious $\mathbb{Z}_2$ symmetries, we will often only give the non-zero components modulo these symmetries. Let GL be the *general linear group*; this is the group of all invertible linear transformations of V. If $\theta \in \otimes^k V^*$ and if $T \in \text{GL}$, we define $T^*\theta \in \otimes^k V^*$ by:

$$(T^*\theta)(v_1, \ldots, v_k) := \theta(Tv_1, \ldots, Tv_k). \tag{1.1.a}$$

Similarly if $\theta \in \otimes^k V^* \otimes V$, we define

$$(T^*\theta)(v_1, \ldots, v_k) := T^{-1}\theta(Tv_1, \ldots, Tv_k). \tag{1.1.b}$$

There is a direct sum decomposition of $V^* \otimes V^*$ into irreducible modules where the structure group is the general linear group:

$$V^* \otimes V^* = \Lambda^2 \oplus S^2$$

as the sum of the *alternating tensors* Λ^2 of rank two and the *symmetric tensors* S^2 of rank two. If $\theta \in V^* \otimes V^*$, this decomposition yields $\theta = \theta_a + \theta_s$ where $\theta_a \in \Lambda^2$ and $\theta_s \in S^2$ are defined by setting:

$$\begin{aligned} \theta_a(x,y) &= \tfrac{1}{2}\{\theta(x,y) - \theta(y,x)\}, \\ \theta_s(x,y) &= \tfrac{1}{2}\{\theta(x,y) + \theta(y,x)\}. \end{aligned} \tag{1.1.c}$$

More generally, let Λ^k and S^k be the space of all *alternating* and *symmetric* tensors of degree k, respectively.

Fix a non-degenerate inner product $\langle\cdot,\cdot\rangle$ of *signature* (p,q) on V. We are in the *Riemannian setting* if $p = 0$ or, equivalently, if $\langle\cdot,\cdot\rangle$ is positive definite. Similarly, we are in the *Lorentzian* setting if $p = 1$. The *neutral setting* $p = q$ also is important. The pair $(V, \langle\cdot,\cdot\rangle)$ is called an *inner product space*. The associated *orthogonal group* $\mathcal{O} = \mathcal{O}(V, \langle\cdot,\cdot\rangle)$ is given by:

$$\mathcal{O} := \{T \in \mathrm{GL} : T^*\langle\cdot,\cdot\rangle = \langle\cdot,\cdot\rangle\}.$$

There is a natural extension of $\langle\cdot,\cdot\rangle$ to $\otimes^k V$ which will play a central role in our development and which we introduce here:

Definition 1.1.1 Let V^k denote the Cartesian product $V \times \cdots \times V$. If $\vec{v} = (v_1, \ldots, v_k)$ and $\vec{w} = (w_1, \ldots, w_k)$ are elements of V^k, the map

$$\vec{v} \times \vec{w} \to \langle v_1, w_1\rangle \ldots \langle v_k, w_k\rangle$$

is a bilinear symmetric map from $V^k \times V^k$ to $\mathbb{R}$ which extends to a symmetric inner product that is the extension of $\langle\cdot,\cdot\rangle$ to $\otimes^k V$. If $\{e_i\}$ is an orthonormal basis for V and if $I = (i_1, \ldots, i_k)$ is a multi-index, let $e_I := e_{i_1} \otimes \cdots \otimes e_{i_k}$. The collection $\{e_I\}_{|I|=k}$ forms a basis for $\otimes^k V$ with

$$\langle e_I, e_K\rangle = \left\{ \begin{array}{cc} 0 & \text{if } I \neq K \\ \langle e_{i_1}, e_{i_1}\rangle \ldots \langle e_{i_k}, e_{i_k}\rangle & \text{if } I = K \end{array} \right\}.$$

Since $\langle e_I, e_I\rangle = \pm 1$, $\langle\cdot,\cdot\rangle$ is non-degenerate on $\otimes^k V$. The orthogonal group $\mathcal{O}$ extends to act naturally on $\otimes^k V$ and preserves this inner product.

We may use $\langle\cdot,\cdot\rangle$ to identify V with V^* and extend $\langle\cdot,\cdot\rangle$ to tensors of all types; the natural action of $\mathcal{O}$ on such tensors then preserves this inner product. For example, let $\varepsilon_{ij} := \langle e_i, e_j\rangle$ give the components of the inner product relative to an arbitrary basis $\{e_i\}$ (which need not be orthonormal) for V. The inverse matrix ε^{ij} then gives the components of the dual inner product on V^* relative to the dual basis $\{e^i\}$ for V^*:

$$\varepsilon^{ij} = \langle e^i, e^j\rangle.$$

The following is a useful identity that will play a central role in many of our calculations:

$$\varepsilon^{ij}\langle v, e_i\rangle e_j = v \quad \text{and} \quad \varepsilon^{ij}\langle e_i, e_j\rangle = m. \tag{1.1.d}$$

If $A \in \otimes^4 V^*$, we define Ricci contractions:

$$\begin{aligned} \rho_{12}(A)_{kl} &:= \varepsilon^{ij} A_{ijkl}, & \rho_{13}(A)_{jl} &:= \varepsilon^{ik} A_{ijkl}, \\ \rho_{14}(A)_{jk} &:= \varepsilon^{il} A_{ijkl}, & \rho_{23}(A)_{il} &:= \varepsilon^{jk} A_{ijkl}, \\ \rho_{24}(A)_{ik} &:= \varepsilon^{jl} A_{ijkl}, & \rho_{34}(A)_{ij} &:= \varepsilon^{kl} A_{ijkl}. \end{aligned} \tag{1.1.e}$$

We set $\rho = \rho_{14}$. These contractions are $\mathcal{O}$ but not GL invariants. Similarly, if $\mathcal{A} \in \otimes^2 V^* \otimes \operatorname{End}(V)$, we define:

$$\rho(\mathcal{A})_{jk} := \mathcal{A}_{ijk}{}^i = \operatorname{Tr}(z \to \mathcal{A}(z,x)y)\,;$$

this contraction does not depend on the inner product. We use Equation (1.1.c) to decompose $\rho = \rho_a + \rho_s$ as the sum of the *alternating Ricci tensor* and the *symmetric Ricci tensor*. The terminology that we will use is motivated by the geometric setting. Therefore, the trace of the Ricci tensor is called the *scalar curvature*; it is given by setting:

$$\tau := \varepsilon^{il}\varepsilon^{jk} A_{ijkl} = \varepsilon^{jk} \mathcal{A}_{ijk}{}^i.$$

Definition 1.1.2 Let $(V, \langle\cdot,\cdot\rangle)$ be an inner product space.

(1) We say that $J_- \in \mathrm{GL}$ is a *complex structure* on V if $J_-^2 = -\operatorname{Id}$; if in addition $J_-^*\langle\cdot,\cdot\rangle = \langle\cdot,\cdot\rangle$, then J_- is said to be a *pseudo-Hermitian complex structure* and the triple $(V, \langle\cdot,\cdot\rangle, J_-)$ is said to be a *pseudo-Hermitian vector space*. Such structures exist if and only if $(V, \langle\cdot,\cdot\rangle)$ has signature (p,q) where both p and q are even. The associated *Kähler form* is given by setting $\Omega_-(x,y) := \langle x, J_- y\rangle$. We shall often let $\Omega = \Omega_-$ when the context is clear.
(2) We say that $J_+ \in \mathrm{GL}$ is a *para-complex structure* if $J_+^2 = \operatorname{Id}$ and if $\operatorname{Tr}(J_+) = 0$. This latter condition is automatic in the complex setting, but must be imposed in the para-complex setting. If $J_+^*\langle\cdot,\cdot\rangle = -\langle\cdot,\cdot\rangle$, then J_+ is said to be a *para-Hermitian complex structure* and the triple $(V, \langle\cdot,\cdot\rangle, J_+)$ is said to be a *para-Hermitian vector space*. Such structures exist only in the *neutral signature* $p = q$. The associated *para-Kähler form* is given by setting $\Omega_+(x,y) := \langle x, J_+ y\rangle$. Again, we shall often set $\Omega = \Omega_+$.

If g is a smooth symmetric non-degenerate bilinear form on the tangent bundle TM of a smooth manifold M, then (M, g) is called a *pseudo-Riemannian manifold.* If J_- is an endomorphism of TM with $J_-^2 = -\operatorname{Id}$, then J_- is said to be an *almost complex structure* on M and the pair (M, J_-) is said to be an *almost complex manifold*; necessarily $m = 2\bar{m}$ is even. The classic integrability result (see [Newlander and Nirenberg (1957)]) is summarized in Theorem 3.4.2. We say that J_- is an *integrable complex structure* and that (M, J_-) is a *complex manifold* if the *Nijenhuis tensor*

$$N_-(x,y) := [x,y] + J_-[J_-x,y] + J_-[x,J_-y] - [J_-x,J_-y] \tag{1.1.f}$$

vanishes or, equivalently, if in a neighborhood of any point of the manifold there are local *holomorphic coordinates* $(x^1,\dots,x^{\bar{m}},y^1,\dots,y^{\bar{m}})$ so that we have $J_-\partial_{x_i} = \partial_{y_i}$ and $J_-\partial_{y_i} = -\partial_{x_i}$. If $J_-^*g = g$, then (M, g, J_-) is called an *almost pseudo-Hermitian manifold*; (M, g, J_-) is said to be a *pseudo-Hermitian manifold* if J_- is an integrable complex structure.

Similarly, following [Cortés et al. (2004)], we say that (M, J_+) is an almost para-complex manifold if J_+ is an endomorphism of TM such that $J_+^2 = \operatorname{Id}$ and $\operatorname{Tr}(J_+) = 0$; necessarily $m = 2\bar{m}$ is even. One says J_+ is an *integrable para-complex structure* if the *para-Nijenhuis tensor*

$$N_+(x,y) := [x,y] - J_+[J_+x,y] - J_+[x,J_+y] + [J_+x,J_+y] \tag{1.1.g}$$

vanishes or, equivalently (see Theorem 3.4.3), if in a neighborhood of any point of the manifold there are local *para-holomorphic coordinates* $(x^1,\dots,x^{\bar{m}},y^1,\dots,y^{\bar{m}})$ so that we have $J_+\partial_{x_i} = \partial_{y_i}$ and $J_+\partial_{y_i} = \partial_{x_i}$. If $J_+^*g = -g$, then (M, g, J_+) is said to be an *almost para-Hermitian manifold*; if J_+ is an integrable para-complex structure, then (M, g, J_+) is said to be a *para-Hermitian manifold.*

The vanishing of $N_\pm$ imposes additional curvature restrictions called the *Gray identity* that will be discussed presently in Theorem 1.9.1 in the complex and in the para-complex settings.

We present a few general purpose references which may provide basic background information in some areas and appologize in advance if your favorite is missing: [Besse (1987)], [Bourbaki (2005)], [Chevalley (1946)], [Cruceanu, Fortuny, and Gadea (1996)], [Eisenhart (1927)], [Eisenhart (1967)], [Evans (1998)], [Ferus, Karcher, and Münzer (1981)], [Frobenius (1877)], [Fukami (1958)], [Fulton and Harris (1991)], [García-Río, Kupeli, and Vázquez-Lorenzo (2002)], [Gilkey (2001)], [Iwahori (1958)], [Kobayashi and Nomizu (1969)], [Newlander and Nirenberg (1957)], [Nomizu (1956)],

[Peter and Weyl (1927)], [Weyl (1921)], [Weyl (1922)], [Weyl (1946)], [Weyl (1988)], and [Yano (1965)]. We also refer to [Hitchin (1982)], [Vaisman (1982)], and [Vaisman (1983)].

1.2 Representation Theory

In Section 1.2, we introduce the associated essential structure groups. Let $(V, \langle\cdot,\cdot\rangle, J_\pm)$ be a para-Hermitian (+) or a pseudo-Hermitian (−) vector space. Let

$$\begin{aligned}
&\mathrm{GL}_\pm := \{T \in \mathrm{GL} : TJ_\pm = J_\pm T\},\\
&\mathrm{GL}_\pm^\star := \{T \in \mathrm{GL} : TJ_\pm = J_\pm T \text{ or } TJ_\pm = -J_\pm T\}, \qquad (1.2.a)\\
&\mathcal{U}_\pm := \mathcal{O} \cap \mathrm{GL}_\pm, \quad \text{and} \quad \mathcal{U}_\pm^\star := \mathcal{O} \cap \mathrm{GL}_\pm^\star\,.
\end{aligned}$$

The group $\mathrm{GL}_\pm$ is the (para)-complex general group and the group $\mathcal{U}_\pm$ is the (para)-unitary group. The groups $\mathrm{GL}_\pm^\star$ and $\mathcal{U}_\pm^\star$ are $\mathbb{Z}_2$ extensions of $\mathrm{GL}_\pm$ and $\mathcal{U}_\pm$, respectively; they permit us to replace $J_\pm$ by $-J_\pm$.

If σ defines the action of a Lie group G on a vector space V, then the pair (V, σ) is said to be a *module for the group* G or, when the particular group in question is clear, simply to be a *module.* We shall also sometimes say that G is the *structure group* of the module or that the module *has structure group* G. If W is a subspace of V, which is invariant under the action of G, then $(W, \sigma|_W)$ will be said to be a submodule of $V, \sigma)$. A *map of modules* or a *module morphism* is a linear map between the underlying vector spaces which commutes with (or equivalently *intertwines*) the group actions. A module is said to be a *general linear module* if $G = \mathrm{GL}$, is said to be an *orthogonal module* if $G = \mathcal{O}$, is said to be a *unitary module* if $G = \mathcal{U}_-$ or if $G = \mathcal{U}_-^\star$, and is said to be a *para-unitary module* if $G = \mathcal{U}_+$ or if $G = \mathcal{U}_-$.

In Section 2.1, we will establish the following result; as a consequence, when describing the decomposition of a module into irreducible modules it suffices for most purposes to simply list the multiplicities with which representations appear for these groups.

Theorem 1.2.1 *Let $G \in \{\mathcal{O}, \mathcal{U}_-, \mathcal{U}_\pm^\star\}$. Then G acts naturally on the tensor algebra $\otimes^k V^*$ via pull-back $(T^*\theta)(v_1, \ldots, v_k) := \theta(Tv_1, \ldots, Tv_k)$.*

(1) No non-trivial submodule of $\otimes^k V^*$ is totally isotropic. Furthermore, the restriction of the natural inner product on $\otimes^k V^*$ induces a non-degenerate inner product on W.

(2) We may decompose any non-trivial submodule ξ of $\otimes^k V^*$ with structure group G as the orthogonal direct sum of irreducible submodules in the form $\xi = \sum_i n_i \xi_i$ where the multiplicities n_i are independent of the particular decomposition chosen.

(3) If $\xi_1 = (V_1, \sigma_1)$ and $\xi_2 = (V_2, \sigma_2)$ are any two inequivalent irreducible submodules of $\otimes^k V^*$, then $V_1 \perp V_2$.

We note that Assertion (1) of Theorem 1.2.1 fails for the group $\mathcal{U}_+$. For this reason we have elected to work with the $\mathbb{Z}_2$ extension $\mathcal{U}_+^\star$. The following modules will play a central role:

Definition 1.2.1 Let $(V, \langle\cdot,\cdot\rangle)$ be an inner product space.

(1) Let $S_0^2 := \{\theta \in S^2 : \varepsilon^{ij}\theta_{ij} = 0\} = \{\theta \in S^2 : \theta \perp \langle\cdot,\cdot\rangle\}$; S_0^2 is an orthogonal module. The elements of S_0^2 are the symmetric tensors of degree two which have trace 0.

(2) Let $(V, \langle\cdot,\cdot\rangle, J_\pm)$ be a para-Hermitian vector space $(+)$ or a pseudo-Hermitian vector space $(-)$. We define the following modules, which have structure group $\mathcal{U}_\pm^\star$, by setting:

$$\begin{aligned} S_+^{2,\mathcal{U}_\pm} &:= \{\theta \in S^2 : J_\pm^*\theta = +\theta\}, & \Lambda_+^{2,\mathcal{U}_\pm} &:= \{\theta \in \Lambda^2 : J_\pm^*\theta = +\theta\},\\ S_-^{2,\mathcal{U}_\pm} &:= \{\theta \in S^2 : J_\pm^*\theta = -\theta\}, & \Lambda_-^{2,\mathcal{U}_\pm} &:= \{\theta \in \Lambda^2 : J_\pm^*\theta = -\theta\}. \end{aligned}$$

There is no linkage between the two sets of signs in the above equation. We now link the signs and define

$$\begin{aligned} S_{0,\mp}^{2,\mathcal{U}_\pm} &:= \{\theta \in S_\mp^{2,\mathcal{U}_\pm} : \theta \perp \langle\cdot,\cdot\rangle\},\\ \Lambda_{0,\mp}^{2,\mathcal{U}_\pm} &:= \{\theta \in \Lambda_\mp^{2,\mathcal{U}_\pm} : \theta \perp \Omega_\pm\}. \end{aligned}$$

(3) If $T \in \mathcal{U}_\pm^\star$ or if $T \in \mathrm{GL}_\pm^\star$, then $TJ_\pm = \chi(T)J_\pm T$ where χ defines a non-trivial representation of $\mathcal{U}_\pm^\star$ and of $\mathrm{GL}_\pm^\star$ into $\mathbb{Z}_2$.

We will establish the following result in Section 2.4 and Section 2.5:

Lemma 1.2.1 *Adopt the notation established above.*

(1) S_0^2 and Λ^2 are irreducible inequivalent orthogonal modules.

(2) $\{S_{0,+}^{2,\mathcal{U}_-}, S_-^{2,\mathcal{U}_-}, \Lambda_-^{2,\mathcal{U}_-}\}$ are irreducible inequivalent modules with structure group $\mathcal{U}_-$.

(3) $\Lambda_{0,+}^{2,\mathcal{U}_-}$ is isomorphic to $S_{0,+}^{2,\mathcal{U}_-}$ as a module with structure group $\mathcal{U}_-$.

(4) $\{S_{0,\mp}^{2,\mathcal{U}_\pm}, S_\pm^{2,\mathcal{U}_\pm}, \Lambda_{0,\mp}^{2,\mathcal{U}_\pm}, \Lambda_\pm^{2,\mathcal{U}_\pm}\}$ are irreducible inequivalent modules with structure group $\mathcal{U}_\pm^\star$.

(5) *The modules* $\Lambda^{2,\mathcal{U}_\pm}_{0,\mp}$ *are isomorphic to the modules* $S^{2,\mathcal{U}_\pm}_{0,\mp} \otimes \chi$ *as modules with structure group* $\mathcal{U}^\star_\pm$.

We are primarily concerned with local theory. Let P_i be points of metric spaces X_i. We say that f is the *germ* of a map from (X_1, P_1) to (X_2, P_2) if f is a continuous map from some neighborhood of P_1 in X_1 to X_2 with $f(P_1) = P_2$. We agree to identify two such maps if they agree on some (possibly) smaller neighborhood of P_1. In a similar fashion, we can talk about the germ of a pseudo-Riemannian manifold, the germ of a connection, and so forth.

1.3 Affine Structures

The results of Section 1.3 will be established in Chapters 3 and 4. An *affine manifold* is a pair (M, ∇) where M is a smooth manifold and where ∇ is an affine connection on the tangent bundle TM or, equivalently, a connection whose torsion tensor vanishes. Further information concerning affine geometry is to be found in [Gilkey, Nikčević, and Simon (2009)] and [Simon, Schwenk-Schellschmidt, and Viesel (1991)]. The associated *curvature operator* $\mathcal{R} \in \otimes^2 T^*M \otimes \operatorname{End}(TM)$ is defined by setting:

$$\mathcal{R}(x, y) := \nabla_x \nabla_y - \nabla_y \nabla_x - \nabla_{[x,y]}.$$

We will show in Lemma 3.1.2 that this tensor satisfies the following identities; the second identity is called the *Bianchi identity*:

$$\begin{aligned} &\mathcal{R}(x, y) = -\mathcal{R}(y, x), \\ &\mathcal{R}(x, y)z + \mathcal{R}(y, z)x + \mathcal{R}(z, x)y = 0. \end{aligned} \tag{1.3.a}$$

A $(1, 3)$ tensor $\mathcal{A} \in \otimes^2 V^* \otimes \operatorname{End}(V)$ satisfying the symmetries given in Equation (1.3.a) is called an *affine algebraic curvature operator*; we shall let $\mathfrak{A} = \mathfrak{A}(V) \subset \otimes^2 V^* \otimes \operatorname{End}(V)$ be the subspace of all such operators. We summarize below the fundamental decomposition of the space of affine curvature operators under the action of the general linear group [Strichartz (1988)]; further details will be presented when we return to this result in Section 4.4:

Theorem 1.3.1 *If* $m \geq 3$, *then* $\mathfrak{A} \approx \{\mathfrak{A} \cap \ker(\rho)\} \oplus \Lambda^2 \oplus S^2$ *as a general linear module where the modules* $\{\mathfrak{A} \cap \ker(\rho)$ *and* Λ^2, S^2 *are inequivalent irreducible general linear modules.*

We list the dimensions of these modules (see [Strichartz (1988)]) for the convenience of the reader; they can also be derived from Theorem 1.4.2.

Theorem 1.3.2

$\dim\{\mathfrak{A}\} = \frac{1}{3}m^2(m^2-1)$	$\dim\{S^2\} = \frac{1}{2}m(m+1)$
$\dim\{\Lambda^2\} = \frac{1}{2}m(m-1)$	$\dim\ker(\rho)\cap\mathfrak{A} = \frac{1}{3}m^2(m^2-4)$

The projection on Λ^2 in Theorem 1.3.1 is provided by ρ_a and the projection on S^2 is provided by ρ_s. We can split these projections as follows. If $\psi \in \Lambda^2$ and if $\phi \in S^2$, define:

$$(\sigma_a(\psi))(x,y)z = -\tfrac{1}{1+m}\{2\psi(x,y)z + \psi(x,z)y - \psi(y,z)x\},$$
$$(\sigma_s(\phi))(x,y)z = \tfrac{1}{1-m}\{\phi(x,z)y - \phi(y,z)x\}.$$

Then $\sigma_a(\psi) \in \mathfrak{A}$ and $\sigma_s(\phi) \in \mathfrak{A}$. Furthermore, we have that:

$$\rho_a\sigma_a = \mathrm{Id}_{\Lambda^2}, \quad \rho_s\sigma_a = 0,$$
$$\rho_a\sigma_s = 0, \qquad \rho_s\sigma_s = \mathrm{Id}_{S^2}\,.$$

Note that the identities $\rho_a\sigma_s = 0$ and $\rho_s\sigma_a = 0$ are immediate since S^2 and Λ^2 are inequivalent irreducible general linear modules. If $\mathcal{A} \in \mathfrak{A}$, define:

$$(\pi_P\mathcal{A})(x,y)z := \mathcal{A}(x,y)z - (\sigma_a(\rho_a)\mathcal{A})(x,y)z - (\sigma_s(\rho_s)\mathcal{A})(x,y)z.$$

The tensor $\pi_P(\mathcal{A})$ is called the *Weyl projective curvature tensor*; π_P is given by orthogonal projection on $\mathfrak{A}\cap\ker(\rho)$ in Theorem 1.3.1. The decomposition of Theorem 1.3.1 motivates the following:

Definition 1.3.1 Let $\mathcal{A} \in \mathfrak{A}$. Let $\rho = \rho(\mathcal{A})$. One says that:

(1) $\mathcal{A}$ is *Ricci symmetric* if and only if $\rho \in S^2$ or, equivalently, if $\rho_a = 0$.
(2) $\mathcal{A}$ is *Ricci anti-symmetric* if and only if $\rho \in \Lambda^2$ or, equivalently, if $\rho_s = 0$.
(3) $\mathcal{A}$ is *Ricci flat* if and only if $\rho = 0$.
(4) $\mathcal{A}$ is *projectively flat* if and only if $\pi_P(\mathcal{A}) = 0$.
(5) $\mathcal{A}$ is *flat* if and only if $\mathcal{A} = 0$.

We say that an affine curvature operator $\mathcal{A} \in \mathfrak{A}$ is *geometrically realizable* if there exists an affine manifold (M, ∇), if there exists a point P of M (which is called the point of realization), and if there exists an isomorphism $\phi : V \to T_PM$ so that $\phi^*\mathcal{R}_P = \mathcal{A}$. The decomposition of $\mathfrak{A}$ as a module over the general linear group has three components so there are eight natural geometric realization questions which are GL equivariant. In Section 4.5, we will discuss the results of [Gilkey and Nikčević (2008)] and the results of [Gilkey, Nikčević, and Westerman (2009)] to establish Theorem 1.3.3 showing, in particular, that the symmetries of Equation (1.3.a) generate the universal symmetries of the curvature operator of an affine connection:

Theorem 1.3.3

(1) Any affine algebraic curvature operator is geometrically realizable by an affine manifold.
(2) Any Ricci symmetric affine algebraic curvature operator is geometrically realizable by a Ricci symmetric affine manifold.
(3) Any Ricci anti-symmetric affine algebraic curvature operator is geometrically realizable by a Ricci anti-symmetric affine manifold.
(4) Any Ricci flat affine algebraic curvature operator is geometrically realizable by a Ricci flat affine manifold.
(5) Any projectively flat affine algebraic curvature operator is geometrically realizable by a projectively flat affine manifold.
(6) Any projectively flat Ricci symmetric affine algebraic curvature operator is geometrically realizable by a projectively flat Ricci symmetric affine manifold.
(7) A non-flat projectively flat Ricci anti-symmetric affine algebraic curvature operator is not geometrically realizable by a projectively flat Ricci anti-symmetric affine manifold.
(8) If $\mathcal{A}$ is flat, then $\mathcal{A}$ is geometrically realizable by a flat affine manifold.

These geometric realizability results can be summarized in the following table; the non-zero components of $\mathcal{A}$ are indicated by $\star$.

Table 1.3.1

$\ker(\rho)$	S^2	Λ^2		$\ker(\rho)$	S^2	Λ^2	
$\star$	$\star$	$\star$	yes	0	$\star$	$\star$	yes
$\star$	$\star$	0	yes	0	$\star$	0	yes
$\star$	0	$\star$	yes	0	0	$\star$	no
$\star$	0	0	yes	0	0	0	yes

Remark 1.3.1 In fact, a bit more is true. Given $\mathcal{A}$, in Section 4.5 we will construct the germ of an affine connection at 0 in V so that the matrix of the Ricci tensor is constant relative to the coordinate frame; this means

that one has that $\rho(\mathcal{R})(\partial_{x_i}, \partial_{x_j}) = \rho(\mathcal{A})(e_i, e_j)$. This result settles other associated realization questions.

Let $\nabla\mathcal{R}(x,y;z)w$ be the *covariant derivative of the curvature operator*:

$$\nabla\mathcal{R}(x,y;z)w := \nabla_z\mathcal{R}(x,y)w - \mathcal{R}(\nabla_z x,y)w - \mathcal{R}(x,\nabla_z y)w - \mathcal{R}(x,y)\nabla_z w.$$

We will show in Lemma 3.1.2 that this has the symmetries:

$$\begin{aligned} &R_{ijk}{}^l{}_{;n} = -R_{jik}{}^l{}_{;n}, \\ &R_{ijk}{}^l{}_{;n} + R_{jki}{}^l{}_{;n} + R_{kij}{}^l{}_{;n} = 0, \\ &R_{ijk}{}^l{}_{;n} + R_{jnk}{}^l{}_{;i} + R_{nik}{}^l{}_{;j} = 0. \end{aligned} \tag{1.3.b}$$

Let $\mathfrak{A}^1 \subset \otimes^3 V^* \otimes \operatorname{End}(V)$ be the subspace of all $(1,4)$ tensors satisfying these relations. In Section 4.5 we will establish the following result:

Theorem 1.3.4 *Let $\mathcal{A} \in \mathfrak{A}$ and let $\mathcal{A}^1 \in \mathfrak{A}^1$. Define an affine connection ∇ on TV by setting*

$$\begin{aligned} \Gamma_{uv}{}^l := &\tfrac{1}{3}(A_{wuv}{}^l + A_{wvu}{}^l)x^w \\ &+\tfrac{5}{24}(A^1_{wuv}{}^l{}_{;n} + A^1_{wvu}{}^l{}_{;n})x^w x^n + \tfrac{1}{24}(A^1_{wun}{}^l{}_{;v} + A^1_{wvn}{}^l{}_{;u})x^w x^n. \end{aligned}$$

Then $\mathcal{R}_{ijk}{}^l(0) = A_{ijk}{}^l$ and $\mathcal{R}_{ijk}{}^l{}_{;n}(0) = A^1_{ijk}{}^l{}_{;n}$.

Affine geometry is a central area of study; we present only a very few references of the many possible [Blaschke (1985)], [Blažić et al. (2006)], [Bokan, Nomizu, and Simon (1990)], [Brozos-Vázquez, Gilkey, and Nikčević (2011)], [Brozos-Vázquez, Gilkey, and Nikčević (2011b)], [Calabi (1982)], [Cortés, Lawn, and Schaefer (2006)], [Gilkey and Nikčević (2008)], [Gilkey, Nikčević, and Simon (2009)], [Itoh (2000)], [Li, Li, and Simon (2004)], [Li et al. (1997)], [Li, Simon, and Zhao (1993)], [Manhart (2003)], [Nomizu and Podestá (1989)], [Nomizu and Sasaki (1993)], [Nomizu and Simon (1992)], [Schirokow and Schirokow (1962)], [Schwenk-Schellschmidt and Simon (2009)], [Simon (1995)], [Simon (2000)], [Simon (2004)], [Simon, Schwenk-Schellschmidt, and Viesel (1991)], [Vrancken, Li, and Simon (1991)], and [Wang (1994)]. We also refer to [Binder (2009)], [Bokan and Nikčević (1994)], and [Pinkall, Schwenk-Schellschmidt, and Simon (1994)].

1.4 Mixed Structures

The results of Section 1.4 will be proved in Chapter 4. We now study an affine structure and a pseudo-Riemannian metric where the given affine connection is not necessarily the Levi-Civita connection of the pseudo-Riemannian metric; thus the two structures are decoupled.

Let $\mathcal{A} \in \mathfrak{A}$. We use the metric to lower the final index and define $A \in \otimes^4 V^*$ by setting:

$$A(x, y, z, w) := \langle \mathcal{A}(x, y)z, w \rangle.$$

The symmetries of Equation (1.3.a) then become:

$$\begin{aligned} &A(x, y, z, w) = -A(y, x, z, w), \\ &A(x, y, z, w) + A(y, z, x, w) + A(z, x, y, w) = 0. \end{aligned} \tag{1.4.a}$$

Again, a curvature decomposition plays a central role. We introduce the following notational conventions:

Definition 1.4.1

$$\begin{aligned} W_6^{\mathcal{O}} &:= \{A \in \mathfrak{A} \cap \ker(\rho) : A_{ijkl} = -A_{ijlk}\}, \\ W_7^{\mathcal{O}} &:= \{A \in \mathfrak{A} \cap \ker(\rho) : A_{ijkl} = A_{ijlk}\}, \\ W_8^{\mathcal{O}} &:= \{A \in \otimes^4 V^* \cap \ker(\rho) : A_{ijkl} = -A_{jikl} = -A_{klij}\}. \end{aligned}$$

Note that $W_6^{\mathcal{O}}$ and $W_7^{\mathcal{O}}$ are submodules of $\mathfrak{A}$ whereas $W_8^{\mathcal{O}} \not\subset \mathfrak{A}$. We also note that $W_6^{\mathcal{O}}$ will play an important role in the $\mathcal{O}$ decomposition of $\mathfrak{R}$ given subsequently in Theorem 1.6.2.

In Section 4.1, we will establish the following result [Bokan (1990)]:

Theorem 1.4.1 *Let $m \geq 4$. We have the following isomorphism decomposing $\mathfrak{A}$ as the direct sum of irreducible and inequivalent orthogonal modules:*

$$\mathfrak{A} \approx \mathbb{R} \oplus 2 \cdot S_0^2 \oplus 2 \cdot \Lambda^2 \oplus W_6^{\mathcal{O}} \oplus W_7^{\mathcal{O}} \oplus W_8^{\mathcal{O}}.$$

Remark 1.4.1 If $m = 3$, we set $W_6^{\mathcal{O}} = W_8^{\mathcal{O}} = 0$. If $m = 2$, then

$$\mathfrak{A} = \mathbb{R} \oplus S_0^2 \oplus \Lambda^2.$$

We shall determine the dimension of these modules subsequently in Theorem 2.4.1 and Corollary 4.1.2; they were first computed in [Bokan (1990)]. We list them here for the convenience of the reader:

Theorem 1.4.2

$\dim\{\mathfrak{R}\} = \frac{1}{12}m^2(m^2-1)$	$\dim\{\mathfrak{A}\} = \frac{1}{3}m^2(m^2-1)$
$\dim\{\mathbb{R}\} = 1$	$\dim\{S_0^2\} = \frac{1}{2}m(m+1)-1$
$\dim\{\Lambda^2\} = \frac{1}{2}m(m-1)$	$\dim\{W_6^{\mathcal{O}}\} = \frac{m(m+1)(m-3)(m+2)}{12}$
$\dim\{W_7^{\mathcal{O}}\} = \frac{(m-1)(m-2)(m+1)(m+4)}{8}$	$\dim\{W_8^{\mathcal{O}}\} = \frac{m(m-1)(m-3)(m+2)}{8}$

Let τ be the scalar curvature and let ρ_0 be the part of the Ricci tensor which has zero trace. Several geometric realization questions, which are natural with respect to the structure group $\mathcal{O}$, can be solved in the real analytic category. As our considerations are local, we shall take $M = V$ and $P = 0$. In Section 4.5, we present results of [Gilkey, Nikčević, and Westerman (2009a)] establishing the following result:

Theorem 1.4.3 *Let g be the germ at $0 \in V$ of a real analytic pseudo-Riemannian metric. Let $\mathcal{A} \in \mathfrak{A}$. There exists the germ of an affine real analytic connection ∇ at $0 \in V$ with:*

(1) $\mathcal{R}_0 = \mathcal{A}$.
(2) ∇ has constant scalar curvature.
(3) If $\mathcal{A}$ is Ricci symmetric, then ∇ is Ricci symmetric.
(4) If $\mathcal{A}$ is Ricci anti-symmetric, then ∇ is Ricci anti-symmetric.
(5) If $\mathcal{A}$ is Ricci traceless, then ∇ is Ricci traceless.

We note there are corresponding results in the C^k category and refer to [Gilkey, Nikčević, and Westerman (2009a)] for further details.

The study of curvature, the covariant derivative of curvature, curvature models, and the spectral geometry of various natural operators associated to the curvature tensor is central in many geometries. We cite only a very few of the many possible references: [Apostolov, Ganchev, and Ivanov (1997)], [Belger and Kowalski (1994)], [Blair (1990)], [Blažić (2006)], [Blažić et al. (2005)], [Blažić et al. (2006)], [Blzaic et al. (2008)], [Bokan (1990)], [Brozos-Vázquez, García-Río, and Gilkey (2008)], [Brozos-Vázquez et al. (2009)], [Brozos-Vázquez et al. (2010)], [Brozos-Vázquez, Gilkey, and Merino (2010)], [Brozos-Vázquez, Gilkey, and Nikčević (2011b)], [Cortés-Ayaso, Díaz-Ramos, and García-Río (2008)], [Deprez, Sekigawa, and Verstraelen (1988)], [De Smedt (1994)], [Díaz-Ramos et al. (2004)], [Díaz-Ramos and García-Río (2004)], [Dunn and Gilkey (2005)], [Falcitelli and Farinola (1994)], [Fiedler (2003)], [Gilkey (1973)], [Gilkey (2001)], [Gilkey, Ivanova, and Zhang (2002)], [Gilkey and Nikčević (2008)], [Gilkey and

Nikčević (2011)], [Gilkey and Nikčević (2011a)], [Gilkey, Nikčević, and Simon (2011)], [Gilkey, Nikčević, and Westerman (2009)], [Gilkey, Nikčević, and Westerman (2009a)], [Gilkey, Park, and Sekigawa (2011)], [Gilkey, Puffini, and Videv (2006)], [Gilkey and Stavrov (2002)], [Gray (1976)], [Higa (1994)], [Martín-Cabrera and Swann (2006)], [Matzeu and Nikčević (1991)], [Nikčević (1992)], [Nikčević (1994)], [Nomizu (1972)], [Ozdeger (2006)], [Sato (1989)], [Sato (2003)], [Sato (2004)], [Schoen (1984)], [Singer and Thorpe (1969)], [Strichartz (1988)], [Tang (2006)], [Tricerri and Vanhecke (1981)], [Tricerri and Vanhecke (1986)], [Vanhecke (1977)], [Vezzoni (2007)], and [Vrancken, Li, and Simon (1991)]. In addition, we refer to [Biswas (2008)], [Boeckx, Kowalski, and Vanhecke (1994)], [Calvino-Louzao et al. (2011)], and [García-Río et al. (2010)], and [Singer (1960)].

1.5 Affine Kähler Structures

The results described here will be established in Section 5.4, in Section 5.5, and in Section 5.6; we refer to [Brozos-Vázquez, Gilkey, and Nikčević (2011)] in the complex setting whereas in the para-complex setting, they are new. Let $J_\pm$ be a (para)-complex structure on V. Set:

$$\begin{aligned}
\mathfrak{K}_\pm^\mathfrak{A} &:= \{\mathcal{A} \in \mathfrak{A} : \mathcal{A}(x,y)J_\pm = J_\pm\mathcal{A}(x,y)\},\\
\mathfrak{K}_{\pm;+}^\mathfrak{A} &:= \{\mathcal{A} \in \mathfrak{K}_\pm^\mathfrak{A} : \mathcal{A}(J_\pm x, J_\pm y) = +\mathcal{A}(x,y)\},\\
\mathfrak{K}_{\pm;-}^\mathfrak{A} &:= \{\mathcal{A} \in \mathfrak{K}_\pm^\mathfrak{A} : \mathcal{A}(J_\pm x, J_\pm y) = -\mathcal{A}(x,y)\}.
\end{aligned} \tag{1.5.a}$$

There are two sets of signs giving rise to four subspaces: $\mathfrak{K}_{+;+}^\mathfrak{A}$, $\mathfrak{K}_{+;-}^\mathfrak{A}$, $\mathfrak{K}_{-;+}^\mathfrak{A}$, and $\mathfrak{K}_{-;-}^\mathfrak{A}$. We shall occasionally use the notation $\mathfrak{K}_{\pm;\delta}^\mathfrak{A}$ where we permit $\delta = \pm$. We may decompose

$$\mathfrak{K}_\pm^\mathfrak{A} = \mathfrak{K}_{\pm;+}^\mathfrak{A} \oplus \mathfrak{K}_{\pm;-}^\mathfrak{A}.$$

Let $\langle\cdot,\cdot\rangle$ be an auxiliary inner product, not necessarily positive definite, used to lower indices and regard $\mathfrak{K}_\pm^\mathfrak{A}$, $\mathfrak{K}_{\pm;+}^\mathfrak{A}$, and $\mathfrak{K}_{\pm;-}^\mathfrak{A}$ as subspaces of $\otimes^4 V^*$. We may now express

$$\begin{aligned}
\mathfrak{K}_\pm^\mathfrak{A} &:= \{A \in \mathfrak{A} : A(x,y,z,w) = \mp A(x,y,J_\pm z, J_\pm w)\},\\
\mathfrak{K}_{\pm;+}^\mathfrak{A} &:= \{A \in \mathfrak{K}_\pm^\mathfrak{A} : A(J_\pm x, J_\pm y, z, w) = A(x,y,z,w)\},\\
\mathfrak{K}_{\pm;-}^\mathfrak{A} &:= \{A \in \mathfrak{K}_\pm^\mathfrak{A} : A(J_\pm x, J_\pm y, z, w) = -A(x,y,z,w)\}.
\end{aligned} \tag{1.5.b}$$

The following result generalizes Theorem 1.3.1 to this setting; it will be established in Section 5.6 by extending results of [Brozos-Vázquez, Gilkey, and Nikčević (2011b)] for the group $\mathrm{GL}_{-}^{\star}$ to the groups $\mathrm{GL}_{+}^{\star}$ and GL_{-}.

Theorem 1.5.1 *If $m \geq 6$, then we have the following isomorphisms decomposing $\mathfrak{K}_{\pm;\delta}^{\mathfrak{A}}$ as the direct sum of irreducible and inequivalent modules with respect to the structure groups $\mathrm{GL}_{\pm}^{\star}$ and GL_{-}:*

$$\mathfrak{K}_{\pm,+}^{\mathfrak{A}} \approx \{\mathfrak{K}_{\pm,+}^{\mathfrak{A}} \cap \ker(\rho)\} \oplus \Lambda_{+}^{2,\mathcal{U}_{\pm}} \oplus S_{+}^{2,\mathcal{U}_{\pm}},$$
$$\mathfrak{K}_{\pm,-}^{\mathfrak{A}} \approx \{\mathfrak{K}_{\pm,-}^{\mathfrak{A}} \cap \ker(\rho)\} \oplus \Lambda_{-}^{2,\mathcal{U}_{\pm}} \oplus S_{-}^{2,\mathcal{U}_{\pm}}.$$

Remark 1.5.1 If $m = 4$, then the corresponding decomposition is obtained by setting $\mathcal{K}_{\pm,\pm}^{\mathfrak{A}} \cap \ker(\rho) = \{0\}$ and deleting it from consideration.

The decomposition of these spaces as modules with structure group $\mathcal{U}_{-}$ in the Hermitian setting is given by [Matzeu and Nikčević (1991)] and by [Nikčević (1992)]. There are four submodules of $\mathfrak{K}_{\pm}^{\mathfrak{A}}$ not corresponding to generalized Ricci tensors. We must examine these.

Definition 1.5.1 Let $(V, \langle\cdot,\cdot\rangle, J_{\pm})$ be a para-Hermitian vector space $(+)$ or a pseudo-Hermitian vector space $(-)$. Let

(1) $W_{\pm,9}^{\mathfrak{A}} := \{A \in \mathfrak{K}_{\pm;\mp}^{\mathfrak{A}} : A(x,y,z,w) = -A(x,y,w,z)\} \cap \ker(\rho)$,
(2) $W_{\pm,10}^{\mathfrak{A}} := \{A \in \mathfrak{K}_{\pm;\mp}^{\mathfrak{A}} : A(x,y,z,w) = A(x,y,w,z)\} \cap \ker(\rho)$,
(3) $W_{\pm,11}^{\mathfrak{A}} := \mathfrak{K}_{\pm,\mp}^{\mathfrak{A}} \cap (W_{\pm,9}^{\mathfrak{A}})^{\perp} \cap (W_{\pm,10}^{\mathfrak{A}})^{\perp} \cap \ker(\rho_{13}) \cap \ker(\rho)$,
(4) $W_{\pm,12}^{\mathfrak{A}} := \mathfrak{K}_{\pm,\pm}^{\mathfrak{A}} \cap \ker(\rho)$,
(5) $\tau_{\pm}^{\mathfrak{A}} := \varepsilon^{il}\varepsilon^{jk} A(e_i, J_{\pm}e_j, e_k, e_l)$.

If we adopt the notation of Definition 1.4.1, we may also express:

$$W_{\pm,9}^{\mathfrak{A}} = \mathfrak{K}_{\pm;\mp}^{\mathfrak{A}} \cap W_6^{\mathcal{O}} \quad \text{and} \quad W_{\pm,10}^{\mathfrak{A}} = \mathfrak{K}_{\pm;\mp}^{\mathfrak{A}} \cap W_7^{\mathcal{O}}.$$

We may also identify $W_{\pm,9}^{\mathfrak{A}}$ with $W_{\pm,3}^{\mathfrak{R}}$ as given in Definition 1.8.1 subsequently. In Section 5.4 and in Section 5.5, we will extend the curvature decomposition which is given in [Matzeu and Nikčević (1991)] and [Nikčević (1992)] from positive definite signatures to more general signatures and also to the para-Hermitian setting to show:

Theorem 1.5.2

(1) We have the following isomorphisms decomposing $\mathfrak{K}_{-}^{\mathfrak{A}}$ as the direct sum of irreducible and inequivalent modules with structure group $\mathcal{U}_{-}$:

(a) If $\dim(V) = 4$, $\mathfrak{K}_{-}^{\mathfrak{A}} \approx 2 \cdot \mathbb{R} \oplus 4 \cdot S_{0,+}^{2,\mathcal{U}_{-}} \oplus \Lambda_{-}^{2,\mathcal{U}_{-}} \oplus S_{-}^{2,\mathcal{U}_{-}} \oplus 2 \cdot W_{-,9}^{\mathfrak{A}}$.

(b) *If* $\dim(V) \geq 6$, $\mathfrak{K}_-^{\mathfrak{A}} \approx 2 \cdot \mathbb{R} \oplus 4 \cdot S_{0,+}^{2,\mathcal{U}_-} \oplus \Lambda_-^{2,\mathcal{U}_-} \oplus S_-^{2,\mathcal{U}_-} \oplus 2 \cdot W_{-,9}^{\mathfrak{A}} \oplus W_{-,11}^{\mathfrak{A}} \oplus W_{-,12}^{\mathfrak{A}}$.

(2) *We have the following isomorphisms decomposing the modules* $\mathfrak{K}_\pm^{\mathfrak{A}}$ *as the direct sum of irreducible and inequivalent modules with structure group* $\mathcal{U}_\pm^\star$:

(a) *If* $m = 4$, $\mathfrak{K}_\pm^{\mathfrak{A}} \approx \mathbb{R} \oplus \chi \oplus 2 \cdot S_{0,\mp}^{2,\mathcal{U}_\pm} \oplus 2 \cdot \Lambda_{0,\mp}^{2,\mathcal{U}_\pm} \oplus \Lambda_\pm^{2,\mathcal{U}_\pm} \oplus S_\pm^{2,\mathcal{U}_\pm} \oplus W_{\pm,9}^{\mathfrak{A}} \oplus W_{\pm,10}^{\mathfrak{A}}$.

(b) *If* $m \geq 6$, $\mathfrak{K}_\pm^{\mathfrak{A}} \approx \mathbb{R} \oplus \chi \oplus 2 \cdot S_{0,\mp}^{2,\mathcal{U}_\pm} \oplus 2 \cdot \Lambda_{0,\mp}^{2,\mathcal{U}_\pm} \oplus \Lambda_\pm^{2,\mathcal{U}_\pm} \oplus S_\pm^{2,\mathcal{U}_\pm} \oplus W_{\pm,9}^{\mathfrak{A}} \oplus W_{\pm,10}^{\mathfrak{A}} \oplus W_{\pm,11}^{\mathfrak{A}} \oplus W_{\pm,12}^{\mathfrak{A}}$.

Remark 1.5.2 We will show subsequently in Lemma 2.5.3 that $S_+^{2,\mathcal{U}_+}$ and $\Lambda_+^{2,\mathcal{U}_+}$ are not irreducible modules for the structure group $\mathcal{U}_+$; Theorem 1.5.2 (2) fails if we replace $\mathcal{U}_+^\star$ by $\mathcal{U}_+$.

We say that $(V, J_\pm, \mathcal{A})$ is a *(para)-Kähler affine curvature model* if $J_\pm$ is a (para)-complex structure on V and if $\mathcal{A} \in \mathfrak{K}_\pm^{\mathfrak{A}}$. Similarly $(M, J_\pm, \nabla)$ is said to be a *(para)-Kähler affine manifold* if $J_\pm$ is a (para)-complex structure on M, if ∇ is an affine connection on TM, and if $\nabla(J_\pm) = 0$. We say that a (para)-Kähler curvature model $(V, J_\pm, A)$ is geometrically realizable if there exists a (para)-Kähler manifold $(M, J_\pm, \nabla)$, a point P in M, and an isomorphism $\Xi : V \to T_PM$ so $\Xi^*\mathcal{R} = \mathcal{A}$ and $\Xi^* J_{\pm,P} = J_\pm$. One then has:

Theorem 1.5.3 *Every (para)-Kähler affine curvature model is geometrically realizable by a (para)-Kähler affine manifold. If* $A \in \mathfrak{K}_{\pm,\pm}^{\mathfrak{A}}$, *the para-Kähler manifold (+) M or pseudo-Kähler manifold (−) M can be chosen so that the curvature belongs to* $\mathfrak{K}_{\pm,\pm}^{\mathfrak{A}}$ *at every point.*

The dimension of these modules was determined by [Matzeu and Nikčević (1991)] in the positive definite setting; we shall simply cite their results for the sake of completeness but shall omit the proof as it plays no role in our development. The dimensions are the same in the para-Hermitian setting or in the pseudo-Hermitian setting; the second column is discussed in Lemma 2.5.1.

Theorem 1.5.4 *Let $m = 2\bar{m} \geq 6$. Then:*

$\dim\{\mathfrak{K}_{\pm}^{\mathfrak{A}}\} = \frac{1}{3}\bar{m}^2(\bar{m}+1)(5\bar{m}-2)$	$\dim\{\mathbb{R}\} = \dim\{\chi\} = 1$
$\dim\{W_{\pm,9}^{\mathfrak{A}}\} = \frac{1}{4}\bar{m}^2(\bar{m}-1)(\bar{m}+3)$	$\dim\{S_{0,\mp}^{2,\mathcal{U}_{\pm}}\} = \bar{m}^2 - 1$
$\dim\{W_{\pm,10}^{\mathfrak{A}}\} = \frac{1}{4}\bar{m}^2(\bar{m}-1)(\bar{m}+3)$	$\dim(\Lambda_{0,\mp}^{2,\mathcal{U}_{\pm}}\} = \bar{m}^2 - 1$
$\dim\{W_{\pm,11}^{\mathfrak{A}}\} = \frac{1}{2}(\bar{m}-1)(\bar{m}+1)(\bar{m}-2)(\bar{m}+2)$	$\dim\{S_{\pm}^{2,\mathcal{U}_{\pm}}\} = \bar{m}^2 + \bar{m}$
$\dim\{W_{\pm,12}^{\mathfrak{A}}\} = \frac{2}{3}\bar{m}^2(\bar{m}-2)(\bar{m}+2)$	$\dim\{\Lambda_{\pm}^{2,\mathcal{U}_{\pm}}\} = \bar{m}^2 - \bar{m}$

1.6 Riemannian Structures

The results of Section 1.6 reflect material that will be presented in Chapter 3, in Chapter 4, and in Chapter 6. One says that $A \in \otimes^4 V^*$ is a *Riemannian algebraic curvature tensor* on V if A satisfies the symmetries of the Riemann curvature tensor (see Lemma 3.3.3):

$$\begin{aligned} &A(x,y,z,w) = -A(y,x,z,w) = A(z,w,x,y),\\ &A(x,y,z,w) + A(y,z,x,w) + A(z,x,y,w) = 0. \end{aligned} \tag{1.6.a}$$

Let $\mathfrak{R} = \mathfrak{R}(V)$ be the space of all such tensors. We have that $\mathfrak{R}$ is invariant under the action of $\mathcal{O}$. Thus by Theorem 1.2.1, $\mathfrak{R} \cap \mathfrak{R}^{\perp} = \{0\}$ and $\langle\cdot,\cdot\rangle$ is a non-degenerate inner product on $\mathfrak{R}$. We say that $(V, \langle\cdot,\cdot\rangle, A)$ is a *curvature model* if $A \in \mathfrak{R}$.

Definition 1.6.1

(1) Let $\phi \in S^2$ be a symmetric bilinear form. Set

$$A_\phi(x,y,z,w) := \phi(x,w)\phi(y,z) - \phi(x,z)\phi(y,w).$$

These tensors arise in the study of hypersurface theory; if ϕ is the second fundamental form of a hypersurface in flat space, then the curvature tensor of the hypersurface is given by A_ϕ.

(2) Let $\psi \in \Lambda^2$ be an anti-symmetric bilinear form. Set

$$A_\psi(x,y,z,w) := \psi(x,w)\psi(y,z) - \psi(x,z)\psi(y,w) - 2\psi(x,y)\psi(z,w).$$

The study of the tensors A_ψ arose in the original instance from the Osserman conjecture and related matters which are contained in [García-Río, Kupeli, and Vázquez-Lorenzo (2002)] and [Gilkey (2001)].

In Section 6.1 we establish a result of [Fiedler (2003)] giving generators for $\mathfrak{R}$ and determine $\dim\{\mathfrak{R}\}$:

Theorem 1.6.1

(1) $\mathfrak{R} = \text{Span}_{\phi \in S^2}\{A_\phi\} = \text{Span}_{\psi \in \Lambda^2}\{A_\psi\}$.

(2) $\dim\{\mathfrak{R}\} = \frac{1}{12}m^2(m^2-1)$.

We adopt the notation of Definition 1.4.1 to define the orthogonal module $W_6^{\mathcal{O}}$. We then have that $W_6^{\mathcal{O}} = \ker(\rho) \cap \mathfrak{R}$. In Section 4.1, we will establish the decomposition [Singer and Thorpe (1969)] of $\mathfrak{R}$ as an orthogonal module:

Theorem 1.6.2 *Let* $\dim(V) \geq 4$. *We have the following isomorphism decomposing* $\mathfrak{R}$ *as the direct sum of irreducible and inequivalent orthogonal modules:*

$$\mathfrak{R} \approx \mathbb{R} \oplus S_0^2 \oplus W_6^{\mathcal{O}}.$$

The projection on S_0^2 in Theorem 1.6.2 is given by the part of the Ricci tensor of zero trace; the projection on $\mathbb{R}$ is given by the scalar curvature τ. The low dimensional setting is a bit different. If $m = 2$, then $\mathfrak{R} \approx \mathbb{R}$ and if $m = 3$, then $\mathfrak{R} \approx \mathbb{R} \oplus S_0^2$.

Let $A \in \mathfrak{R}$. We say that the curvature model $(V, \langle \cdot, \cdot \rangle, A)$ is *geometrically realizable* if there exists a pseudo-Riemannian manifold (M, g), if there exists a point P of M, and if there exists an isomorphism $\phi : V \to T_PM$ so that $\phi^* g_P = \langle \cdot, \cdot \rangle$ and $\phi^* R_P = A$.

The Weyl conformal curvature tensor W is the projection of A on $\ker(\rho)$ in Theorem 1.6.2 (see Equation (6.2.a)); we say a curvature model or a pseudo-Riemannian manifold is *conformally flat* if and only if $W = 0$. In Section 6.4, we establish results [Brozos-Vázquez et al. (2009)] dealing with geometric realizations by pseudo-Riemannian manifolds with constant scalar curvature:

Theorem 1.6.3

(1) Any curvature model is geometrically realizable by a pseudo-Riemannian manifold of constant scalar curvature.

(2) Any conformally flat curvature model is geometrically realizable by a conformally flat pseudo-Riemannian manifold of constant scalar curvature.

1.7 Weyl Geometry I

The results of Section 1.7 will be proved in Chapter 6. Again, we consider a mixed structure. Consider a triple $\mathcal{W} := (M, g, \nabla)$ where g is a pseudo-Riemannian metric on a smooth manifold M of dimension m and where ∇ is an affine connection on TM. We say that $\mathcal{W}$ is a *Weyl manifold* if the following identity is satisfied:

$$\nabla g = -2\phi \otimes g \quad \text{for some} \quad \phi \in C^\infty(T^*M). \tag{1.7.a}$$

This notion is conformally invariant. If $\mathcal{W} = (M, g, \nabla)$ is a Weyl manifold and if $f \in C^\infty(M)$, then $\tilde{\mathcal{W}} := (M, e^{2f}g, \nabla)$ is again a Weyl manifold where $\tilde{\phi} := \phi - df$. Let ∇^g be the Levi-Civita connection determined by the metric g. There exists a conformally equivalent metric $\tilde{g}$ locally so that $\nabla = \nabla^{\tilde{g}}$ if and only if $d\phi = 0$; if $d\phi = 0$, such a conformally equivalent metric exists globally if and only if $[\phi] = 0$ in de Rham cohomology; this means that $\phi = df$ for some smooth globally defined function f defined on M.

Weyl geometry fits in between affine and Riemannian geometry. Let (M, g) be a pseudo-Riemannian manifold. Since ∇^g is an affine connection and since $\nabla g = 0$, the triple (M, g, ∇^g) is a Weyl manifold. There are, however, examples with $d\phi \neq 0$ so Weyl geometry is more general than Riemannian geometry or even conformal Riemannian geometry. Every Weyl manifold gives rise to an underlying affine and an underlying Riemannian manifold; Equation (1.7.a) provides the link between these two structures. If (M, g, ∇) is a Weyl manifold, there is an extra curvature symmetry we shall establish in Theorem 6.5.1:

Theorem 1.7.1 *Let (M, g, ∇) be a Weyl manifold. Let R be the curvature of the Weyl connection ∇. Then:*

$$R(x, y, z, w) + R(x, y, w, z) = \tfrac{2}{m}\{\rho(R)(y, x) - \rho(R)(x, y)\}g(z, w).$$

We define the subspace of *Weyl curvature tensors* $\mathfrak{W}$ by imposing the relations of Equation (1.4.a) and the relation of Theorem 1.7.1. This means that:

$$\begin{aligned}\mathfrak{W} :&= \big\{A \in \mathfrak{A} : A(x, y, z, w) + A(x, y, w, z)\\ &= \tfrac{2}{m}[\rho(A)(y, x) - \rho(A)(x, y)]g(z, w)\big\}.\end{aligned}$$

We will establish in Section 6.5 the decomposition of [Higa (1993)] and [Higa (1994)] of $\mathfrak{W}$ as an orthogonal module; note that the decomposition of $\mathfrak{R}$ as an orthogonal module is given by Theorem 1.6.2:

Theorem 1.7.2 *If $m \geq 4$, then there is an orthogonal module direct sum decomposition:*

$$\mathfrak{W} \approx \mathfrak{R} \oplus \Lambda^2.$$

We say that a tensor $A \in \mathfrak{W}$ is *geometrically realizable* by a Weyl manifold $\mathcal{W} = (M, g, \nabla)$ if there exists a point $P \in M$ and an isomorphism $\phi : V \to T_P M$ so that $\phi^* g_P = \langle \cdot, \cdot \rangle$ and $\phi^* R_P = A$. In Section 6.5 we will establish the following result [Gilkey, Nikčević, and Simon (2011)] showing that the relations of Equation (1.4.a) and Theorem 1.7.1 generate the universal symmetries of the curvature tensor in Weyl geometry:

Theorem 1.7.3 *Every $A \in \mathfrak{W}$ is geometrically realizable by a Weyl manifold with constant scalar curvature.*

The following is an interesting illustration of the extent to which the geometric category is determined by the algebraic setting. The following useful result characterizes *trivial Weyl manifolds*:

Theorem 1.7.4 *Let $\mathcal{W} = (M, g, \nabla)$ be a Weyl manifold with $H^1(M;\mathbb{R}) = 0$. The following assertions are equivalent. If any is satisfied, then we say that $\mathcal{W}$ is* trivial.

(1) $d\phi = 0$.
(2) $\nabla = \nabla^{\tilde{g}}$ *for some $\tilde{g}$ in the conformal class defined by g.*
(3) $\nabla = \nabla^{\tilde{g}}$ *for some pseudo-Riemannian metric $\tilde{g}$.*
(4) $R_P(\nabla) \in \mathfrak{R}$ *for every $P \in M$.*
(5) ∇ *is Ricci symmetric.*

There are many references that deal with Weyl geometries; we list only a few as follows for further information in this important area: [Alexandrov and Ivanov (2003)], [Bokan, Gilkey, and Simon (1997)], [Bonneau (1998)], [Calderbank and Pedersen (2000)], [Canfes (2006)], [Dunajski, Mason, and Tod (2001)], [Dunajski and Tod (2002)], [Folland (1970)], [Gilkey and Nikčević (2011)], [Gilkey, Nikčević, and Simon (2011)], [Higa (1993)], [Higa (1994)], [Itoh (2000)], [Jones and Tod (1985)], [Matsuzoe (2001)], [Matzeu (2002)], [Miritzis (2004)], [Narita and Satou (2004)], [Oprea (2005)], [Ozdeger (2006)], [Pedersen and Swann (1991)], [Pedersen, Poon, and Swann (1993)], [Pedersen and Tod (1993)], [Scholz (2009)].

1.8 Almost Pseudo-Hermitian Geometry

We now discuss the decomposition of $\mathfrak{R}$ as a module with structure group $\mathcal{U}_-$ or with structure group $\mathcal{U}_\pm^\star$. This result was given by [Tricerri and Vanhecke (1981)] in the positive definite setting; we will extend the decomposition to the remaining geometries in Section 7.1. Let $(V, \langle\cdot,\cdot\rangle, J_\pm)$ be a para-Hermitian vector space $(+)$ or a pseudo-Hermitian vector space $(-)$. Define:

$$\begin{aligned}\rho_{J_\pm}(x,y) &:= \varepsilon^{il} A(e_i, x, J_\pm y, J_\pm e_l),\\ \tau_{J_\pm} &:= \varepsilon^{il}\varepsilon^{jk} A(e_i, e_j, J_\pm e_k, J_\pm e_l).\end{aligned} \tag{1.8.a}$$

If $T \in \otimes^4 V^*$, define the *Gray symmetrizer* by setting:

$$\begin{aligned}\mathcal{G}_\pm(T)(x,y,z,w) := {}& T(x,y,z,w) + T(J_\pm x, J_\pm y, J_\pm z, J_\pm w)\\ &\pm T(J_\pm x, J_\pm y, z, w) \pm T(J_\pm x, y, J_\pm z, w) \pm T(J_\pm x, y, z, J_\pm w)\\ &\pm T(x, J_\pm y, J_\pm z, w) \pm T(x, J_\pm y, z, J_\pm w) \pm T(x, y, J_\pm z, J_\pm w).\end{aligned} \tag{1.8.b}$$

If σ is a permutation, let $(\sigma^* T)(x_1, x_2, x_3, x_4) = T(x_{\sigma(1)}, x_{\sigma(2)}, x_{\sigma(3)}, x_{\sigma(4)})$. The Gray symmetrizer is invariant under permuting the factors:

$$\sigma^* \mathcal{G}_\pm = \mathcal{G}_\pm \sigma^*. \tag{1.8.c}$$

We also consider the following modules:

Definition 1.8.1

(1) $\mathfrak{R}_+^{\mathcal{U}_\pm} := \{A \in \mathfrak{R} : A(J_\pm x, J_\pm y, J_\pm z, J_\pm w) = A(x,y,z,w)\}$.

(2) $\mathfrak{R}_-^{\mathcal{U}_\pm} := \{A \in \mathfrak{R} : A(J_\pm x, J_\pm y, J_\pm z, J_\pm w) = -A(x,y,z,w)\}$.

(3) $\mathfrak{G}_\pm := \mathfrak{R} \cap \ker(\mathcal{G}_\pm)$.

(4) $\mathfrak{K}_\pm^{\mathfrak{R}} := \{A \in \mathfrak{R} : A(x,y,z,w) = \mp A(J_\pm x, J_\pm y, z, w)\}$.

(5) $W_{\pm,3}^{\mathfrak{R}} := \mathfrak{K}_\pm^{\mathfrak{R}} \cap \ker(\rho)$.

(6) $W_{\pm,6}^{\mathfrak{R}} := \{\mathfrak{K}_\pm^{\mathfrak{R}}\}^\perp \cap \mathfrak{G}_\pm \cap \mathfrak{R}_+^{\mathcal{U}_\pm} \cap \ker(\rho \oplus \rho_{J_\pm})$.

(7) $W_{\pm,7}^{\mathfrak{R}} := \{A \in \mathfrak{R} : A(J_\pm x, y, z, w) = A(x, y, J_\pm z, w)\}$.

(8) $W_{\pm,10}^{\mathfrak{R}} := \mathfrak{R}_-^{\mathcal{U}_\pm} \cap \ker(\rho \oplus \rho_{J_\pm})$.

In the notation of Definition 1.5.1, $W_{\pm,9}^{\mathfrak{A}} = W_{\pm,3}^{\mathfrak{R}}$.

Theorem 1.8.1 *Let* $\dim(V) \geq 8$. *Let* $(V, \langle\cdot,\cdot\rangle, J_\pm)$ *be a para-Hermitian vector space* $(+)$ *or a pseudo-Hermitian vector space* $(-)$. *We have an orthogonal direct sum decomposition* $\mathfrak{R} = W^{\mathfrak{R}}_{\pm,1} \oplus \cdots \oplus W^{\mathfrak{R}}_{\pm,10}$ *into irreducible modules with structures group* $\mathcal{U}_-$ *or with structure group* $\mathcal{U}^\star_\pm$. *One has:*

$$W^{\mathfrak{R}}_{\pm,1} \approx W^{\mathfrak{R}}_{\pm,4} \approx \mathbb{R},\ W^{\mathfrak{R}}_{\pm,2} \approx W^{\mathfrak{R}}_{\pm,5} \approx S^{2,\mathcal{U}_\pm}_{0,\mp},\ W^{\mathfrak{R}}_{\pm,8} \approx S^{2,\mathcal{U}_\pm}_{\pm},\ W^{\mathfrak{R}}_{\pm,9} \approx \Lambda^{2,\mathcal{U}_\pm}_{\pm}.$$

Except for the isomorphisms $W^{\mathfrak{R}}_{\pm,1} \approx W^{\mathfrak{R}}_{\pm,4}$ *and* $W^{\mathfrak{R}}_{\pm,2} \approx W^{\mathfrak{R}}_{\pm,5}$, *these are inequivalent modules for the structure groups* $\mathcal{U}_-$ *and* $\mathcal{U}^\star_\pm$.

Remark 1.8.1 If $m = 4$, we set $W^{\mathfrak{R}}_{\pm,5} = W^{\mathfrak{R}}_{\pm,6} = W^{\mathfrak{R}}_{\pm,10} = \{0\}$ and if $m = 6$, we set $W^{\mathfrak{R}}_{\pm,6} = \{0\}$ to derive the corresponding decomposition. The two trivial factors $W^{\mathfrak{R}}_{\pm,1} \oplus W^{\mathfrak{R}}_{\pm,4} \approx 2\cdot\mathbb{R}$ are detected by $\tau \oplus \tau_{J_\pm}$. The module $W^{\mathfrak{R}}_{\pm,2} \oplus W^{\mathfrak{R}}_{\pm,8} \oplus W^{\mathfrak{R}}_{\pm,9}$ if $m = 4$, and the module $W^{\mathfrak{R}}_{\pm,2} \oplus W^{\mathfrak{R}}_{\pm,5} \oplus W^{\mathfrak{R}}_{\pm,8} \oplus W^{\mathfrak{R}}_{\pm,9}$ if $m \geq 6$ is detected by $\rho \oplus \rho_{J_\pm}$. We will identify the submodules $W^{\mathfrak{R}}_{\pm,i}$ for $i = 1, 2, 4, 5, 8, 9$ which arise from the Ricci tensors very explicitly in Chapter 7 and postpone until that time a more detailed description. We shall show in Lemma 2.5.3 that $S^{2,\mathcal{U}_+}_{+}$ is not an irreducible module for the group $\mathcal{U}_+$. Thus Theorem 1.8.1 fails if we replace $\mathcal{U}^\star_+$ by $\mathcal{U}_+$.

Let $\nu_{\pm,i} := \dim\{W^{\mathfrak{R}}_{\pm,i}\}$. The discussion of Section 7.1 together with analytic continuation shows that $\nu_{+,i} = \nu_{-,i}$ is independent of the signature of the inner product; these constants only depend on the dimension of V. In the positive definite setting, these dimensions were determined previously (see [Tricerri and Vanhecke (1981)]); we shall simply cite their results for the sake of completeness and shall omit the proof as it plays no role in our development.

Theorem 1.8.2 *Let* $m = 2\bar{m}$.

	$m=4$	$m=6$	$m\geq 8$		$m=4$	$m=6$	$m\geq 8$
ν_1	1	1	1	ν_2	3	8	$\bar{m}^2-1$
ν_3	5	27	$\frac{1}{4}\bar{m}^2(\bar{m}-1)(\bar{m}+3)$	ν_4	1	1	1
ν_6	0	0	$\frac{1}{4}\bar{m}^2(\bar{m}+1)(\bar{m}-3)$	ν_5	0	8	$\bar{m}^2-1$
ν_7	2	12	$\frac{1}{6}\bar{m}^2(\bar{m}^2-1)$	ν_8	6	12	$\bar{m}^2+\bar{m}$
ν_{10}	0	30	$\frac{1}{3}2\bar{m}^2(\bar{m}^2-4)$	ν_9	2	6	$\bar{m}^2-\bar{m}$

One says that $(V, \langle\cdot,\cdot\rangle, J_\pm, A)$ is an *almost para-Hermitian curvature model* $(+)$ or an *almost pseudo-Hermitian curvature model* $(-)$ if $A \in \mathfrak{R}$, and if $(V, \langle\cdot,\cdot\rangle, J_\pm)$ is a para-Hermitian vector space (+) or a pseudo-Hermitian vector space (−). The notion of geometric realizability in these contexts is defined similarly; we say that $(M, g, J_\pm)$ is an almost para-Hermitian manifold (+) or an almost pseudo-Hermitian manifold (−) is a geometric realization of $(V, \langle\cdot,\cdot\rangle, J_\pm, A)$ at a point P of M if there is an isomorphism ϕ from V to T_PM so that:

$$\Phi^*\langle\cdot,\cdot\rangle = g_P, \quad \Phi^*J_\pm = J_\pm(P), \quad \text{and} \quad \Phi^*A = R_P.$$

We focus our attention on the scalar curvature. In Section 7.4, we present results of [Brozos-Vázquez et al. (2009)] to establish:

Theorem 1.8.3 *Let $m \geq 4$. Any almost para-Hermitian curvature model (+) and any almost pseudo-Hermitian curvature model (−) is geometrically realizable by an almost para-Hermitian manifold (+) or an almost pseudo-Hermitian manifold (−) with τ and $\tau_{J_\pm}$ constant.*

The study of Hermitian, almost Hermitian, para-Hermitian, and almost para-Hermitian geometry is a central area. We cite the following references; these are only a few amongst the many possibilities [Apostolov, Ganchev, and Ivanov (1997)], [Balas and Gauduchon (1985)], [Bejan (1989)], [Brozos-Vázquez, García-Río, and Gilkey (2008)], [Brozos-Vázquez et al. (2011)], [Brozos-Vázquez et al. (2010)], [Brozos-Vázquez et al. (2009a)], [del Río and Simanca (2003)], [Díaz-Ramos, García-Río, and Vázquez-Lorenzo (2006)], [Falcitelli and Farinola (1994)], [Falcitelli, Farinola, and Salmon (1994)], [Gadea and Masque (1991)], [Gadea and Oubiña (1992)], [Gray (1965)], [Gray (1969)], [Gray (1969a)], [Gray (1976)], [Gray and Hervella (1980)], [Kim (2007)], [Kirchberg (2004)], [Martín-Cabrera (2005)], [Martín-Cabrera and Swann (2004)], [Martín-Cabrera and Swann (2006)], [Matzeu and Nikčević (1991)], [Nikčević (1992)], [Sato (1989)], [Sato (2003)], [Sato (2004)], [Tang (2006)], [Tricerri and Vanhecke (1981)], [Vanhecke (1977)], and [Vezzoni (2007)]. In addition, we refer to [Butruille (2007)] and [del Río and Simanca (2003)].

1.9 The Gray Identity

The results of Section 1.9 will be proved in Chapter 3 and in Chapter 7. The curvature tensor of a para-Hermitian manifold or of a pseudo-Hermitian manifold has an additional symmetry given below in Theorem 1.9.1; it is quite striking that a geometric integrability condition imposes an additional algebraic symmetry on the curvature tensor.

In Section 3.5, we will follow the discussion in [Brozos-Vázquez et al. (2009a)] and [Brozos-Vázquez et al. (2010)]. We first extend a result of [Gray (1976)] in the positive definite case to more general signatures and to the para-Hermitian setting. Let $(V, \langle\cdot,\cdot\rangle, J_\pm)$ be a para-Hermitian vector space (+) or a pseudo-Hermitian vector space (−). Let $\mathcal{G}_\pm$ be the Gray symmetrizer defined in Equation (1.8.b).

Theorem 1.9.1 *If the complex curvature model $\mathfrak{C} := (V, \langle\cdot,\cdot\rangle, J_\pm, A)$ is geometrically realizable by a para-Hermitian manifold (+) or by a pseudo-Hermitian manifold (−), then*

$$\mathcal{G}_\pm(A) = 0.$$

We say that a curvature model $(V, \langle\cdot,\cdot\rangle, J_\pm, A)$ is a *para-Hermitian curvature model* (+) or a pseudo-Hermitian curvature model (−) if $\mathcal{G}_\pm(A) = 0$. We will establish the following result in Section 7.3.

Theorem 1.9.2 *Any para-Hermitian curvature model (+) or pseudo-Hermitian curvature model (−) is geometrically realizable by a para-Hermitian manifold (+) or pseudo-Hermitian manifold (−) with τ and $\tau_{J_\pm}$ constant.*

The universal symmetries of the curvature tensor of a para-Hermitian manifold (+) or of a pseudo-Hermitian manifold (−) are generated by the identity $\mathcal{G}_\pm(R) = 0$ and by the usual curvature symmetries given in Equation (1.6.a). This result emphasizes the difference between the class of almost para-Hermitian manifolds or almost pseudo-Hermitian manifolds and the class of para-Hermitian manifolds or pseudo-Hermitian manifolds.

The para-Hermitian geometric realization or the pseudo-Hermitian geometric realization in Theorem 1.9.2 can be chosen so that $d\Omega_\pm(P) = 0$. Thus imposing the (para)-Kähler identity $d\Omega_\pm(P) = 0$ at a single point imposes no additional curvature restrictions. If $d\Omega_\pm = 0$ globally, then the manifold is said to be *almost (para)-Kähler*. This is a very rigid structure, see for example the discussion in [Tang (2006)], and there are additional curvature restrictions. Thus Theorem 1.9.2 also emphasizes the difference between $d\Omega_\pm$ vanishing at a single point and $d\Omega_\pm$ vanishing globally.

We shall establish the following algebraic characterization in Section 7.3:

Theorem 1.9.3 $\mathfrak{G}_\pm \perp W^{\mathfrak{R}}_{\pm,7}$ *and* $\mathfrak{R} = \mathfrak{G}_\pm \oplus W^{\mathfrak{R}}_{\pm,7}$.

1.10 Kähler Geometry in the Riemannian Setting I

In Section 7.5, we will report on results of [Brozos-Vázquez, Gilkey, and Merino (2010)]. We begin with a classic result that we shall establish in Section 3.6; we also refer to Section 7.7 for related results.

Theorem 1.10.1 *Let $(M, g, J_\pm)$ be an almost para-Hermitian manifold $(+)$ or an almost pseudo-Hermitian manifold $(-)$.*

(1) The following assertions are equivalent. If either is satisfied, then $(M, g, J_\pm)$ is said to be a (para)-Kähler manifold.

(a) $\nabla J_\pm = 0$.
(b) $J_\pm$ *is an integrable complex structure and* $d\Omega_\pm = 0$.
(c) $\nabla\Omega_\pm = 0$.

(2) If $\nabla\Omega_\pm = 0$, *then*

(a) $d\Omega_\pm = 0$ *and* $\delta\Omega_\pm = 0$.
(b) $J_\pm\mathcal{R}(x, y) = \mathcal{R}(x, y)J_\pm$.
(c) $R(J_\pm x, J_\pm y, z, w) = \mp R(x, y, z, w)$.

We say that $(V, \langle\cdot,\cdot\rangle, J_\pm, A)$ is a *(para)-Kähler curvature model* if $(V, \langle\cdot,\cdot\rangle, J_\pm)$ is a para-Hermitian vector space $(+)$ or a pseudo-Hermitian vector space $(-)$ and if $A \in \mathfrak{K}_\pm^{\mathfrak{R}}$. The Gray identity is then necessarily satisfied. Note that $\tau = \tau_{J_\pm}$ in the (para)-Kähler setting. We will establish the following associated geometrical realization result in Section 7.5:

Theorem 1.10.2 *Any (para)-Kähler curvature model is geometrically realizable by a (para)-Kähler manifold of constant scalar curvature.*

Theorem 1.10.1 and Theorem 1.10.2 provide necessary and sufficient conditions for a curvature model to be geometrically realizable by a (para)-Kähler manifold. We will prove the following result in Section 7.1:

Theorem 1.10.3 *If* $m \geq 4$, *then* $\mathfrak{K}_\pm^{\mathfrak{R}} = W_{\pm,1}^{\mathfrak{R}} \oplus W_{\pm,2}^{\mathfrak{R}} \oplus W_{\pm,3}^{\mathfrak{R}}$.

The curvature tensors of Hermitian or Kähler manifolds satisfy linear identities. On the other hand, there are examples where one has relations rather than identities. For example, if the metric in question is positive definite, then the invariants τ and τ_{J_-} of an almost Kähler manifold $(d\Omega_- = 0)$ satisfy $\tau_{J_-} - \tau = \frac{1}{2}|\nabla J_-|^2$ and thus the curvature lies in the half-space defined by the relation $\tau_{J_-} \geq \tau$. We refer to [Davidov et al. (2007)] for further details concerning almost Kähler manifolds in both the Riemannian and the higher signature settings. The geometry of such manifolds is discussed further in Section 7.7.

Kähler, almost Kähler, and nearly Kähler geometry (and analogues in the para complex setting) together with affine analogues play a central role in differential geometry. Here are just a few references for further reading in this important area: [Abbena (1984)], [Apostolov, Armstrong, and Drăghici (2002)], [Balas and Gauduchon (1985)], [Blair (1990)], [Brozos-Vázquez et al. (2011)], [Brozos-Vázquez, Gilkey, and Merino (2010)], [Brozos-Vázquez, Gilkey, and Nikčević (2011)], [Brozos-Vázquez, Gilkey, and Nikčević (2011b)], [Cordero, Fernández, and de León (1985)], [Cortés, Lawn, and Schaefer (2006)], [Davidov et al. (2007)], [Deprez, Sekigawa, and Verstraelen (1988)], [Fino (2005)], [Ganchev and Mihova (2008)], [Ganchev and Mihova (2008a)], [Gilkey (1973)], [Gilkey and Nikčević (2011)], [Gilkey and Nikčević (2011a)], [Hitchin et al. (1987)], [Kamada (1999)], [Kirchberg (2004)], [Koto (1960)], [Moroianu and Ornea (2008)], [Nagy (2002)], [Oguro and Sekigawa (2004)], [Sato (2003)], [Sekigawa (1987)], [Sekigawa and Vanhecke (1990)], and [Watson (1983)].

1.11 Curvature Kähler–Weyl Geometry

The results described here (see [Gilkey and Nikčević (2011)]) will be established in Section 7.6. We say that $(M, g, \nabla, J_\pm)$ is an *almost (para)-complex Weyl manifold* if (M, g, ∇) is a Weyl manifold and if $J_\pm$ is an almost para-Hermitian or an almost pseudo-Hermitian structure on (M, g). If one has that $\nabla(J_\pm) = 0$, the structure is said to be a *Kähler–Weyl structure* in either the para complex $(+)$ or in the complex $(-)$ settings. Necessarily $J_\pm$ is an integrable (para)-complex structure by Lemma 3.4.1. [Pedersen, Poon, and Swann (1993)] used results contained in [Vaisman (1982)] and [Vaisman (1983)] to establish the following generalization of Theorem 1.7.4 in the Riemannian setting; the extension to the higher signature setting and to the para-Kähler setting is immediate.

Theorem 1.11.1 *Let $H^1(M; \mathbb{R}) = 0$ and let $m \geq 6$. Let $(g, \nabla, J_\pm)$ give M a Kähler–Weyl structure. This means that $J_\pm^* g = \mp g$, $\nabla J_\pm = 0$, and $\nabla g = -2\phi \otimes g$. Then the underlying Weyl structure is trivial; this means that $d\phi = 0$.*

We introduce some additional notation. Set:

$$\mathfrak{K}_\pm^{\mathfrak{W}} := \mathfrak{K}_\pm^{\mathfrak{A}} \cap \mathfrak{W} = \{\mathcal{A} \in \mathfrak{W} : \mathcal{A}(x, y) J_\pm = J_\pm \mathcal{A}(x, y)\}.$$

If $(M, g, \nabla, J_\pm)$ is a Kähler–Weyl manifold, then ∇ is an affine Kähler connection and one has an additional curvature symmetry since $R \in \mathfrak{K}_\pm^{\mathfrak{W}}$. Conversely, we shall say that $(M, g, \nabla, J_\pm)$ is a *(para)-Kähler curvature Weyl manifold* if (M, g, ∇) is a Weyl manifold, if $(M, g, J_\pm)$ is an almost para-Hermitian manifold (+) or an almost pseudo-Hermitian manifold (−), and if $\mathcal{R}(P) \in \mathfrak{K}_\pm^{\mathfrak{W}}(T_pM, g, J_\pm)$ for all $P \in M$. We will see presently in Remark 1.12.1 that there exist curvature Kähler–Weyl manifolds which are not Kähler–Weyl manifolds.

The following result gives a curvature condition in the (para)-complex setting. It ensures that the Weyl structure is trivial and extends Theorem 1.11.1 to this context; it can be regarded as a generalization of Theorem 1.7.4 to the (para)-complex category:

Theorem 1.11.2 *Let $H^1(M;\mathbb{R}) = 0$ and let $m \geq 6$. Any curvature Kähler–Weyl structure on M is trivial.*

Theorem 1.7.2 decomposes $\mathfrak{W} \approx \mathfrak{R} \oplus \Lambda^2$ as an orthogonal module. The decomposition of $\mathfrak{R}$ as a module with structure group $\mathcal{U}_-$ or with structure group $\mathcal{U}_\pm^\star$ is given in Lemma 2.5.2; there is an orthogonal direct sum decomposition into inequivalent irreducible modules

$$\Lambda^2 \approx \left\{ \begin{array}{lll} \mathbb{R}\cdot\Omega_- \oplus \Lambda_{0,+}^{2,\mathcal{U}_-} \oplus \Lambda_-^{2,\mathcal{U}_-} & \text{as modules for } \mathcal{U}_- & \text{and } \mathcal{U}_-^\star \\ \mathbb{R}\cdot\Omega_+ \oplus \Lambda_{0,-}^{2,\mathcal{U}_+} \oplus \Lambda_+^{2,\mathcal{U}_+} & \text{as modules for } \mathcal{U}_+^\star & \end{array} \right\}.$$

The following decompositions are then an immediate consequence of these results and of Theorem 1.8.1:

Theorem 1.11.3 *Let $(V, \langle\cdot,\cdot\rangle, J_\pm)$ be a para-Hermitian vector space (+) of dimension $m \geq 8$ or a pseudo-Hermitian vector space (−) of dimension $m \geq 8$. We have the following isomorphism decomposing $\mathfrak{W}$ as the direct sum of irreducible modules with structure groups $\mathcal{U}_-$ and $\mathcal{U}_\pm^\star$:*

$$\begin{array}{l} \mathfrak{W} = W_{\pm,1}^{\mathfrak{R}} \oplus \cdots \oplus W_{\pm,10}^{\mathfrak{R}} \oplus W_{\pm,11}^{\mathfrak{W}} \oplus W_{\pm,12}^{\mathfrak{W}} \oplus W_{\pm,13}^{\mathfrak{W}}, \quad \textit{where} \\ W_{\pm,11}^{\mathfrak{W}} \approx \chi, \quad W_{\pm,12}^{\mathfrak{W}} \approx \Lambda_{0,\mp}^{2,\mathcal{U}_\pm}, \quad W_{\pm,13}^{\mathfrak{W}} \approx \Lambda_\pm^{2,\mathcal{U}_\pm}. \end{array}$$

Except for the isomorphisms $W_{\pm,1}^{\mathfrak{R}} \approx W_{\pm,4}^{\mathfrak{R}}$, $W_{\pm,2}^{\mathfrak{R}} \approx W_{\pm,5}^{\mathfrak{R}}$, $W_{\pm,9}^{\mathfrak{R}} \approx W_{\pm,13}^{\mathfrak{W}}$, these are inequivalent modules for the structure group $\mathcal{U}_\pm^\star$. As modules with structure group $\mathcal{U}_-$, we also have the isomorphism

$$W_{-,1}^{\mathfrak{R}} \approx W_{-,4}^{\mathfrak{R}} \approx W_{-,11}^{\mathfrak{W}}, \quad W_{-,2}^{\mathfrak{R}} \approx W_{-,5}^{\mathfrak{R}} \approx W_{-,12}^{\mathfrak{W}}.$$

Remark 1.11.1 If $m = 4$, then we set $W^{\mathfrak{R}}_{\pm,5} = W^{\mathfrak{R}}_{\pm,6} = W^{\mathfrak{R}}_{\pm,10} = \{0\}$ and if $m = 6$, then we set $W^{\mathfrak{R}}_{\pm,6} = \{0\}$ to derive the corresponding decomposition of $\mathfrak{W}$.

Let $\psi \in \Lambda^2$ and let $S \in \Lambda^{2,\mathcal{U}_\pm}_\pm$. We define

$$\{\sigma^{\mathfrak{W}}\psi\}_{ijkl} := 2\psi_{ij}\varepsilon_{kl} + \psi_{ik}\varepsilon_{jl} - \psi_{jk}\varepsilon_{il} - \psi_{il}\varepsilon_{jk} + \psi_{jl}\varepsilon_{ik},$$

and we define

$$\begin{aligned}\Psi_\pm(S)(x,y,z,w) : = \; & 2\langle x, J_\pm y\rangle S(z, J_\pm w) + 2\langle z, J_\pm w\rangle S(x, J_\pm y) \\ & + \langle x, J_\pm z\rangle S(y, J_\pm w) + \langle y, J_\pm w\rangle S(x, J_\pm z) \\ & - \langle x, J_\pm w\rangle S(y, J_\pm z) - \langle y, J_\pm z\rangle S(x, J_\pm w).\end{aligned}$$

We shall introduce notation subsequently that expresses

$$\sigma^{\mathfrak{W}} = (\sigma_4 - \sigma_5).$$

We will show in Theorem 6.5.3 that $\sigma^{\mathfrak{W}}$ provides the imbedding of Λ^2 in $\mathfrak{W}$ given in Theorem 1.7.2. We will also show in Lemma 7.2.1 that $\Psi_\pm$ is an isomorphism from $\Lambda^{2,\mathcal{U}_\pm}_\pm$ to $W^{\mathfrak{R}}_{\pm,9}$. Consequently

$$\begin{aligned} W^{\mathfrak{R}}_{\pm,9} &= \Psi_\pm(\Lambda^{2,\mathcal{U}_\pm}_\pm), & W^{\mathfrak{W}}_{\pm,11} &= \sigma^{\mathfrak{W}}(\Omega)\cdot\mathbb{R}, \\ W^{\mathfrak{W}}_{\pm,12} &= \sigma^{\mathfrak{W}}(\Lambda^{2,\mathcal{U}_\pm}_{0,\mp}), & W^{\mathfrak{W}}_{\pm,13} &= \sigma^{\mathfrak{W}}(\Lambda^{2,\mathcal{U}_\pm}_\pm). \end{aligned}$$

Theorem 1.11.2 will follow from Theorem 1.7.4 and from the following purely algebraic result:

Theorem 1.11.4 *If $(V, \langle\cdot,\cdot\rangle, J_\pm)$ is a para-Hermitian vector space $(+)$ of dimension $m \geq 6$ or is a pseudo-Hermitian vector space $(-)$ of dimension $m \geq 6$, then $\mathfrak{K}^{\mathfrak{W}}_\pm = \mathfrak{K}^{\mathfrak{R}}_\pm$.*

The situation is very different if $m = 4$. Theorem 1.11.1 is shown to fail in this setting by [Calderbank and Pedersen (2000)]. But much more is true. We summarize below the situation in dimension $m = 4$ and refer to [Gilkey and Nikčević (2011a)] for further details as we shall not establish these results in this book.

Theorem 1.11.5

(1) Let $(V, \langle\cdot,\cdot\rangle, J_\pm)$ be a para-Hermitian vector space $(+)$ of dimension $m = 4$ or a pseudo-Hermitian vector space $(-)$ of dimension $m = 4$.

(a) $\mathfrak{K}^{\mathfrak{W}}_\pm = \mathfrak{K}^{\mathfrak{R}}_\pm \oplus (\sigma^{\mathfrak{W}} \mp 3\Psi_\pm)(\Lambda^{2,\mathcal{U}_\pm}_\pm) \oplus W^{\mathfrak{W}}_{\pm,12}$.

(b) *Every $A \in \mathfrak{K}_{\pm}^{\mathfrak{W}}$ is geometrically realizable by a Kähler–Weyl manifold (+).*

(2) *Every para-Hermitian manifold and every pseudo-Hermitian manifold of dimension 4 admits a unique Kähler–Weyl structure where $\phi = \frac{1}{2} J_{\pm}^{*} \delta \Omega_{\pm}$.*

1.12 The Covariant Derivative of the Kähler Form I

In Section 1.12, we report results contained in [Brozos-Vázquez et al. (2011)] which generalize earlier results contained in [Gray and Hervella (1980)] and [Gadea and Masque (1991)]. Let $\nabla\Omega_{\pm}$ be the covariant derivative of the Kähler form of $(M, g, J_{\pm})$. We will show in Lemma 7.7.2 that the following symmetries are satisfied:

$$\nabla\Omega_{\pm}(x, y; z) = -\nabla\Omega_{\pm}(y, x; z) = \pm\nabla\Omega_{\pm}(J_{\pm}x, J_{\pm}y; z). \qquad (1.12.a)$$

We therefore define:

$$\mathfrak{H}_{\pm} := \Lambda_{\pm}^{2,\mathcal{U}_{\pm}} \otimes V^{*}.$$

In Section 7.7, we will establish the following geometric realization result. It shows that Equation (1.12.a) generates the universal symmetries satisfied by $\nabla\Omega_{\pm}$ and provides a rich family of examples. It is striking that we can fix the metric and only vary the almost (para)-complex structure; in particular, we could take the background structures to be flat.

Theorem 1.12.1 *Let $(M, g, J_{\pm})$ be a background almost para-Hermitian manifold (+) or a background pseudo-Hermitian manifold (−) and let P be a point of M. Let $H_{\pm} \in \mathfrak{H}_{\pm}(T_P M, g_P, J_{\pm,P})$. Then there exists a new almost para-Hermitian structure (+) or a new pseudo-Hermitian structure (−) $\tilde{J}_{\pm}$ on M with $J_{\pm}(P) = \tilde{J}_{\pm}(P)$ so that*

$$\nabla\Omega_{\pm}(M, g, \tilde{J}_{\pm})(P) = H_{\pm}.$$

Remark 1.12.1 We may take $(M, g, J_{\pm})$ to be the flat (para)-Kähler torus. We take $H_{\pm} \neq 0$ and construct $(M, g, \tilde{J}_{\pm})$. Then $(M, g, \tilde{J}_{\pm}, \nabla^{g})$ is a (para)-Kähler curvature Weyl manifold. Furthermore, it is not a (para)-Kähler manifold. By taking suitable product structures, we can construct examples where the underlying metric is not flat.

We consider the following subspace:

$$U_{\pm,3} := \{H_\pm \in \mathfrak{H}_\pm : H_\pm(x,y;z) = \mp H_\pm(x, J_\pm y; J_\pm z)\}.$$

If $(M, g, J_\pm)$ is a para-Hermitian manifold $(+)$ or is a pseudo-Hermitian manifold $(-)$ (in other words if $J_\pm$ is an integrable (para)-complex structure), then $\nabla\Omega_\pm \in U_{\pm,3}$ as we shall see presently in Lemma 7.7.4. Conversely:

Theorem 1.12.2 *Let $(M, g, J_\pm)$ be a background para-Hermitian manifold $(+)$ or a background pseudo-Hermitian manifold $(-)$ and let P be a point of M. Let $H_\pm$ in $U_{\pm,3}(T_PM, g_P, J_{\pm,P})$. Then there exists a new para-Hermitian metric $(+)$ or a new pseudo-Hermitian metric $(-)$ $\tilde{g}$ on M with $\tilde{g}(P) = g(P)$ so that $\nabla\Omega_\pm(M, \tilde{g}, J_\pm)(P) = H_\pm$.*

Theorem 1.12.1 and Theorem 1.12.2 are global results. It is necessary to have an initial background structure as not every manifold admits a para-Hermitian structure or a pseudo-Hermitian structure of a given signature.

These results are based on a decomposition of $\mathfrak{H}_\pm$ extending the decomposition given in [Gray and Hervella (1980)] in the positive definite context; the corresponding decomposition of $\mathfrak{H}_+$ as a module for the group $\mathcal{U}_+$ is given in [Gadea and Masque (1991)]. We introduce the following notation:

Definition 1.12.1 Let $(V, \langle\cdot,\cdot\rangle, J_\pm)$ be a para-Hermitian vector space $(+)$ or a pseudo-Hermitian vector space $(-)$.

(1) If $H \in \otimes^3 V^*$, define $\tau_1(H) \in V^*$ by contracting the final two indices: $(\tau_1 H)(x) := \varepsilon^{ij} H(x, e_i; e_j)$.

(2) If $\kappa \in \mathrm{GL}$ and if $\phi \in V^*$, define $\sigma_\kappa(\phi) \in \otimes^3 V^*$ by setting: $\sigma_\kappa(\phi)(x,y;z) := \phi(\kappa x)\langle y,z\rangle - \phi(\kappa y)\langle x,z\rangle + \phi(x)\langle \kappa y,z\rangle - \phi(y)\langle \kappa x,z\rangle$.

(3) $W_{\pm,1}^{\mathfrak{H}} := \{H \in \mathfrak{H}_\pm : H(x,y;z) + H(x,z;y) = 0\}$.

(4) $W_{\pm,2}^{\mathfrak{H}} := \{H \in \mathfrak{H}_\pm : H(x,y;z) + H(y,z;x) + H(z,x;y) = 0\}$.

(5) $W_{\pm,3}^{\mathfrak{H}} := U_{\pm,3} \cap \ker(\tau_1)$.

(6) $W_{\pm,4}^{\mathfrak{H}} := \mathrm{Range}(\sigma_{J_\pm})$.

Theorem 1.12.3 *Let $(V, \langle\cdot,\cdot\rangle, J_\pm)$ be a para-Hermitian vector space $(+)$ of dimension $m \geq 6$ or a pseudo-Hermitian vector space $(-)$ of dimension $m \geq 6$. We have the following orthogonal direct sum decompositions of $\mathfrak{H}_\pm$ into irreducible inequivalent modules with structure group $\mathcal{U}_\pm^\star$:*

$$\mathfrak{H}_\pm = W_{\pm,1}^{\mathfrak{H}} \oplus W_{\pm,2}^{\mathfrak{H}} \oplus W_{\pm,3}^{\mathfrak{H}} \oplus W_{\pm,4}^{\mathfrak{H}}.$$

The decomposition of Theorem 1.12.3 is also a $\mathcal{U}_-$ decomposition and was derived by [Gray and Hervella (1980)] in the positive definite setting. The decomposition of [Gadea and Masque (1991)] as a module with structure group $\mathcal{U}_+$ is a finer decomposition as the modules $W_{i,+}^{\mathfrak{H}}$ need not be irreducible modules over the group $\mathcal{U}_+$. If $\dim(V) = 4$, then we set $W_{\pm,1}^{\mathfrak{H}} = W_{\pm,3}^{\mathfrak{H}} = \{0\}$ to obtain the corresponding decompositions. The dimensions $\nu_{\pm,i}^{\mathfrak{H}} := \dim\{W_{\pm,i}^{\mathfrak{H}}\}$ of these modules are independent of the signature and are the same in both the complex and in the para-complex settings. They were computed in the positive definite setting by [Gray and Hervella (1980)]:

Theorem 1.12.4 *Let $m = 2\bar{m}$.*

$\nu_{\pm,1}^{\mathfrak{H}} = \frac{1}{3}\bar{m}(\bar{m}-1)(\bar{m}-2)$	$\nu_{\pm,2}^{\mathfrak{H}} = \frac{2}{3}\bar{m}(\bar{m}-1)(\bar{m}+1)$
$\nu_{\pm,3}^{\mathfrak{H}} = \bar{m}(\bar{m}+1)(\bar{m}-2)$	$\nu_{\pm,4}^{\mathfrak{H}} = m$
$\dim\{\mathfrak{H}_\pm\} = 2\bar{m}^2(\bar{m}-1)$	

In Theorem 1.12.1, we show that every element of $\mathfrak{H}_\pm$ can be obtained by perturbing the almost (para)-complex structure on a given almost para-Hermitian manifold (+) or on a given almost pseudo-Hermitian manifold (−). In Theorem 1.12.2, we show that every element of $U_{\pm,3}$ can be obtained by perturbing the metric of a para-Hermitian manifold or of a pseudo-Hermitian manifold. Thus the focus is on geometrically realizing every element in a suitable context. One can, however, focus instead on the classes defined by the representation theory. We restrict our attention at this point to the complex setting. Let ξ be one of the sixteen submodules of $\mathfrak{H}_-$ for the group $\mathcal{U}_-^\star$ if $m \geq 6$ or one of the four submodules of $\mathfrak{H}_-$ for the group $\mathcal{U}_-^\star$ if $m = 4$. We say that (M, g, J_-) is a *manifold corresponding to the representation* ξ if $\nabla\Omega_-$ belongs to ξ for every point of the manifold and if ξ is minimal with this property. This gives rise to the celebrated sixteen classes of almost Hermitian manifolds (in the positive definite setting) [Gray and Hervella (1980)]. Many of these classes have extensively investigated geometrical meanings. For example:

(1) $\xi = \{0\}$ is the class of Kähler manifolds.
(2) $\xi = W_{1,-}$ is the class of nearly Kähler manifolds.
(3) $\xi = W_{2,-}$ is the class of almost Kähler manifolds.
(4) $\xi = W_{3,-}$ is the class of Hermitian semi-Kähler manifolds.
(5) $\xi = W_{1,-} \oplus W_{2,-}$ is the class of quasi-Kähler manifolds.
(6) $\xi = W_{3,-} \oplus W_{4,-} = U_{3,-}$ is the class of pseudo-Hermitian manifolds.

(7) $\xi = W_{1,-} \oplus W_{2,-} \oplus W_{3,-}$ is the class of semi-Kähler manifolds.
(8) $\xi = \mathfrak{H}_-$ is the class of almost pseudo-Hermitian manifolds.

In the positive definite setting all these classes are non-trivial [Gray and Hervella (1980)]:

Theorem 1.12.5 *Let $p = 0$, let q be even, and let ξ be a submodule of $\mathfrak{H}_-$. Then there exists an almost Hermitian manifold corresponding to the representation ξ of signature $(0, q)$.*

We shall generalize this result to the indefinite setting in Section 7.7; we shall suppose $m \geq 10$ to simplify the discussion as we wish to apply Theorem 1.12.5 rather than construct radically new examples:

Theorem 1.12.6 *Let $(2\bar{p}, 2\bar{q})$ with $2\bar{p} + 2\bar{q} \geq 10$ be given. Let ξ be a submodule of $\mathfrak{H}_-$. Then there exists a manifold corresponding to the representation ξ of signature $(2\bar{p}, 2\bar{q})$.*

1.13 Hyper-Hermitian Geometry

In Section 1.13, we summarize results contained in [Brozos-Vázquez et al. (2009)] and [De Smedt (1994)] and work in progress; we shall omit the proofs. We refer to [Apostolov, Ganchev, and Ivanov (1997)], [Blair (1990)], [Kim (2007)], [Sato (2004)], and [Vezzoni (2007)] for additional material on hyper-Hermitian geometry.

Let V be a real vector space which has dimension $r < \infty$. We say that $\mathcal{J} = \{J_1, J_2, J_3\}$ is a *hyper-complex structure* on V or, equivalently, a *quaternion structure* if the maps J_i are linear maps of V satisfying the *quaternion identities*:

$$J_i J_j + J_j J_i = -2\delta_{ij} \quad \text{and} \quad J_1 J_2 = J_3.$$

Similarly an *almost hyper-complex structure* on a smooth manifold M is a triple $\mathcal{J} = \{J_1, J_2, J_3\}$ of almost complex structures satisfying the quaternion identities given above.

The canonical example is $\mathbb{H}^{\bar{m}}$ for $m = 4\bar{m}$ where $\mathbb{H}$ are the quaternions and where $J_1^{\mathbb{H}} = i$, $J_2^{\mathbb{H}} = j$, $J_3^{\mathbb{H}} = k$ are given by quaternion multiplication; if $P \in M$ and if $\phi : (M, P) \to (\mathbb{H}, 0)$ is the germ of a coordinate system on M, we say the coordinate system is *hyper-holomorphic* if $\phi^* J_\mu^{\mathbb{H}} = J_\mu$ for $\mu = 1, 2, 3$. Let N_{J_i} be the Nijenhuis tensors corresponding to the complex structures J_i. One can establish the following extension of Theorem 3.4.2 [Newlander and Nirenberg (1957)] in this setting:

Theorem 1.13.1 *Let $(M, \mathcal{J})$ be an almost hyper-complex manifold. The following assertions are equivalent. If any is satisfied, then $(M, \mathcal{J})$ is said to be a* hyper-complex manifold *and $\mathcal{J}$ is said to be an* integrable hyper-complex structure*:*

(1) $N_{J_i} = 0$ for $i = 1, 2, 3$.
(2) M is covered by hyper-holomorphic coordinate charts.

Let $\langle \cdot, \cdot \rangle$ be a positive definite inner product on V. A hyper-complex structure $\mathcal{J}$ on V is said to be *hyper-Hermitian* if each complex structure J_i is Hermitian. Let $\mathcal{U}(V, \langle \cdot, \cdot \rangle, J_i)$ be corresponding unitary groups. The *symplectic group* $\mathcal{S} = \mathcal{S}(V, \langle \cdot, \cdot \rangle, \mathcal{J})$ is the associated structure group. It is given by:

$$\mathcal{S} := \{T \in \mathcal{O}(V, \langle \cdot, \cdot \rangle) : TJ_i = J_i T \quad \text{for} \quad i = 1, 2, 3\} = \cap_{i=1}^{3} \mathcal{U}(V, \langle \cdot, \cdot \rangle, J_i).$$

An almost hyper-complex structure $\mathcal{J}$ on a Riemannian manifold (M, g) is said to be *almost hyper-Hermitian* if each J_i is almost Hermitian; $\mathcal{J}$ is said to be *hyper-Hermitian* if $\mathcal{J}$ is an integrable hyper-complex structure. We say that $(V, \langle \cdot, \cdot \rangle, \mathcal{J}, A)$ is a *hyper-Hermitian curvature model* if $A \in \mathfrak{R}(V)$ and if $\mathcal{J}$ is a hyper-Hermitian complex structure on $(V, \langle \cdot, \cdot \rangle)$. We extend the invariant τ_J of Equation (1.8.a) to this setting by defining:

$$\tau_{\mathcal{J}} := \tau_{J_1} + \tau_{J_2} + \tau_{J_3}.$$

Theorem 1.8.3 can be generalized to this setting to this setting [Brozos-Vázquez et al. (2009)]:

Theorem 1.13.2 *Any hyper-Hermitian curvature model is geometrically realizable by an almost hyper-Hermitian manifold with τ and $\tau_{\mathcal{J}}$ constant.*

What is interesting is that Theorem 1.9.2 does not generalize to the hyper-Hermitian setting; an additional curvature restriction is imposed. Let $\mathcal{G}_{J_i}$ be the Gray symmetrizers of Equation (1.8.b) defined by the Hermitian complex structures J_i for $i = 1, 2, 3$. Denote the simultaneous Kähler component which has zero trace by:

$$W_{3,\mathcal{J}}^{\mathfrak{R}} := W_{3,J_1}^{\mathfrak{R}} \cap W_{3,J_2}^{\mathfrak{R}} \cap W_{3,J_3}^{\mathfrak{R}}.$$

The module $W_{3,\mathcal{J}}^{\mathfrak{R}}$ is an irreducible module with structure group $\mathcal{S}$ [De Smedt (1994)]. Let $\pi_{3,\mathcal{J}}^{\mathfrak{R}}$ denote orthogonal projection on $W_{3,\mathcal{J}}^{\mathfrak{R}}$. The following represents work in progress; details are available upon request:

Theorem 1.13.3

(1) There are local hyper-holomorphic coordinates centered at a point P of a hyper-Hermitian manifold so $g = \delta + O(|x|^2)$ if and only if $\pi_{3,\mathcal{J}}^{\mathfrak{R}} R_P = 0$.
(2) Let $\mathfrak{H}$ be a hyper-Hermitian curvature model with $\pi_{3,\mathcal{J}}^{\mathfrak{R}} A = 0$ and $\mathcal{G}_{J_i} A = 0$ for $i = 1, 2, 3$. Then there exists a hyper-Hermitian manifold which geometrically realizes $\mathfrak{H}$.

The results of [Tricerri and Vanhecke (1981)] have been extended from the Hermitian to the hyper-Hermitian setting [De Smedt (1994)]. We summarize these results as follows:

Theorem 1.13.4 *If $m \geq 16$, we may decompose*

$$\mathfrak{R} = \{\oplus_{i=1}^{7} \mathfrak{B}_i\} \oplus \{\oplus_{i=1}^{10} \mathfrak{C}_i\} \oplus \{\oplus_{i=1}^{6} \mathfrak{D}_i\} \oplus \{\oplus_{i=1}^{14} \mathfrak{F}_i\}$$

as the orthogonal direct sum of thirty-seven irreducible modules for the group $\mathcal{S}$; the decomposition if $m = 4$, $m = 8$, or $m = 12$ is obtained by omitting certain factors in this decomposition.

(1) The modules $\{\mathfrak{B}_1, ..., \mathfrak{B}_7\}$ are isomorphic to $\mathbb{R}$. They are detected by various scalar invariants.
(2) The modules $\{\mathfrak{C}_1, ..., \mathfrak{C}_{10}\}$ are isomorphic to submodules of S^2 and the modules $\{\mathfrak{D}_1, ..., \mathfrak{D}_6\}$ are isomorphic to submodules of Λ^2. They are detected by various tensors that are analogous to the Ricci tensor.
(3) The modules $\{\mathfrak{F}_1, ..., \mathfrak{F}_{14}\}$ are in the kernel of all the tensors given in (1) and (2) above.

Hyper-Hermitian geometry, hyper-Kähler geometry, almost hyper-Hermitian geometry, and almost hyper-Kähler geometry appears in many contexts. We cite below just a few references in this area: [Barberis, Dotti, and Fino (2006)], [Bredthauer (2007)], [Burdík, Krivonos, and Scherbakov (2006)], [Calderbank and Tod (2001)], [De Smedt (1994)], [Hitchin et al. (1987)], [Kamada (1999)], and [Kath and Olbrich (2007)]. Quaternion geometry also plays an important role and we refer to: [Ivanov and Zamkovoy (2005)], [Martín-Cabrera and Swann (2004)], and [Pedersen, Poon, and Swann (1993)].

Chapter 2

Representation Theory

In Chapter 2, we present some basic results concerning representation theory. We will use these results subsequently. We discuss both general theory and theory as it relates to the orthogonal, unitary, and para-unitary groups in arbitrary signature. We refer to [Bourbaki (2005)], [Chevalley (1946)], and [Fulton and Harris (1991)] for more classical theory. We shall be interested primarily in the indefinite setting where perhaps the classical references are more sparse.

Let G be a Lie group. In Section 2.1, we give a brief introduction to the theory of modules. In Section 2.2, we discuss the theory of quadratic invariants as it relates to representation theory. In Section 2.3, we discuss Weyl's theorem on the linear invariants of the orthogonal group as well as related results for the unitary and para-unitary groups. In Section 2.4, we decompose $\otimes^2 V^*$ and $\Lambda^2 \otimes S^2$ as orthogonal modules; this will play a crucial role in our analysis of $\mathfrak{A}$ as an orthogonal module in Section 4.1 as we shall see subsequently. In Section 2.5, we study $\otimes^2 V^*$ as a unitary and as a para-unitary module. In Section 2.6 we present some standard material for compact Lie groups. Unless otherwise noted, all vector spaces are assumed to have finite dimension.

2.1 Modules for a Group G

Section 2.1 is devoted to the proof of Theorem 1.2.1. Let G be a Lie group. A pair $\xi := (V, \sigma)$ is said to be a *module for a group* G if V is a real vector space and if σ is a smooth group homomorphism from G to GL; ξ is also said to be a *representation of* G. If $v \in V$ and if $g \in G$, we set $g \cdot v := \sigma(g)v$. We will often identify ξ with V and suppress the role of σ if the action of G is clear.

Let $\langle\cdot,\cdot\rangle$ be a non-degenerate inner product of signature (p,q) on a vector space V of dimension $m=p+q$. Before proceeding further with our analysis, we recall some relatively elementary facts concerning indefinite geometry. We say a subspace W of V is *spacelike/timelike/totally isotropic* if the restriction of the inner product $\langle\cdot,\cdot\rangle$ to W is positive definite/negative definite/zero.

Lemma 2.1.1 *Let $\langle\cdot,\cdot\rangle$ be a non-degenerate inner product on V. We can find an orthogonal direct sum decomposition $V=V_+\oplus V_-$ where V_+ is spacelike and where V_- is timelike. If J_- is a pseudo-Hermitian complex structure on $(V,\langle\cdot,\cdot\rangle)$, we can choose the decomposition to be J_- invariant; if J_+ is a para-Hermitian complex structure on $(V,\langle\cdot,\cdot\rangle)$, we can choose the decomposition so $J_+V_\pm=V_\mp$.*

Proof. Let $0\neq v\in V$. Since $\langle\cdot,\cdot\rangle$ is non-degenerate, we can choose $w\in V$ so $\langle v,w\rangle=1$. If $\langle v,v\rangle=\langle w,w\rangle=0$, we consider $x=v+w$; otherwise we set $x=v$ or $x=w$ to find $x\in V$ with $\langle x,x\rangle\neq 0$. Set $e_1:=x/\sqrt{|\langle x,x\rangle|}$. Then $\langle e_1,e_1\rangle=\pm1$. Set $W_2:=e_1^\perp$. The restriction of $\langle\cdot,\cdot\rangle$ to W_2 is non-degenerate. Thus we may choose $e_2\in W_2$ so $e_1\perp e_2$ and $\langle e_2,e_2\rangle=\pm1$. Continuing in this fashion constructs an orthonormal basis $\{e_1,...,e_n\}$ for V so that $\langle e_i,e_i\rangle=\pm1$ and $\langle e_i,e_j\rangle=0$ for $i\neq j$. We set

$$V_+:=\mathrm{Span}\{e_i:\langle e_i,e_i\rangle=+1\},$$
$$V_-:=\mathrm{Span}\{e_i:\langle e_i,e_i\rangle=-1\}.$$

Suppose J_- is a pseudo-Hermitian complex structure on V. We argue as above to choose $e_1\in V$ so $\langle e_1,e_1\rangle=\pm1$. Set $e_2:=J_-e_1$. Then $\langle e_1,e_2\rangle=0$ and $\langle e_1,e_1\rangle=\langle e_2,e_2\rangle$. We then set $W_3:=\mathrm{Span}\{e_1,e_2\}^\perp$ and iterate the process to find a basis $\{e_1,J_-e_2=e_1,\dots,e_{n-1},e_m=J_-e_{m-1}\}$ and thereby construct vector spaces V_+ and V_- invariant under the action of J_-. Finally, suppose J_+ is a para-Hermitian complex structure. As before, set $e_2=J_+e_1$. As $\langle e_1,e_1\rangle=-\langle e_2,e_2\rangle$, we may choose the notation so that $\langle e_1,e_1\rangle=+1$ and $\langle e_2,e_2\rangle=-1$. We proceed as above and set

$$V_+:=\mathrm{Span}\{e_1,e_3,...\}\text{ and }V_-:=\mathrm{Span}\{e_2,e_4,...\}.\qquad\square$$

The metric of Definition 1.1.1 on $\otimes^kV$ is non-degenerate but it is indefinite. We identify $\otimes^kV=\otimes^kV^*$ to extend $\langle\cdot,\cdot\rangle$ to this setting as well. The following is a crucial fact concerning that metric which establishes Assertion (1) of Theorem 1.2.1.

Lemma 2.1.2 *Use* Definition 1.1.1 *to extend the given inner product* $\langle\cdot,\cdot\rangle$ *to* $\otimes^k V^*$*. Let* W *be a non-trivial subspace of* $\otimes^k V^*$ *which is invariant under the action of the group* G *where* G *belongs to* $\{\mathcal{O},\mathcal{U}_-,\mathcal{U}_\pm^\star\}$*. Then the restriction of* $\langle\cdot,\cdot\rangle$ *to* W *is non-degenerate. In particular,* W *is not totally isotropic.*

Proof. We first suppose $G = \mathcal{O}$. Apply Lemma 2.1.1 to find an orthogonal direct sum decomposition

$$V = V_+ \oplus V_- \tag{2.1.a}$$

where V_+ is spacelike and V_- is timelike. Let $g = \pm\,\mathrm{Id}$ on $V_\pm$; $g \in \mathcal{O}$. Let $\{e_1, \ldots, e_p\}$ be an orthonormal basis for V_- and let $\{e_{p+1}, \ldots, e_m\}$ be an orthonormal basis for V_+. If $I = (i_1, \ldots, i_k)$ is a multi-index, set $e^I := e^{i_1} \otimes \cdots \otimes e^{i_k}$. Then

$$g^* e^I = \langle e^{i_1}, e^{i_1}\rangle \ldots \langle e^{i_k}, e^{i_k}\rangle e^I = \langle e^I, e^I\rangle e^I = \pm e^I.$$

Thus if $g^* w = w$, then w is a spacelike vector in $\otimes^k V^*$ while if $g^* w = -w$, then w is a timelike vector in $\otimes^k V^*$. Let W be a non-trivial $\mathcal{O}$-invariant subspace of $\otimes^k V^*$. Since $g \in \mathcal{O}$, g preserves W by assumption. Thus we may decompose $W = W_+ \oplus W_-$ into the ± 1 eigenspaces of g. Since g acts orthogonally, $W_+ \perp W_-$. Since W_+ is spacelike and W_- is timelike, the induced inner product on W is non-degenerate.

Next let J_- be a pseudo-Hermitian structure on $(V, \langle\cdot,\cdot\rangle)$. Let $G = \mathcal{U}_-$. By Lemma 2.1.1, the decomposition of Equation (2.1.a) can be chosen to be J_- invariant and thus the involution $g \in \mathcal{U}_-$. The remainder of the argument is the same. Since $\mathcal{U}_-^\star$ contains $\mathcal{U}_-$, we obtain the same result for this larger group as well. Finally, let J_+ be a para-Hermitian structure on $(V, \langle\cdot,\cdot\rangle)$. Choose the decomposition of Equation (2.1.a) so $J_+ : V_\pm \to V_\mp$. We then have $g \in \mathcal{U}_+^\star - \mathcal{U}_+$ and the same argument pertains. □

Remark 2.1.1 Let $G \in \{\mathcal{O},\mathcal{U}_-,\mathcal{U}_\pm^\star\}$ and let W be a submodule of $\otimes^k V^*$ for the group G. Set

$$W^\perp := \{\theta \in \otimes^k V^* : \langle\theta, \phi\rangle = 0\}.$$

Then $W^\perp$ also is a submodule of $\otimes^k V^*$ with structure group G. Since the module $W \cap W^\perp$ is a totally isotropic submodule of $\otimes^k V^*$ for the group G, $W \cap W^\perp = \{0\}$. We may decompose $\otimes^k V^* = W \oplus W^\perp$; the orthogonal projection π_W on W is then given by the first factor in this decomposition. Alternatively, since the induced inner product on W is non-degenerate, we may choose an orthonormal basis $\{e_1^-, ..., e_p^-, e_1^+, \ldots, e_q^+\}$ for W where the vectors e_i^- are timelike and the vectors e_j^+ are spacelike. If $\psi \in \otimes^k V^*$, then

$$\pi_W(\psi) = -\langle\psi, e_1^-\rangle e_1^- - \cdots - \langle\psi, e_p^-\rangle e_p^- + \langle\psi, e_1^+\rangle e_1^+ + \cdots + \langle\psi, e_q^+\rangle e_q^+.$$

Remark 2.1.2 Following the notation of Section 1.13, let $\mathcal{J}$ be a hyper-complex structure on V. Let $\langle\cdot,\cdot\rangle$ be a non-degenerate bilinear form on V invariant under the action of $\mathcal{J}$; we do not assume $\langle\cdot,\cdot\rangle$ to be positive definite. Let $\mathcal{S}$ be the associated symplectic group:

$$\mathcal{S} := \{T \in \mathcal{O}(V, \langle\cdot,\cdot\rangle) : TJ_i = J_iT \quad \text{for} \quad i = 1, 2, 3\}.$$

The same argument as that given to establish Lemma 2.1.2 extends to the symplectic group in arbitrary signature; thus the results of Section 2.1 also hold valid in the hyper-pseudo-Hermitian setting.

A *hyper-para-Hermitian structure* on V is a pair $\{J_1, J_2\}$ of linear maps of V so that $J_1^*\langle\cdot,\cdot\rangle = \langle\cdot,\cdot\rangle$, $J_2\langle\cdot,\cdot\rangle = -\langle\cdot,\cdot\rangle$, $J_1^2 = -\operatorname{Id}$, $J_2^2 = \operatorname{Id}$, and $J_1J_2 + J_2J_1 = 0$; necessarily $\langle\cdot,\cdot\rangle$ has neutral signature and $\dim(V)$ is divisible by 4. One sets

$$\tilde{\mathcal{S}}^\star := \{T \in \mathcal{O}(V, \langle\cdot,\cdot\rangle) : TJ_1 = J_1T, \ TJ_2 = \pm J_2T\}.$$

Again, Lemma 2.1.2 extends to $\tilde{\mathcal{S}}^\star$ and the results of Section 2.1 hold valid in the hyper-para-Hermitian setting. We have omitted details to simplify the exposition.

Lemma 2.1.2 fails for $\mathcal{U}_+$; this is the crucial fact that will require us to deal with $\mathcal{U}_+^\star$ rather than with $\mathcal{U}_+$ in much of what follows when discussing para-complex geometry.

Lemma 2.1.3 *Let J_+ be a para-Hermitian structure on $(V, \langle\cdot,\cdot\rangle)$. For any k, we may decompose $\otimes^k V^* = \oplus_\varepsilon W_\varepsilon$ as the direct sum of totally isotropic $\mathcal{U}_+$ invariant subspaces.*

Proof. Let $m = 2\bar{m}$. Since $J_+^2 = \operatorname{Id}$, we may decompose $V = V_+ \oplus V_-$ into the ± 1 eigenspaces of J_+. Since $J_+^*\langle\cdot,\cdot\rangle = -\langle\cdot,\cdot\rangle$, the spaces $V_\pm$ are totally isotropic and thus $\dim(V_+) = \dim(V_-) = \bar{m}$. Let $\varepsilon = (\varepsilon_1, \ldots, \varepsilon_k)$ be a choice of signs. We set

$$W_\varepsilon := V_{\varepsilon_1}^* \otimes \cdots \otimes V_{\varepsilon_k}^* \subset \otimes^k V^*.$$

These spaces are totally isotropic subspaces of $\otimes^k V^*$ which are invariant under $\mathcal{U}_+$ and which give rise to a direct sum decomposition of $\otimes^k V^*$. □

We say that a representation ξ is *irreducible* if the only subspaces of V which are invariant under the action of G are $\{0\}$ and V itself. Let ξ_1 and ξ_2 be modules with structure group G. We let

$$\begin{aligned}\mathrm{Hom}(\xi_1, \xi_2) &:= \mathrm{Hom}_G(V_1, V_2)\\ &:= \{T \in \mathrm{Hom}(V_1, V_2) : T(g \cdot v_1) = g \cdot Tv_1\}\end{aligned}$$

be the linear space of *intertwining operators.* We say that ξ_1 and ξ_2 are *isomorphic modules with structure group G* if there exists a $T \in \mathrm{Hom}(\xi_1, \xi_2)$ providing a linear isomorphism between V_1 and V_2.

Lemma 2.1.4 *Two irreducible modules ξ_1 and ξ_2 with structure group G are isomorphic if and only if* $\dim\{\mathrm{Hom}(\xi_1, \xi_2)\} > 0$.

Proof. Let $\xi_1 = (V_1, \sigma_1)$ and $\xi_2 = (V_2, \sigma_2)$ be irreducible modules with structure group G with $\dim\{\mathrm{Hom}(\xi_1, \xi_2)\} > 0$. Choose an element $0 \neq T$ in $\mathrm{Hom}(\xi_1, \xi_2)$. Then $\mathrm{Range}(T)$ is a non-trivial subspace of V_2 which is invariant under the action of G. Since ξ_2 is irreducible and since $\mathrm{Range}(T) \neq \{0\}$, $\mathrm{Range}(T) = V_2$ so T is surjective. Since ξ_1 is irreducible and since $\ker(T) \neq V_1$, $\ker(T) = \{0\}$ so T is injective. Thus $\dim\{\mathrm{Hom}(\xi_1, \xi_2)\} > 0$ implies ξ_1 and ξ_2 are isomorphic. Conversely, of course, if there exists a isomorphism between ξ_1 and ξ_2 which is equivariant with respect to the action of G; this means that there is a linear isomorphism T from V_1 to V_2 satisfying $gT = Tg$. In this setting, necessarily $\dim\{\mathrm{Hom}(\xi_1, \xi_2)\} > 0$. □

We say that a module $\xi = (V, \sigma)$ with structure group G is *completely reducible* if there are irreducible modules $\xi_i = (V_i, \sigma_i)$ with structure group G so that

$$V = V_1 \oplus \cdots \oplus V_k \quad \text{and} \quad \sigma_i = \sigma|_{V_i}.$$

There exist Lie groups with not completely reducible representations:

Example 2.1.1 Let $G = \mathbb{R}$ and let

$$\sigma(x) := \begin{pmatrix} 1 & x \\ 0 & 1 \end{pmatrix}$$

define a representation from $\mathbb{R}$ to $\mathrm{GL}(\mathbb{R}^2)$ so that

$$\sigma(x)\begin{pmatrix} a \\ b \end{pmatrix} = \begin{pmatrix} a + bx \\ b \end{pmatrix}.$$

The only subspaces of $\mathbb{R}^2$ which are invariant under the action of G are

$$\{0\}, \qquad \mathbb{R}^2, \qquad V_1 := \mathbb{R} \cdot \begin{pmatrix} 1 \\ 0 \end{pmatrix}.$$

Thus this representation is not completely reducible; it contains a proper invariant subspace but does not contain a complementary invariant subspace. This provides an example where Lemma 2.1.5 below fails. The natural projection π defines a natural short exact sequence of modules with structure group G not splitting in the category of modules with structure group G:

$$0 \to V_1 \to \mathbb{R}^2 \xrightarrow{\pi} \mathbb{R}^2/V_1 \to 0.$$

Lemma 2.1.2 shows that if $G \in \{\mathcal{O}, \mathcal{U}_-, \mathcal{U}_\pm^\star\}$ and if ξ is a submodule of $\otimes^k V^*$ with structure group G, then ξ is not totally isotropic and the following sequence of Lemmas applies to ξ. Thus the remaining assertions of Theorem 1.2.1 will follow from this discussion. We begin our study with:

Lemma 2.1.5 *Let $\xi = (V, \sigma)$ be a module with structure group G admitting a non-degenerate inner product $\langle \cdot, \cdot \rangle$ which is invariant under the action of G so that no non-trivial submodule of V is totally isotropic. Then:*

(1) There is an orthogonal direct sum decomposition $\xi = \eta_1 \oplus \cdots \oplus \eta_\ell$ where the η_i are irreducible and where $V_{\eta_i} \perp V_{\eta_j}$ for $i \neq j$.

(2) If ξ_1 and ξ_2 are any two irreducible non-isomorphic submodules of ξ, then $V_{\xi_1} \perp V_{\xi_2}$.

(3) Let $0 \to \xi_1 \to \xi \to \xi_2 \to 0$ be a short exact sequence of modules for the group G. Then ξ is isomorphic to $\xi_1 \oplus \xi_2$.

Remark 2.1.3 Let W be a non-trivial invariant subspace of ξ. Then $W^\perp$ also is a non-trivial invariant subspace of ξ so $W \cap W^\perp$ is a totally isotropic subspace of ξ and hence $W \cap W^\perp = \{0\}$. Thus under the assumptions of Lemma 2.1.5, the restriction of $\langle \cdot, \cdot \rangle$ to any non-trivial submodule of ξ is non-degenerate.

Proof. If ξ is irreducible, Assertion (1) of Lemma 2.1.5 is immediate so we suppose ξ is reducible. Choose a minimal invariant subspace V_1 of V so that $0 \neq V_1$ and $V_1 \neq V$. Let $\sigma_1 = \sigma|_{V_1}$. Then $V_1^\perp$ also is a subspace of V which is invariant under the action of G. Since $V_1 \cap V_1^\perp$ is a totally isotropic subspace of V, we have $V_1 \cap V_1^\perp = \{0\}$. Thus we may split $V = V_1 \oplus V_1^\perp$ as the orthogonal direct sum of two non-trivial subspaces which are invariant under the group action. If $(V_1^\perp, \sigma|_{V_1^\perp})$ is irreducible, we are done. Otherwise, we iterate the construction. This process does not continue indefinitely because we have $\dim(V) > \dim(V_1^\perp) > \dots$ are positive integers. Assertion (1) follows.

Let π_2 denote orthogonal projection on V_{ξ_2}. Since the inner product is invariant under the action of the group G, $\pi_2(gv) = g\pi_2(v)$ for all $g \in G$ so we may regard $\pi_2 \in \text{Hom}(\xi_1, \xi_2)$. If V_{ξ_1} is not perpendicular to V_{ξ_2}, then $\pi_2 V_{\xi_1} \neq 0$ and so $\text{Hom}(\xi_1, \xi_2) \neq \{0\}$. This is false by Lemma 2.1.4. Assertion (2) now follows.

Let $0 \to \xi_1 \to \xi \xrightarrow{\pi} \xi_2 \to 0$ be a short exact sequence of modules with structure group G. Since ξ contains no totally isotropic invariant subspaces, the intersection $\xi_1^\perp \cap \xi_1 = \{0\}$. Thus $V = \xi_1 \oplus \xi_1^\perp$. Because $\xi_1 = \ker(\pi)$, the projection $\pi : \xi_1^\perp \to \xi_2$ is 1-1 and onto. Consequently, π provides a natural module isomorphism between the modules $\xi_1^\perp$ and ξ_2. Assertion (3) now follows. □

By Lemma 2.1.5, we may decompose $\xi = \eta_1 \oplus \dots \oplus \eta_\ell$ as the orthogonal direct sum of irreducible representations. Of course, the components η_i can be isomorphic. We group the isomorphic factors to express

$$\xi = n_1\xi_1 + \dots + n_k\xi_k,$$

where n_1 factors of the η_i are isomorphic to ξ_1, where n_2 factors of the η_i are isomorphic to ξ_2 and so forth and where ξ_i is not isomorphic to ξ_j for $i \neq j$. The following assertion shows the multiplicities are independent of the particular decomposition chosen:

Lemma 2.1.6 *Let ξ be a module for the group G admitting a non-degenerate inner product which is invariant under the action of G and which has no non-trivial totally isotropic subspaces which are invariant under the action of G. Decompose $\xi = \eta_1 + \dots + \eta_l$ as the direct sum of irreducible modules for the group G. If η is any irreducible module for the group G, let $\mu_\eta(\xi)$ be the number of factors η_i in the decomposition which are isomorphic to η. Then*

$$\mu_\eta(\xi) = \frac{\dim\{\text{Hom}(\eta, \xi)\}}{\dim\{\text{Hom}(\eta, \eta)\}}.$$

In particular, the multiplicity is independent of the particular decomposition chosen.

Proof. Group the isomorphic irreducible modules to express $\xi = \sum_i n_i\xi_i$. It is immediate that

$$\dim\{\text{Hom}(\eta, \xi)\} = \sum_i n_i \dim\{\text{Hom}(\eta, \xi_i)\}.$$

By Lemma 2.1.4, this vanishes if η is not isomorphic to ξ_i for any i. Lemma 2.1.6 now follows. □

Lemma 2.1.7 *Let ξ be a module for the group G admitting a non-degenerate invariant inner product which is invariant under the group G with no non-trivial totally isotropic submodules.*

(1) Assume that $\xi_1 = (V_1, \sigma|V_1)$ appears with multiplicity one in ξ. Let π_1 denote orthogonal projection on V_1. Let $\eta = (W, \sigma_W)$ be any submodule of ξ. Then either $\pi_1 W = \{0\}$ or $V_1 \subset W$.
(2) If every module occurs with multiplicity one, then any decomposition of $\xi = \xi_1 \oplus \cdots \oplus \xi_\ell$ as a direct sum of irreducible modules is an orthogonal decomposition.

Proof. Assume ξ_1 appears with multiplicity one. The map π (orthogonal projection of V on V_1) is a map which is invariant under the action of G. Consequently $\pi_1 W \neq 0$ implies that $0 \neq \pi_1|_W \in \operatorname{Hom}_G(W, V_1)$ so ξ_1 appears with multiplicity at least one in $(W, \sigma|_W)$. Suppose V_1 is not contained in W. Then π_1 is also a non-trivial map from $W^\perp$ to V_1 so ξ_1 appears with multiplicity at least one in $(W^\perp, \sigma|_{W^\perp})$. This means ξ_1 appears with multiplicity at least two in ξ; this is contrary to the given assumption. This contradiction shows we have $V_1 \subset W$ as desired. This establishes Assertion (1); Assertion (2) follows from Assertion (1). □

2.2 Quadratic Invariants

In Section 2.2, we shall present some basic tests for irreducibility. Let ξ be a module for the group G admitting a non-degenerate inner product invariant under the action of G $\langle\cdot,\cdot\rangle$. Let $\operatorname{Hom}^{\mathrm{sa}}(\xi, \xi)$ be the subspace of self-adjoint maps from ξ to ξ which are equivariant with respect to the G action:

$$\operatorname{Hom}^{\mathrm{sa}}(\xi, \xi) := \{T \in \operatorname{Hom}(\xi, \xi) : \langle T v_1, v_2 \rangle = \langle v_1, T v_2 \rangle\}.$$

Let $\mathcal{I}(\xi) = \mathcal{I}^G(\xi)$ be the set of linear maps from V to $\mathbb{R}$ which are equivariant with respect to the action of G and let $\mathcal{I}_2^G(\xi)$ be the vector space of symmetric bilinear forms on V which are invariant under the action of G. These are the spaces of linear and quadratic invariants, respectively. By Lemma 2.1.2, if G is one of the groups $\{\mathcal{O}, \mathcal{U}_-, \mathcal{U}_\pm^\star\}$ and if ξ is a submodule of $\otimes^k V^*$, then the natural inner product on $\otimes^k V^*$ is non-degenerate on ξ and hence $\dim\{\mathcal{I}_2^G(\xi)\} > 0$. If η is irreducible, then the integer $\dim\{\text{Hom}(\eta,\eta)\}$ again plays a crucial role just as it did in Lemma 2.1.6.

If θ is a bilinear form on V, define a linear map T_θ of V_ξ characterized by:

$$\theta(v_1, v_2) = \langle v_1, T_\theta v_2\rangle. \tag{2.2.a}$$

Lemma 2.2.1 *Let ξ be a module for the group G admitting a non-degenerate inner product $\langle\cdot,\cdot\rangle$ which is invariant under the action of G with no non-trivial totally isotropic submodules.*

(1) The map $\theta \to T_\theta$ identifies $\mathcal{I}(\xi\otimes\xi)$ with $\text{Hom}(\xi,\xi)$ and $\mathcal{I}_2^G(\xi)$ with $\text{Hom}^{\text{sa}}(\xi,\xi)$.

(2) Suppose $\xi = \eta_1 \oplus \cdots \oplus \eta_l$ where the modules η_i are not necessarily irreducible. Then $\text{Hom}^{\text{sa}}(\xi,\xi) = \oplus_i \text{Hom}^{\text{sa}}(\eta_i,\eta_i) \oplus_{i<j} \text{Hom}(\eta_i,\eta_j)$.

(3) Suppose that each η_i decomposes as n_i copies of irreducible modules ξ_i where ξ_i is not isomorphic to ξ_j for $i \neq j$. Then

$\dim\{\text{Hom}(\eta_i,\eta_j)\} = 0$ for $i \neq j$ and

$\dim \mathcal{I}_2^G(\xi) = \sum_i n_i \dim\{\text{Hom}^{\text{sa}}(\xi_i,\xi_i)\} + \frac{1}{2} n_i(n_i-1)\dim\{\text{Hom}(\xi_i,\xi_i)\}$.

Proof. Linear maps from $V\otimes V$ to $\mathbb{R}$ are bilinear forms on V, in other words elements of $V^*\otimes V^*$. Given such a θ, there exists T satisfying the relations of Equation (2.2.a); conversely, given T, we can define a corresponding bilinear form θ. Since

$$\theta(gv_1, gv_2) = \langle gv_1, Tgv_2\rangle = \langle v_1, g^*Tgv_2\rangle = \langle v_1, g^{-1}Tgv_2\rangle,$$

θ is a linear invariant if and only if $g^{-1}Tg = T$; this means that T is equivariant with respect to the action of G. This identifies $\mathcal{I}(\xi\otimes\xi)$ with $\text{Hom}(\xi,\xi)$. We have

$$\langle v_1, Tv_2\rangle - \langle v_2, Tv_1\rangle = \theta(v_1,v_2) - \theta(v_2,v_1).$$

Thus T is self-adjoint with respect to $\langle\cdot,\cdot\rangle$ if and only if θ is symmetric and $\mathcal{I}_2^G(\xi)$ can be identified with $\text{Hom}^{\text{sa}}(\xi,\xi)$. Assertion (1) follows.

We argue as follows to prove Assertion (2). Decompose $v\in V_\xi$ in the form $v = (v_1,\ldots,v_l)$ for $v_i \in V_{\eta_i}$. Let θ be a bilinear form on V which is invariant under the action of G. Set $\theta_{ij}(v,w) = \theta(v_i,w_j)$. This defines a family of bilinear forms on V_ξ determining θ which are invariant under the group action. Let $\theta^t(x,y) = \theta(y,x)$. If $i = j$, we need $\theta_{ii}^t = \theta_{ii}$ to ensure that θ is symmetric; such elements correspond to elements of $\text{Hom}^{\text{sa}}(\eta_i,\eta_i)$.

On the other hand, if $i < j$, then the form θ_{ij} need not be symmetric; all that is required is that θ_{ij} is invariant under the action of G as we can then define $\theta_{ji} := \theta_{ij}^t$. Such elements therefore correspond to arbitrary elements of $\mathrm{Hom}(\eta_i, \eta_j)$. This establishes Assertion (2).

Suppose that $\xi = \eta_1 \oplus \cdots \oplus \eta_l$ where η_i and η_j decompose into inequivalent irreducible modules for $i \neq j$. Then by Lemma 2.1.4, $\mathrm{Hom}(\eta_i, \eta_j) = 0$. If we suppose that $\eta_i = n_i \xi_i$, we can then apply Assertions (1) and (2) to η_i to complete the proof of Assertion (3). □

We can now give a criteria for irreducibility which is quite useful:

Lemma 2.2.2 *Let $\xi = (V, \sigma)$ be a module for the group G admitting a non-degenerate inner product which is invariant under the action of G with no non-trivial totally isotropic submodules.*

(1) Suppose $\xi = \sum_i n_i \xi_i$ decomposes as a sum of modules ξ_i for the group G. Then

$$\dim\{\mathcal{I}_2^G(\xi)\} \geq \sum_i \tfrac{1}{2} n_i(n_i + 1).$$

(2) If equality holds in Assertion (1), then each ξ_i is irreducible and ξ_i is not isomorphic to ξ_j for $i \neq j$.

(3) If η is a proper submodule of ξ, then restriction induces a surjective map from $\mathcal{I}_2^G(\xi) \to \mathcal{I}_2^G(\eta) \to 0$. This means that every quadratic invariant of η can be extended to a quadratic invariant of ξ and consequently that $\dim\{\mathcal{I}_2^G(\eta)\} < \dim\{\mathcal{I}_2^G(\xi)\}$.

(4) If $\dim\{\mathcal{I}_2^G(\xi \oplus \xi)\} \leq 3$, *then ξ is irreducible and* $\dim\{\mathrm{Hom}(\xi, \xi)\} = 1$.

(5) Assume that $\dim\{\mathrm{Hom}(\xi, \xi)\} = 1$. *If $\eta = (W, (\sigma \oplus \sigma)|_W)$ is a proper non-trivial submodule of $\xi \oplus \xi$, then there exist real numbers a_1 and a_2 with $(a_1, a_2) \neq (0, 0)$ so that we can parametrize W in the form:*

$$W = \{(a_1 v, a_2 v)\}_{v \in V}.$$

Remark 2.2.1 In Lemma 2.5.4 we shall exhibit an irreducible module ξ for the group $\mathcal{U}_-$ with $\dim\{\mathrm{Hom}(\xi, \xi)\} \geq 2$. Thus $\dim\{\mathcal{I}_2^{\mathcal{U}_-}(\xi \oplus \xi)\} \geq 4$. Consequently, the inequality of Assertion (1) in Lemma 2.2.2 can be strict.

Proof. Since the restriction of $\langle \cdot, \cdot \rangle$ to any non-trivial submodule is non-trivial by Remark 2.1.3, $\dim\{\mathrm{Hom}^{\mathrm{sa}}(\eta, \eta)\} \geq 1$ for any non-trivial submodule η of ξ. Assertion (1) now follows from Lemma 2.2.1. Suppose $n_1 \geq 1$

and $n_2 \geq 1$. We observe that

$$(n_1 + n_2)(n_1 + n_2 + 1) - n_1(n_1 + 1) - n_2(n_2 + 1) = 2n_1n_2 > 0.$$

Since each ξ_i is totally reducible as the direct sum of modules for the group G, Assertion (2) now follows from Assertion (1) and from the estimate given above. If η is any submodule of ξ, then $\xi = \eta \oplus \eta^\perp$. If $\theta \in \mathcal{I}_2^G(\eta)$, we extend θ to ξ to be zero on $\eta^\perp$. Assertion (3) now follows.

Let ξ satisfy $\dim\{\mathcal{I}_2^G(\xi \oplus \xi)\} \leq 3$. By Assertion (2), ξ is irreducible. We use Lemma 2.2.1 to complete the proof of Assertion (4) by computing:

$$\begin{aligned} 3 = \dim\{\mathcal{I}_2^G(\xi \oplus \xi)\} &= 2\dim\{\operatorname{Hom}^{\text{sa}}(\xi, \xi)\} + \dim\{\operatorname{Hom}(\xi, \xi)\} \\ &\geq 2 \cdot 1 + 1 = 3. \end{aligned}$$

Assume that $\dim\{\operatorname{Hom}(\xi, \xi)\} = 1$. Let $\eta = (W, (\sigma \oplus \sigma)|_W)$ be a proper submodule of $\xi \oplus \xi$. Let π_1 and π_2 denote the natural projections on $V \oplus 0$ and on $0 \oplus V$, respectively. If $\pi_2 W = 0$, then $W = \{(v, 0)\}_{v \in V}$; similarly, if $\pi_1 W = 0$, then $W = \{(0, v)\}_{v \in V}$. Thus to complete the proof of Assertion (5), we shall assume that $\pi_1 W \neq 0$ and $\pi_2 W \neq 0$. Since η is a proper submodule, it follows η is abstractly isomorphic to ξ. Thus π_1 and π_2 are isomorphisms. Let $T = \pi_2 \circ \pi_1^{-1}$. We may then parametrize $W = \{(v, Tv)\}_{v \in V}$. Since $\dim\{\operatorname{Hom}(\xi, \xi)\} = 1$, $T = a \operatorname{Id}$ for some a. This shows $W = \{(v, av)\}_{v \in V}$ and establishes Assertion (5). □

2.3 Weyl's Theory of Invariants

We say $\psi : \otimes^k V^* \to \mathbb{R}$ is a *linear orthogonal invariant* if ψ is a linear map and if

$$\psi(g \cdot w) = \psi(w) \quad \forall g \in \mathcal{O}, \forall w \in \otimes^k V^*.$$

We can construct such maps as follows. Let $k = 2\ell$ and let $\pi \in \operatorname{Perm}(2\ell)$ be a permutation of the integers from 1 to 2ℓ. Define

$$\psi_\pi(v^1, \dots, v^{2\ell}) := \langle v^{\pi(1)}, v^{\pi(2)} \rangle \cdots \langle v^{\pi(2\ell-1)}, v^{\pi(2\ell)} \rangle. \tag{2.3.a}$$

Note that the inner product of Definition 1.1.1 arises in this way when viewed as a linear orthogonal invariant on $\otimes^{2k} V^*$ for a suitable choice of the permutation π. We show ψ_π is an orthogonal invariant by computing

$$\begin{aligned} \psi_\pi(gv^1, \dots, gv^{2\ell}) &= \langle gv^{\pi(1)}, gv^{\pi(2)} \rangle \cdots \langle gv^{\pi(2\ell-1)}, gv^{\pi(2\ell)} \rangle \\ &= \langle v^{\pi(1)}, v^{\pi(2)} \rangle \cdots \langle v^{\pi(2\ell-1)}, v^{\pi(2\ell)} \rangle = \psi_\pi(v^1, \dots, v^{2\ell}). \end{aligned}$$

Since ψ_π is a multi-linear map, it extends naturally to a linear orthogonal invariant mapping $\otimes^{2\ell}V$ to $\mathbb{R}$. We refer to [Weyl (1946)] (see Theorem 2.9.A on page 53 and the discussion on page 66 for the extension from the positive definite to the indefinite setting) for the proof of the following result:

Theorem 2.3.1 *The space of linear orthogonal invariants of $\otimes^{2k}V^*$ is spanned by the maps $\psi_\pi(v^1,\ldots,v^{2\ell}) := \langle v^{\pi(1)}, v^{\pi(2)}\rangle \ldots \langle v^{\pi(2\ell-1)}, v^{\pi(2\ell)}\rangle$.*

Definition 2.3.1 Let $\{e_i\}$ be an orthonormal basis for V used to contract indices in pairs. Let $I = (i_1, i_2, i_3, i_4)$ be a multi-index. We set

$$\varepsilon_I := \varepsilon_{i_1 i_1}\varepsilon_{i_2 i_2}\varepsilon_{i_3 i_3}\varepsilon_{i_4 i_4} = \pm 1.$$

Let $S = (\nu_1, \nu_2, \nu_3, \nu_4)(\mu_1, \mu_2, \mu_3, \mu_4)$ be a string of eight symbols where each index $(1, 2, 3, 4)$ appears exactly twice in S. We use such a string to construct an invariant $\mathcal{I}(S)$ of $\otimes^4 V^*$ setting:

$$\mathcal{I}(S)(A) := \sum_{i_1=1}^{4}\sum_{i_2=1}^{4}\sum_{i_3=1}^{4}\sum_{i_4=1}^{4} \varepsilon_I A_{i_{\nu_1} i_{\nu_2} i_{\nu_3} i_{\nu_4}} A_{i_{\mu_1} i_{\mu_2} i_{\mu_3} A_{\mu_4}}.$$

Thus, for example,

$$\tau^2 := \mathcal{I}\{(1,2,2,1)(3,4,4,3)\} = \sum_{i_1=1}^{m}\sum_{i_2=1}^{m}\sum_{i_3=1}^{m}\sum_{i_4=1}^{m} \varepsilon_I A_{i_1 i_2 i_2 i_1} A_{i_3 i_4 i_4 i_3},$$

$$|A|^2 := \mathcal{I}\{(1,2,3,4)(1,2,3,4)\} = \sum_{i_1=1}^{m}\sum_{i_2=1}^{m}\sum_{i_3=1}^{m}\sum_{i_4=1}^{m} \varepsilon_I A_{i_1 i_2 i_3 i_4} A_{i_1 i_2 i_3 i_4}.$$

Clearly we can permute the symbols without changing the invariant. Let $\mathcal{S}$ be the set of all such strings; Theorem 2.3.1 then yields

$$\mathcal{I}_2^{\mathcal{O}}(\otimes^4 V^*) = \mathrm{Span}_{S\in\mathcal{S}}\{\mathcal{I}(S)\}.$$

By considering strings of shorter or greater length, similar spanning sets for $\mathcal{I}_2^{\mathcal{O}}(\otimes^k V^*)$ can be constructed for other values of k. This formalism will play an important role in our subsequent discussion.

There is a convenient formalism for describing such elements. Let $\{e_i\}$ be an orthonormal basis for V; set $\varepsilon_i := \langle e_i, e_i\rangle = \pm 1$. If $\Theta \in \otimes^k V^*$, we may expand $\Theta = \Theta_{i_1\ldots i_k} e^{i_1} \otimes \cdots \otimes e^{i_k}$ where, as always, we sum over repeated indices. Set $\varepsilon_I := \varepsilon_{i_1} \ldots \varepsilon_{i_k}$. If π is the identity permutation and if $k = 2\ell$, then

$$\psi_\pi(\Theta) = \varepsilon_I \Theta_{i_1 i_1 i_2 i_2 \ldots i_\ell i_\ell}.$$

More generally, ψ_π is defined similarly for other permutations; such invariants are said to be given by *contractions of indices.* The following result [Weyl (1946)] (see Theorem 2.17.A on page 75) plays a prominent role in the discussion of [Gilkey, Park, and Sekigawa (2011)]:

Theorem 2.3.2 *Let $(V, \langle\cdot,\cdot\rangle)$ be an inner product of dimension m. Every relation among scalar products is an algebraic consequence of the relations:*

$$\det\begin{pmatrix} \langle v^0, w^0\rangle & \langle v^0, w^1\rangle & \dots & \langle v^0, w^m\rangle \\ \langle v^1, w^0\rangle & \langle v^1, w^1\rangle & \dots & \langle v^1, w^m\rangle \\ \dots & \dots & \dots & \dots \\ \langle v^m, w^0\rangle & \langle v^m, w^1\rangle & \dots & \langle v^m, w^m\rangle \end{pmatrix} = 0.$$

We have, for example:

Remark 2.3.1 The space of orthogonally invariant linear maps from $\otimes^4 V^*$ to $\mathbb{R}$ is generated by the maps: $\{\varepsilon_i\varepsilon_j\Theta_{iijj}, \varepsilon_i\varepsilon_j\Theta_{ijij}, \varepsilon_i\varepsilon_j\Theta_{ijji}\}$. Thus, in particular, the space of linear invariants on $\otimes^4 V^*$ has dimension at most 3; these maps are linearly independent as long as $m \geq 2$.

Remark 2.3.2 Let ξ be an orthogonal submodule of $\otimes^k V^*$. Since we may decompose $\otimes^k V^* = \xi \oplus \xi^\perp$, every element of $\theta \in \mathcal{I}_2^{\mathcal{O}}(\xi)$ extends to an element $\tilde\theta \in \mathcal{I}_2^{\mathcal{O}}(\otimes^k V^*)$. We can polarize $\tilde\theta$ to define a corresponding linear orthogonal invariant of $\otimes^{2k} V^*$. This shows that every θ can be written in terms of contractions of indices.

We turn now to the (para)-complex setting. Let $\kappa_{0,ij}$ be the components of the inner product on V^*. If $(V, \langle\cdot,\cdot\rangle, J_\pm)$ is a para-Hermitian vector space $(+)$ or a pseudo-Hermitian vector space $(-)$, let $\kappa_{1,ij}$ be the components of the (para)-Kähler form:

$$\kappa_{0,ij} := \langle e_i, e_j\rangle \quad \text{and} \quad \kappa_{1,ij} := \langle e_i, J_\pm e_j\rangle.$$

Raise indices to define κ_0^{ij} and κ_1^{ij}. Let $\pi \in \operatorname{Perm}(2k)$ and let $\vec a$ be a sequence of 0's and 1's. If $\Theta \in \otimes^k V^*$, define:

$$\psi_{\pi,\vec a}(\Theta) := \kappa_{a_1}^{i_{\pi(1)} i_{\pi(2)}} \dots \kappa_{a_k}^{i_{\pi(2k-1)} i_{\pi(2k)}} \Theta_{i_1 \dots i_k} \Theta_{i_{k+1} \dots i_{2k}}.$$

Let $n(\vec a)$ be the number of times $a_i = 1$; this is the number of contractions involving the (para)-Kähler form. Theorem 2.3.1 extends to this context to become:

Lemma 2.3.1 *Let $(V, \langle\cdot,\cdot\rangle, J_\pm)$ be a para-Hermitian vector space (+) or a pseudo-Hermitian vector space (−).*

(1) If ξ is a submodule of $\otimes^k V^*$ with structure group $\mathcal{U}_-$, then

$$\mathcal{I}_2^{\mathcal{U}_-}(\xi) = \text{Span}\{\psi_{\pi,\vec{a}}\}.$$

(2) If ξ is a submodule of $\otimes^k V^*$ with structure group $\mathcal{U}_\pm^\star$, then

$$\mathcal{I}_2^{\mathcal{U}_\pm^\star}(\xi) = \text{Span}_{n(\vec{a})\equiv 0 \bmod 2}\{\psi_{\pi,\vec{a}}\}.$$

Proof. Let $G \in \{\mathcal{U}_-, \mathcal{U}_\pm^\star\}$. Since by Lemma 2.2.2 the restriction map $\mathcal{I}_2^G(\otimes^k V^*) \to \mathcal{I}_2^G(\xi)$ is surjective, we may assume without loss of generality that $\xi = \otimes^k V^*$.

Suppose first that $G = \mathcal{U}_-$. The inner product and the Kähler form are invariant under the action of $\mathcal{U}_-$. Consequently, $\psi_{\pi,\vec{a}} \in \mathcal{I}_2^{\mathcal{U}_-}(\otimes^k V^*)$ are quadratic invariants. If $\langle\cdot,\cdot\rangle$ is positive definite, then the desired result follows from the results contained in [Fukami (1958)] and [Iwahori (1958)] so our task is to extend this result from the definite to the indefinite setting. We complexify to set:

$$V_{\mathbb{C}} := V \otimes_{\mathbb{R}} \mathbb{C}.$$

We extend $\langle\cdot,\cdot\rangle$ to $V_{\mathbb{C}}$ as a symmetric complex bilinear form $\langle\cdot,\cdot\rangle_{\mathbb{C}}$ on $V_{\mathbb{C}}$ and we extend J_- to a complex linear endomorphism $J_{-,\mathbb{C}}$ of $V_{\mathbb{C}}$. Let $\text{GL}_{\mathbb{C}}$ be the set of all complex linear maps of $V_{\mathbb{C}}$ and let

$$\mathcal{U}_{-,\mathbb{C}} := \{T \in \text{GL}_{\mathbb{C}} : T^*\langle\cdot,\cdot\rangle = \langle\cdot,\cdot\rangle, \quad TJ_{-,\mathbb{C}} = J_{-,\mathbb{C}}T\}.$$

Let $\mathcal{U}_{-,\mathbb{C}}^0$ be the connected component of the identity in $\mathcal{U}_{-,\mathbb{C}}$ and let $\mathcal{U}_-^0$ be the connected component of the identity in $\mathcal{U}_-$;

$$\mathcal{U}_-^0 \subset \mathcal{U}_{-,\mathbb{C}}^0 \cap \text{GL}_{\mathbb{R}}(V).$$

Let $\langle\cdot,\cdot\rangle$ have signature $(2p, 2q)$. Let

$$\{e_1^-, J_-e_1^-, \dots, e_q^-, J_-e_q^-, f_1^+, J_-f_1^+, \dots, f_p, J_-f_p^+\}$$

be an orthonormal basis for $(V, \langle\cdot,\cdot\rangle)$ where the superscript "−" indicates timelike vectors and the superscript "+" indicates spacelike vectors. Let

$$W := \text{Span}_{\mathbb{R}}\{\sqrt{-1}e_1^-, \sqrt{-1}J_-e_1^-, \dots, \sqrt{-1}e_p^-, \sqrt{-1}J_-e_p^-, \\ f_1^+, J_-f_1^+, \dots, f_q^+, J_-f_q^+\}.$$

Let $\langle\cdot,\cdot\rangle_W$ be the restriction of $\langle\cdot,\cdot\rangle_{\mathbb{C}}$ to W; this is a positive definite inner product. Since $J_{-,\mathbb{C}}W = W$, we may restrict $J_{-,\mathbb{C}}$ to define a positive definite Hermitian complex structure $J_{-,W}$ on $(W, \langle\cdot,\cdot\rangle_W)$. Let $\mathcal{U}_{-,W}$ be the associated unitary group; $\mathcal{U}_{-,W}$ is a connected Lie group and

$$\mathcal{U}_{-,W} \subset \mathcal{U}^0_{-,\mathbb{C}} \cap \mathrm{GL}_{\mathbb{R}}(W).$$

Let $f \in \mathcal{I}_2^{\mathcal{U}_-}(\otimes^k V^*)$. We regard f as a symmetric bilinear form on $\otimes^k V^*$ and extend f to be a complex linear symmetric bilinear form $f_{\mathbb{C}}$ on $\otimes^k V^*_{\mathbb{C}}$ which is holomorphic. We consider the associated Lie algebras:

$$\begin{aligned} \mathfrak{u} &:= \{u \in \mathrm{Hom}_{\mathbb{R}}(V,V) : u + u^* = 0\}, \\ \mathfrak{u}_{\mathbb{C}} &:= \{u \in \mathrm{Hom}_{\mathbb{C}}(V_{\mathbb{C}},V_{\mathbb{C}}) : u + u^* = 0\}, \end{aligned}$$

where the adjoint u^* is defined relative to the bilinear form $\langle\cdot,\cdot\rangle_{\mathbb{C}}$. We use the exponential map to define coordinates on $\mathcal{U}_-$ and $\mathcal{U}_{-,\mathbb{C}}$ near Id. The action of $\mathfrak{u}_{\mathbb{C}}$ on $\otimes^k V^*_{\mathbb{C}}$ is holomorphic. Since $f_{\mathbb{C}}$ is a holomorphic function and since $f_{\mathbb{C}}$ is invariant under the action of $\exp(u)$ for all $u \in \mathfrak{u}$, the identity theorem for holomorphic functions implies that $f_{\mathbb{C}}$ is invariant under the action of $\exp(u)$ for all $u \in \mathfrak{u}_{\mathbb{C}}$. Restricting to $\mathcal{U}_{-,W}$ yields $f_W \in \mathcal{I}_2^{\mathcal{U}_{-,W}}(\otimes^k W) \otimes \mathbb{C}$. Thus since Lemma 2.3.1 holds in the positive definite setting, by considering the real and imaginary parts of f_W, we may conclude that there are complex constants $c_{\pi,\vec{a}}$ satisfying:

$$f_W = \sum_{\pi,\vec{a}} c_{\pi,\vec{a}} \psi_{\pi,\vec{a}}. \tag{2.3.b}$$

This identity holds for f_W viewed as a map from $\otimes^k W$ to $\mathbb{C}$. It therefore holds as a map from $\otimes^k(W \otimes \mathbb{C})$ to $\mathbb{C}$ as well. We restrict to $\otimes^k V^*$ and take the real part to obtain Assertion (1). Since the Kähler form Ω_- changes sign under the action of $\mathcal{U}^\star_- - \mathcal{U}_-$, elements of $\mathcal{I}_2^{\mathcal{U}^\star_-}$ must involve contractions involving an even number of occurrences of the Kähler form. Assertion (2) in the pseudo-Hermitian setting now follows from Assertion (1).

We pass to the para-Hermitian setting. Choose an orthonormal basis

$$\{e_1^+, J_+e_1^+, \ldots, e_p^+, J_+e_p^+\}$$

for V so the vectors $\{e_i^+\}$ are spacelike and the vectors $\{J_+e_i^+\}$ are timelike. As before, we complexify. We set $J_- = \sqrt{-1}J_+$ and take

$$W := \mathrm{Span}_{\mathbb{R}}\{e_1^+, J_-e_1^+, \ldots, e_p^+, J_-e_p^+\} \subset V_{\mathbb{C}}.$$

Let $(\cdot,\cdot)$ be the restriction of $\langle\cdot,\cdot\rangle$ to W; this inner product is positive definite and J_- is a Hermitian complex structure on $(W,(\cdot,\cdot))$. Let f_V belong to $\mathcal{I}_2^{\mathcal{U}_+}(\otimes^k V^*)$. As before, extend f to a holomorphic function $f_{\mathbb{C}}$ on $\otimes^k V_{\mathbb{C}}^*$. Let f_W be the restriction of $f_{\mathbb{C}}$ to $\otimes^k W$; a similar argument using exponential coordinates and analytic continuation shows that we have $f_W \in \mathcal{I}_2^{\mathcal{U}_-,W}(\otimes^k W)$ and consequently, we may express

$$f_W = c_{\pi,\vec{a}}\psi_{\pi,\vec{a}}.$$

Consequently, $f = c_{\pi,\vec{a}}\psi_{\pi,\vec{a}}$ on $\otimes^k V^*$. Assertion (2) now follows from Assertion (1). □

We now extend Definition 2.3.1 to the para-Hermitian and the pseudo-Hermitian setting.

Definition 2.3.2 Let $(V,\langle\cdot,\cdot\rangle,J_\pm)$ be a para-Hermitian vector space $(+)$ or a pseudo-Hermitian vector space $(-)$. A slight additional amount of formality is required as we must consider invariants including $J_\pm$. Instead of considering just a single orthonormal basis, we consider four distinct orthonormal bases $\{e^1_{i_1},e^2_{i_2},e^3_{i_3},e^4_{i_4}\}$ for V; we index these bases by the indices $\{i_1,i_2,i_3,i_4\}$, respectively, for $1\le i_\mu\le m$ with $1\le\mu\le 4$. We use these bases to contract indices in pairs and use $J_\pm$ as well. We consider a string S of eight symbols to construct an invariant $\mathcal{I}(S)$. Thus, for example, the string $(1,2,J_\pm 2,1)(3,4,J_\pm 4,3)$, consisting of two quadruples, is shorthand for the invariant

$$\begin{aligned}&\mathcal{I}((1,2,J_\pm 2,1)(3,4,J_\pm 4,3))\\ &:=\sum_{i_1=1}^m\sum_{i_2=1}^m\sum_{i_3=1}^m\sum_{i_4=1}^m \varepsilon_I A(e^1_{i_1},e^2_{i_2},J_\pm e^2_{i_2},e^1_{i_1})A(e^3_{i_3},e^4_{i_4},J_\pm e^4_{i_4},e^3_{i_3}).\end{aligned}$$

Let $\mathcal{S}(J_\pm)$ be the set of all such strings. If $S\in\mathcal{S}(J_\pm)$, let $n(S)$ be the number of times $J_\pm$ appears in S. Lemma 2.3.1 yields:

$$\begin{aligned}\mathcal{I}_2^{\mathcal{U}_-}(\otimes^4V^*)&=\operatorname{Span}_{S\in\mathcal{S}(J_-)}\{\mathcal{I}(S)\},\\ \mathcal{I}_2^{\mathcal{U}_\pm^*}(\otimes^4V^*)&=\operatorname{Span}_{S\in\mathcal{S}(J_\pm),n(S)\equiv 0 \bmod 2}\{\mathcal{I}(S)\}.\end{aligned}$$

Clearly we can permute the bases. We can also replace the basis $\{e^1_{i_1}\}$ by the basis $\{J_\pm e^1_{i_1}\}$. Since the metric changes sign in the para-Hermitian setting, we have

$$\begin{aligned}\mathcal{I}\{(-,1,-,J_\pm 1,-)\}&=\mp\mathcal{I}\{(-,J_\pm 1,-,J_\pm J_\pm 1,-)\}\\ &=-\mathcal{I}\{(-,J_\pm 1,-,1,-)\}\end{aligned}$$

where the symbol "–" simply means we do not know exactly where the remaining indices appear nor if they are decorated suitably.

2.4 Some Orthogonal Modules

We apply Lemma 2.2.2 to establish some useful results. We set

$$\begin{aligned}
S_0^2 &:= \{\theta \in \otimes^2 V^* : \theta_{ij} = \theta_{ji}, \quad \tau(\theta) = 0\},\\
\Lambda^2 &:= \{\theta \in \otimes^2 V^* : \theta_{ij} = -\theta_{ji}\},\\
W_7^{\mathcal{O}} &:= \{A \in \mathfrak{A} \cap \ker(\rho) : A_{ijkl} = A_{ijlk}\},\\
\tilde{W}_7^{\mathcal{O}} &:= \{A \in \otimes^4 V^* \cap \ker(\rho) : A_{ijkl} = -A_{jikl} = A_{ijlk},\\
&\qquad\qquad A_{kjil} + A_{ikjl} - A_{ljik} - A_{iljk} = 0\},\\
W_8^{\mathcal{O}} &:= \{A \in \otimes^4 V^* \cap \ker(\rho) : A_{ijkl} = -A_{jikl} = -A_{klij}\}.
\end{aligned}$$

Theorem 2.4.1 *Let $m \geq 4$.*

(1) $\otimes^2 V^* = \Lambda^2 \oplus S_0^2 \oplus \mathbb{R} \cdot \langle \cdot, \cdot \rangle$ as an orthogonal module.

(2) $\Lambda^2(\Lambda^2) \approx W_8^{\mathcal{O}} \oplus \Lambda^2$ as an orthogonal module.

(3) $\Lambda^2 \otimes S^2 \approx S_0^2 \oplus 2 \cdot \Lambda^2 \oplus \tilde{W}_7^{\mathcal{O}} \oplus W_8^{\mathcal{O}}$ as an orthogonal module.

(4) $\{\mathbb{R}, \Lambda^2, S_0^2, \tilde{W}_7^{\mathcal{O}}, W_8^{\mathcal{O}}\}$ are inequivalent irreducible orthogonal modules.

(5) $W_7^{\mathcal{O}} = \tilde{W}_7^{\mathcal{O}}$.

(6) We have

(a) $\dim\{\Lambda^2\} = \frac{1}{2}m(m-1)$.

(b) $\dim\{S_0^2\} = \frac{1}{2}m(m+1) - 1$.

(c) $\dim\{W_8^{\mathcal{O}}\} = \frac{m(m-1)(m-3)(m+2)}{8}$.

(d) $\dim\{W_7^{\mathcal{O}}\} = \frac{(m-1)(m-2)(m+1)(m+4)}{8}$.

Remark 2.4.1 We refer to Corollary 4.1.2 where $\dim\{W_6^{\mathcal{O}}\}$ is computed.

Proof. We first establish Assertion (1). If $\theta \in \otimes^2 V^*$, we may decompose $\theta = \theta_a + \theta_s + \tau \cdot \langle \cdot, \cdot \rangle$ as the sum of an alternating tensor, a symmetric tensor of trace 0, and a multiple of the inner product. Since the dimension $m \geq 2$, Λ^2, S_0^2, and $\mathbb{R}$ are non-trivial representations. By Remark 2.3.1, $\dim\{\mathcal{I}_2^{\mathcal{O}}(V^* \otimes V^*)\} \leq 3$. Assertion (1) now follows from Lemma 2.2.2; the dimensions of these modules are easily computed; Assertions (6a) and (6b) now follow. We may also conclude that Λ^2, S_0^2, and $\mathbb{R}$ are inequivalent and irreducible orthogonal modules.

Next we turn our attention to Assertion (2). Since

$$W_8^{\mathcal{O}} = \ker(\rho) \cap \Lambda^2(\Lambda^2),$$

the Ricci tensor defines a four term short exact sequence:

$$0 \to W_8^{\mathcal{O}} \to \Lambda^2(\Lambda^2) \xrightarrow{\rho} \Lambda^2.$$

To show that ρ is surjective, we define a splitting. Choose an orthonormal basis $\{e_i\}$ for V. If $\theta \in \Lambda^2$, set:

$$\sigma(\theta)_{ijkl} := \varepsilon_{il}\theta_{jk} - \varepsilon_{jl}\theta_{ik} - \varepsilon_{ik}\theta_{jl} + \varepsilon_{jk}\theta_{il}.$$

It is immediate that $\sigma(\theta) \in \Lambda^2(\Lambda^2)$. We compute:

$$\rho(\sigma(\theta))_{jk} = \varepsilon^{il}\{\varepsilon_{il}\theta_{jk} - \varepsilon_{jl}\theta_{ik} - \varepsilon_{ik}\theta_{jl} + \varepsilon_{jk}\theta_{il}\} = (m-2)\theta_{jk}.$$

Thus ρ is surjective and Assertion (2) follows from Lemma 2.1.5. We may now establish Assertion (6c) by computing:

$$\begin{aligned}\dim\{W_8^{\mathcal{O}}\} &= \dim\{\Lambda^2(\Lambda^2)\} - \dim\{\Lambda^2\}\\ &= \tfrac{1}{2}\{\tfrac{1}{2}m(m-1)\}\{\tfrac{1}{2}m(m-1)-1\} - \tfrac{1}{2}m(m-1)\\ &= \tfrac{1}{8}m(m-1)(m-3)(m+2).\end{aligned}$$

We use the formalism of Definition 2.3.1 to study $\mathcal{I}_2^{\mathcal{O}}(\Lambda^2(\Lambda^2))$. We examine the possible cases:

(1) Both monomials in the string contain two indices. This leads to the invariant $\mathcal{I}\{(1,2,1,2)(3,4,3,4)\}$. This expression vanishes since $\Theta(a,b,c,d) = -\Theta(c,d,a,b)$ for $\Theta \in \Lambda^2(\Lambda^2)$.
(2) Each monomial in the string contains three indices. This leads to the invariant $\psi_1 := \mathcal{I}\{(1,2,1,3)(4,2,4,3)\}$.
(3) Each monomial in the string contains all four of the indices. This leads to two possibilities:
    (a) $\psi_2 := \mathcal{I}\{(1,2,3,4)(1,2,3,4)\}$.
    (b) We permute $1 \leftrightarrow 3$ and $2 \leftrightarrow 4$ to see $\mathcal{I}\{(1,2,3,4)(1,3,2,4)\} = \mathcal{I}\{(3,4,1,2)(3,1,4,2)\}$ $= -\mathcal{I}\{(1,2,3,4)(1,3,2,4)\}$ so this invariant is zero.

This shows that $\dim\{\mathcal{I}_2^{\mathcal{O}}(\Lambda^2(\Lambda^2))\} \le 2$. By Assertion (6c), $W_8^{\mathcal{O}}$ is non-trivial for $m \ge 4$. The fact that Λ^2 and $W_8^{\mathcal{O}}$ are inequivalent irreducible orthogonal modules now follows from Lemma 2.2.2. This establishes Assertion (2) and part of Assertion (4).

We now prove Assertion (3). Let $\Theta \in \Lambda^2 \otimes S^2$. Define

$$(\Xi_2\Theta)_{ijkl} := \Theta_{kjil} + \Theta_{ikjl} - \Theta_{ljik} - \Theta_{iljk}.$$

This defines an $\mathcal{O}$ equivariant map

$$0 \to \ker(\Xi_2) \to \Lambda^2 \otimes S^2 \xrightarrow{\Xi_2} \Lambda^2(\Lambda^2).$$

We wish to show this map is surjective. In the proof of Assertion (2), we showed $\Lambda^2(\Lambda^2) = W_8^{\mathcal{O}} \oplus \Lambda^2$ and that the Ricci tensor provided the projection on Λ^2. Since $m \geq 4$, we may consider the element:

$$\Theta := (e^1 \otimes e^2 - e^2 \otimes e^1) \otimes (e^3 \otimes e^1 + e^1 \otimes e^3) \in \Lambda^2 \otimes S^2.$$

The non-zero components of Θ are determined by $\Theta_{1213} = 1$. Consequently $\Xi_2(\Theta)_{ijkl}$ is only non-zero when $\{i, j, k, l\}$ is a rearrangement of $\{1, 1, 2, 3\}$. Since $\{e_i\}$ is an orthonormal basis, contracting with respect to ε then yields:

$$\Xi_2(\Theta)_{1231} = \Theta_{3211} + \Theta_{1321} - \Theta_{1213} - \Theta_{1123} = -1,$$
$$\rho(\Xi_2\Theta)_{23} = -\varepsilon_{11}.$$

Thus the component of $\Xi_2(\Theta)$ in Λ^2 is non-zero. We have shown that $\Lambda^2(\Lambda^2) = \Lambda^2 \oplus W_8^{\mathcal{O}}$ is a decomposition into inequivalent irreducible modules. Consequently,

$$\Lambda^2 \subset \text{Range}\{\Xi_2\}.$$

We clear the previous notation and consider the element:

$$\Theta := (e^3 \otimes e^2 - e^2 \otimes e^3) \otimes (e^1 \otimes e^4 + e^4 \otimes e^1) \in \Lambda^2 \otimes S^2.$$

Clearly $(\Xi_2\Theta)_{ijkl} = 0$ unless $\{i, j, k, l\}$ is a permutation of $\{1, 2, 3, 4\}$. In particular, $\rho(\Xi_2\Theta) = 0$ so $\Xi_2\Theta \in W_8^{\mathcal{O}}$. We compute:

$$(\Xi_2\Theta)_{1234} = \Theta_{3214} + \Theta_{1324} - \Theta_{4213} - \Theta_{1423} = 1.$$

Consequently $\Xi_2\Theta$ has a non-zero component in $W_8^{\mathcal{O}}$ as desired; this implies that $W_8^{\mathcal{O}} \subset \text{Range}(\Xi_2)$ so we have a short exact sequence:

$$0 \to \ker(\Xi_2) \to \Lambda^2 \otimes S^2 \xrightarrow{\Xi_2} \Lambda^2 \oplus W_8^{\mathcal{O}} \to 0$$

and consequently by Lemma 2.1.5,

$$\Lambda^2 \otimes S^2 = \ker(\Xi_2) \oplus \Lambda^2 \oplus W_8^{\mathcal{O}}. \tag{2.4.a}$$

Next, we consider the Ricci tensor. If $\Theta \in \Lambda^2 \otimes S^2$, then

$$\tau(\rho(\Theta)) = \varepsilon^{il}\varepsilon^{jk}\Theta_{ijkl} = -\varepsilon^{il}\varepsilon^{jk}\Theta_{jilk} = -\tau(\rho(\Theta))$$

and thus, since $\tau = 0$, ρ takes values in $\Lambda^2 \oplus S_0^2$. By definition, we have $\tilde{W}_7^{\mathcal{O}} = \ker(\rho) \cap \ker(\Xi_2)$. Thus we have a short exact sequence:

$$0 \to \tilde{W}_7^{\mathcal{O}} \to \ker(\Xi_2) \xrightarrow{\rho} \Lambda^2 \oplus S_0^2.$$

We wish to show ρ is surjective. We clear the previous notation. Let ε be a real parameter. Set

$$\Theta := (e^1 \otimes e^2 - e^2 \otimes e^1) \otimes (e^1 \otimes e^1 + \varepsilon e^2 \otimes e^2).$$

This is an example of dimension 2. Since $\Lambda^2(\Lambda^2(\mathbb{R}^2)) = \Lambda^2(\mathbb{R}) = 0$, necessarily $\Theta \in \ker(\Xi_2)$. We compute:

$$\rho(\Theta)_{12} = -\varepsilon\varepsilon_{22}, \quad \rho(\Theta)_{21} = \varepsilon_{11}.$$

Taking $\varepsilon = -\varepsilon_{11}\varepsilon_{22}$ yields an element of S_0^2 while taking $\varepsilon = \varepsilon_{11}\varepsilon_{22}$ yields an element of Λ^2. Since, by Assertion (1), Λ^2 and S_0^2 are inequivalent and irreducible orthogonal modules, ρ is surjective. Consequently, we may decompose

$$\ker(\Xi_2) = \tilde{W}_7^{\mathcal{O}} \oplus \Lambda^2 \oplus S_0^2.$$

Combining this decomposition with Equation (2.4.a) yields the decomposition given in Assertion (3). We use the decomposition of Assertion (3) together with our previous computations to see:

$$\begin{aligned}\dim\{\tilde{W}_7^{\mathcal{O}}\} &= \dim\{\Lambda^2 \otimes S^2\} - 2\dim\{\Lambda^2\} - \dim\{S_0^2\} - \dim\{W_8^{\mathcal{O}}\} \\ &= \tfrac{1}{8}(m-1)(m+1)(m-2)(m+4) \neq 0.\end{aligned}$$

We show that the decomposition of Assertion (3) is into inequivalent irreducible modules and establish Assertion (4) by examining the space of quadratic invariants. We clear the previous notation. We use the formalism of Definition 2.3.1 once again to construct a spanning set for the space of quadratic invariants $\mathcal{I}_2^{\mathcal{O}}(\Lambda^2 \otimes S^2)$:

(1) General remarks. We can permute the indices and we can use the given $\mathbb{Z}_2$ symmetries. Thus in $\mathcal{I}\{(i_1, i_2, i_3, i_4)(j_1, j_2, j_3, j_4)\}$ we may assume $i_1 = 1 < i_2 = 2$, $i_3 \leq i_4$, $j_1 < j_2$, and $j_3 \leq j_4$.

(2) Two indices appear in each monomial. This leads to the invariant $\mathcal{I}\{(1,2,1,2)(*,*,*,*)\} = -\mathcal{I}\{(2,1,2,1)(*,*,*,*)\}$ $= -\mathcal{I}\{(1,2,1,2)(*,*,*,*)\}$ so this vanishes.

(3) Three indices appear in each monomial. This yields:
   (a) $\psi_1 = \mathcal{I}\{(1,2,1,3)(2,4,3,4)\}$.
   (b) $\psi_2 = \mathcal{I}\{(1,2,1,3)(3,4,2,4)\}$.
   (c) $\psi_3 = \mathcal{I}\{(1,2,1,3)(2,3,4,4)\}$.
   (d) $\psi_4 = \mathcal{I}\{(1,2,3,3)(1,2,4,4)\}$.

(4) Each index appears in each monomial. This yields:
   (a) $\psi_5 = ((1,2,3,4)(1,2,3,4))$.
   (b) $\psi_6 = ((1,2,3,4)(1,3,2,4))$.
   (c) $\mathcal{I}\{(1,2,3,4)(3,4,1,2)\} = -\mathcal{I}\{(1,2,4,3)(4,3,1,2)\}$ $= -\mathcal{I}\{(1,2,3,4)(3,4,1,2)\}$ so this invariant vanishes.

This shows that

$$\dim\{\mathcal{I}_2^{\mathcal{O}}(\Lambda^2 \otimes S^2)\} \leq 6.$$

Assertions (3) and (4) now follow from Lemma 2.2.2 and from this calculation.

We now establish Assertion (5). Let $A \in W_7^{\mathcal{O}}$. We check that $A \in \tilde{W}_7^{\mathcal{O}}$ by using the Bianchi identity to verify that the defining relation holds:

$$A_{kjil} + A_{ikjl} - A_{ljik} - A_{iljk} = A_{ijkl} - A_{ijlk} = 0.$$

Thus $W_7^{\mathcal{O}} \subset \tilde{W}_7^{\mathcal{O}}$. Let $\{e_i\}$ be a basis for V. Since $\dim(V) \geq 4$, we can define $A \in \mathfrak{A}$ whose non-zero components are:

$$A_{1234} = A_{1243} = A_{1324} = A_{1342} = A_{1423} = A_{1432} = 1,$$
$$A_{2134} = A_{2143} = A_{3124} = A_{3142} = A_{4123} = A_{4132} = -1.$$

Clearly $\rho(A) = 0$ and $A_{ijkl} = A_{ijlk}$. We also verify A satisfies the Bianchi identity. Thus $W_7^{\mathcal{O}} \neq \{0\}$. Since $W_7^{\mathcal{O}}$ and $\tilde{W}_7^{\mathcal{O}}$ are orthogonal modules and $\tilde{W}_7^{\mathcal{O}}$ is irreducible, the desired equality $W_7^{\mathcal{O}} = \tilde{W}_7^{\mathcal{O}}$ follows. □

We conclude with a general linear module decomposition:

Theorem 2.4.2 *If $m \geq 2$, then $\otimes^2 V^* = \Lambda^2 \oplus S^2$ decomposes as the direct sum of two irreducible general linear modules.*

Proof. Since Λ^2 is an irreducible orthogonal module, it is necessarily an irreducible general linear module. Furthermore, either S^2 is an irreducible general linear module, or S^2 splits as $S_0^2 \oplus \mathbb{R} \cdot \langle \cdot, \cdot \rangle$ as a general linear module.

Since $\mathbb{R}\cdot\langle\cdot,\cdot\rangle$ is not preserved by GL, this latter possibility is ruled out and Theorem 2.4.2 follows. □

2.5 Some Unitary Modules

We adopt the notation of Definition 1.2.1.

Lemma 2.5.1 *Let $(V,\langle\cdot,\cdot\rangle,J_\pm)$ be a para-Hermitian vector space (+) of dimension $m=2\bar{m}$ or a pseudo-Hermitian vector space (−) of dimension $m=2\bar{m}$. Then:*

$\dim\{S^{2,\mathcal{U}_\pm}_\pm\}=\bar{m}^2+\bar{m}$	$\dim\{\Lambda^{2,\mathcal{U}_\pm}_\pm\}=\bar{m}^2-\bar{m}$
$\dim\{S^{2,\mathcal{U}_\pm}_{0,\mp}\}=\bar{m}^2-1$	$\dim\{\Lambda^{2,\mathcal{U}_\pm}_{0,\mp}\}=\bar{m}^2-1$

Proof. Let $\{e_1,J_\pm e_1,\ldots,e_{\bar{m}},J_\pm e_{\bar{m}}\}$ be a basis for V. Let

$$\xi\circ\eta:=\tfrac{1}{2}(\xi\otimes\eta+\eta\otimes\xi)\quad\text{and}\quad\xi\wedge\eta=\tfrac{1}{2}(\xi\otimes\eta-\eta\otimes\xi).$$

Let $1\le i\le\bar{m}$ and $1\le j<k\le\bar{m}$. We have:

$$\begin{aligned}
S^{2,\mathcal{U}_\pm}_\pm&=\operatorname{Span}\{e_i\circ e_i\pm J_\pm e_i\circ J_\pm e_i,\quad e_i\circ J_\pm e_i,\quad e_j\circ e_k\pm J_\pm e_j\circ J_\pm e_k,\\
&\qquad e_j\circ J_\pm e_k+J_\pm e_j\circ e_k\},\\
\Lambda^{2,\mathcal{U}_\pm}_\pm&=\operatorname{Span}\{e_j\wedge e_k\pm J_\pm e_j\wedge J_\pm e_k,\quad e_j\wedge J_\pm e_k+J_\pm e_j\wedge e_k\},\\
S^{2,\mathcal{U}_\pm}_\mp&=\operatorname{Span}\{e_i\circ e_i\mp J_\pm e_i\circ J_\pm e_i,\quad e_j\circ e_k\mp J_\pm e_j\circ J_\pm e_k,\\
&\qquad e_j\circ J_\pm e_k-J_\pm e_j\circ e_k)\},\\
\Lambda^{2,\mathcal{U}_\pm}_\mp&=\operatorname{Span}\{e_i\wedge J_\pm e_i,\quad e_j\wedge e_k\mp J_\pm e_j\wedge J_\pm e_k,\\
&\qquad e_j\wedge J_\pm e_k-J_\pm e_j\wedge e_k\}.
\end{aligned}$$

This permits us to estimate

$$\begin{aligned}
&\dim\{S^{2,\mathcal{U}_\pm}_-\}\le\bar{m}+\bar{m}+\tfrac{1}{2}\bar{m}(\bar{m}-1)+\tfrac{1}{2}\bar{m}(\bar{m}-1)=\bar{m}^2+\bar{m},\\
&\dim\{\Lambda^{2,\mathcal{U}_\pm}_-\}\le\tfrac{1}{2}\bar{m}(\bar{m}-1)+\tfrac{1}{2}\bar{m}(\bar{m}-1)=\bar{m}^2-\bar{m},\\
&\dim\{S^{2,\mathcal{U}_\pm}_+\}\le\bar{m}+\tfrac{1}{2}\bar{m}(\bar{m}-1)+\tfrac{1}{2}\bar{m}(\bar{m}-1)=\bar{m}^2,\\
&\dim\{\Lambda^{2,\mathcal{U}_\pm}_+\}\le\bar{m}+\tfrac{1}{2}\bar{m}(\bar{m}-1)+\tfrac{1}{2}\bar{m}(\bar{m}-1)=\bar{m}^2,\qquad(2.5.a)\\
&4\bar{m}^2=\dim\{V^*\otimes V^*\},\\
&\quad=\dim\{S^{2,\mathcal{U}_\pm}_-\}+\dim\{\Lambda^{2,\mathcal{U}_\pm}_-\}+\dim\{S^{2,\mathcal{U}_\pm}_+\}+\dim\{\Lambda^{2,\mathcal{U}_\pm}_+\}\\
&\quad\le 4\bar{m}^2.
\end{aligned}$$

Consequently, all the inequalities which are present in Equation (2.5.a) are in fact equalities. We complete the proof by verifying:

$$\dim\{S_{0,\mp}^{2,\mathcal{U}_\pm}\} = \dim\{S_{\mp}^{2,\mathcal{U}_\pm}\} - 1, \quad \dim\{\Lambda_{0,\mp}^{2,\mathcal{U}_\pm}\} = \dim\{\Lambda_{\mp}^{2,\mathcal{U}_\pm}\} - 1. \quad \square$$

Let χ be the non-trivial representation of $\mathcal{U}_\pm^\star$ into $\mathbb{Z}_2$ specified previously in Definition 1.2.1. Lemma 1.2.1 will follow from the following result:

Lemma 2.5.2

(1) We have the following orthogonal direct sum decompositions into irreducible modules with structure groups $\mathcal{U}_-$, $\mathcal{U}_-^\star$, or $\mathcal{U}_+^\star$:

(a) $\Lambda^2 = \Lambda_{0,\mp}^{2,\mathcal{U}_\pm} \oplus \mathbb{R}\cdot\Omega_\pm \oplus \Lambda_\pm^{2,\mathcal{U}_\pm}$.

(b) $S^2 = S_{0,\mp}^{2,\mathcal{U}_\pm} \oplus \mathbb{R}\cdot\langle\cdot,\cdot\rangle \oplus S_\pm^{2,\mathcal{U}_\pm}$.

(c) $\otimes^2 V^* = \Lambda_{0,\mp}^{2,\mathcal{U}_\pm} \oplus \mathbb{R}\cdot\Omega_\pm \oplus \Lambda_\pm^{2,\mathcal{U}_\pm} \oplus S_{0,\mp}^{2,\mathcal{U}_\pm} \oplus \mathbb{R}\cdot\langle\cdot,\cdot\rangle \oplus S_\pm^{2,\mathcal{U}_\pm}$.

(2) We have $\mathbb{R}\cdot\Omega_\pm \approx \mathbb{R}\cdot\langle\cdot,\cdot\rangle$ and $\Lambda_{0,+}^{2,\mathcal{U}_-} \approx S_{0,+}^{2,\mathcal{U}_-}$ as modules for the group $\mathcal{U}_-$. Otherwise, these modules are inequivalent.

(3) $\Lambda_{0,\mp}^{2,\mathcal{U}_\pm} \approx S_{0,\mp}^{2,\mathcal{U}_\pm} \otimes \chi$ as a module for the group $\mathcal{U}_\pm^\star$.

(4) $\dim\{\mathcal{I}_2^{\mathcal{U}_-}(V^*\otimes V^*)\} = 8$ and $\dim\{\mathcal{I}_2^{\mathcal{U}_\pm^\star}(V^*\otimes V^*)\} = 6$.

Proof. By Lemma 2.5.1, the modules in Assertion (1) are non-trivial for $m \geq 4$. We examine the space of quadratic invariants of $\otimes^2 V^*$. We modify the notation of Definition 2.3.2 to examine the space of quadratic invariants; we consider strings of the form $(*,*)(*,*)$. We will stratify the space of invariants by the number of times that $J_\pm$ appears. We can permute the basis. We can also replace $\mathcal{I}\{(-,J_\pm 1,-,1,-)\}$ by $-\mathcal{I}\{(-,1,-,J_\pm 1,-)\}$.

(1) $J_\pm$ does not appear.

(a) $\psi_1 := \mathcal{I}\{(1,1)(2,2)\}$.

(b) $\psi_2 := \mathcal{I}\{(1,2)(1,2)\}$.

(c) $\psi_3 := \mathcal{I}\{(1,2)(2,1)\}$.

(2) $J_\pm$ appears once.

(a) $\psi_4 := \mathcal{I}\{(1,J_\pm 1)(2,2)\}$.

(b) $\psi_5 := \mathcal{I}\{(1,2)(2,J_\pm 1)\}$.

(c) $\mathcal{I}\{(1,2)(1,J_\pm 2)\} = -\mathcal{I}\{(1,J_\pm 2)(1,2)\}$ so this invariant does not appear.

(3) $J_\pm$ appears twice.

(a) $\psi_6 := \mathcal{I}\{(1,J_\pm 1)(2,J_\pm 2)\}$.

(b) $\psi_7 := \mathcal{I}\{(1,2)(J_\pm 1,J_\pm 2)\}$.

(c) $\psi_8 := \mathcal{I}\{(1,2)(J_\pm 2,J_\pm 1)\}$.

Following the discussion in Section 2.3, this shows that:

$$\dim \mathcal{I}_2^{\mathcal{U}_-}(V^* \otimes V^*) \leq 8 \quad \text{and} \quad \dim \mathcal{I}_2^{\mathcal{U}_\pm^*}(V^* \otimes V^*) \leq 6. \tag{2.5.b}$$

There are six factors in the decomposition for $\otimes^2 V^*$ as a module for the group $\mathcal{U}_\pm^*$. Consequently, these factors are inequivalent and irreducible by Lemma 2.2.2. This establishes Assertions (1) and (2) in this setting.

Since Ω_- is invariant under the action of $\mathcal{U}_-$, $\mathbb{R} \cdot \Omega_-$ and $\mathbb{R} \cdot \langle \cdot, \cdot \rangle$ are isomorphic trivial modules for the group $\mathcal{U}_-$ of dimension 1. If $\theta \in S_{0,\mp}^{2,\mathcal{U}_\pm}$, we define $T\theta(x, y) := \theta(x, J_\pm y)$. Since $\theta \in S_{0,\mp}^{2,\mathcal{U}_\pm}$, $J_\pm^* \theta = \mp \theta$ so

$$T\theta(y, x) = \theta(y, J_\pm x) = \mp\theta(J_\pm y, J_\pm J_\pm x) = -\theta(x, J_\pm y) = -T\theta(x, y).$$

This shows $T\theta \in \Lambda^2$. We compute:

$$\begin{aligned} J_\pm^* T\theta(x, y) &= T\theta(J_\pm x, J_\pm y) = \theta(J_\pm x, J_\pm J_\pm y) = \mp\theta(x, J_\pm y) \\ &= \mp T\theta(x, y). \end{aligned}$$

This shows $T\theta \in \Lambda_\mp^{2,\mathcal{U}_\pm}$. We show $T\theta \perp \Omega_\pm$ by computing the contraction:

$$\begin{aligned} \langle \Omega_\pm, T\theta \rangle &= \varepsilon^{ij}\varepsilon^{kl}\langle e_i, J_\pm e_k \rangle \theta(e_j, J_\pm e_l) \\ &= \mp\varepsilon^{ij}\varepsilon^{kl}\langle e_i, e_k \rangle \theta(e_j, e_l) = \mp\langle \{\langle \cdot, \cdot \rangle\}, \theta \rangle \\ &= 0. \end{aligned}$$

This defines a $\mathcal{U}_\pm$ equivariant map from $S_{0,\mp}^{2,\mathcal{U}_\pm}$ to $\Lambda_{0,\mp}^{2,\mathcal{U}_\pm}$. The inverse of this map is defined by setting $T\omega(x, y) := \pm\omega(x, J_\pm y)$ for $\omega \in \Lambda_{0,\mp}^{2,\mathcal{U}_\pm}$. Thus these two modules are isomorphic. This shows that

$$\otimes^2 V^* = \mathbb{R} \oplus \chi \oplus S_{0,\mp}^{2,\mathcal{U}_\pm} \oplus \Lambda_{0,\mp}^{2,\mathcal{U}_\pm} \oplus S_\pm^{2,\mathcal{U}_\pm} \oplus \Lambda_\pm^{2,\mathcal{U}_\mp}.$$

Assertions (1) and (2) for the group $\mathcal{U}_-$ now follow from Lemma 2.2.2 and from the estimate

$$\dim\{\mathcal{I}_2^{\mathcal{U}_-}(\otimes^2 V^*)\} \leq 8.$$

Assertion (3) follows by observing that $Tg = -gT$ if $g \in \mathcal{U}^\star - \mathcal{U}$; Assertion (4) follows from the arguments given above. □

We observe that Lemma 2.5.2 fails for the group $\mathcal{U}_+$:

Lemma 2.5.3 *$S_+^{2,\mathcal{U}_+}$ and $\Lambda_+^{2,\mathcal{U}_+}$ are not irreducible modules with structure group $\mathcal{U}_+$.*

Proof. Since $J_+^2 = \mathrm{Id}$ and $\mathrm{Tr}(J_+) = 0$, we can decompose $V = V_+ \oplus V_-$ into the ± 1 eigenvalues of J_+ where $\dim(V_+) = \dim(V_-) = \frac{1}{2}\dim(V)$. This decomposes

$$\begin{aligned}
&S^2 \approx S^2(V_+^*) \oplus S^2(V_-^*) \oplus \{V_+^* \otimes V_-^*\},\\
&S_+^{2,\mathcal{U}_+} \approx S^2(V_+^*) \oplus S^2(V_-^*), \quad S_-^{2,\mathcal{U}_+} \approx V_+^* \otimes V_-^*,\\
&\Lambda^2 \approx \Lambda^2(V_+^*) \oplus \Lambda^2(V_-^*) \oplus \{V_+^* \otimes V_-^*\},\\
&\Lambda_+^{2,\mathcal{U}_+} \approx \Lambda^2(V_+^*) \oplus \Lambda^2(V_-^*), \quad \Lambda_-^{2,\mathcal{U}_+} \approx V_+^* \otimes V_-^*.
\end{aligned}$$ □

We continue our examination of these modules:

Lemma 2.5.4

(1) $\dim\{\mathrm{Hom}_{\mathcal{U}_-}(S_-^{2,\mathcal{U}_-}, S_-^{2,\mathcal{U}_-})\} = \dim\{\mathrm{Hom}_{\mathcal{U}_-}(\Lambda_-^{2,\mathcal{U}_-}, \Lambda_-^{2,\mathcal{U}_-})\} = 2.$

(2) $\dim\{\mathrm{Hom}_{\mathcal{U}_-}(S_{0,+}^{2,\mathcal{U}_-}, S_{0,+}^{2,\mathcal{U}_-})\} = \dim\{\mathrm{Hom}_{\mathcal{U}_-}(\Lambda_{0,+}^{2,\mathcal{U}_-}, \Lambda_{0,+}^{2,\mathcal{U}_-})\} = 1.$

(3) $\dim\{\mathrm{Hom}_{\mathcal{U}_\pm^\star}(S_\pm^{2,\mathcal{U}_-}, S_\pm^{2,\mathcal{U}_-})\} = \dim\{\mathrm{Hom}_{\mathcal{U}_\pm^\star}(\Lambda_\pm^{2,\mathcal{U}_-}, \Lambda_\pm^{2,\mathcal{U}_-})\} = 1.$

(4) $\dim\{\mathrm{Hom}_{\mathcal{U}_\pm^\star}(S_{0,\mp}^{2,\mathcal{U}_\pm^\star}, S_{0,\mp}^{2,\mathcal{U}_\pm})\} = \dim\{\mathrm{Hom}_{\mathcal{U}_\pm^\star}(\Lambda_{0,\mp}^{2,\mathcal{U}_\pm}, \Lambda_{0,\mp}^{2,\mathcal{U}_\pm})\} = 1.$

Proof. If $\theta \in V^* \otimes V^*$, let $T\theta(x, y) := \theta(x, J_- y)$. Then T is a $\mathcal{U}_-$ equivariant map of $V^* \otimes V^*$. Consequently by Lemma 2.5.2, T preserves $S_-^{2,\mathcal{U}_-}$ and $\Lambda_-^{2,\mathcal{U}_-}$. Since $T^* = -\mathrm{Id}$, and since the identity is always an equivariant map, we have

$$\begin{aligned}
&\dim\{\mathrm{Hom}_{\mathcal{U}_-}(S_-^{2,\mathcal{U}_-}, S_-^{2,\mathcal{U}_-})\} \geq 2, \ \dim\{\mathrm{Hom}_{\mathcal{U}_-}(\Lambda_-^{2,\mathcal{U}_-}, \Lambda_-^{2,\mathcal{U}_-})\} \geq 2,\\
&\dim\{\mathrm{Hom}_{\mathcal{U}_-}(S_{0,+}^{2,\mathcal{U}_-}, S_{0,+}^{2,\mathcal{U}_-})\} \geq 1, \ \dim\{\mathrm{Hom}_{\mathcal{U}_-}(\Lambda_{0,+}^{2,\mathcal{U}_-}, \Lambda_{0,+}^{2,\mathcal{U}_-})\} \geq 1,\\
&\dim\{\mathrm{Hom}_{\mathcal{U}_\pm^\star}(S_\pm^{2,\mathcal{U}_\pm}, S_\pm^{2,\mathcal{U}_\pm})\} \geq 1, \ \dim\{\mathrm{Hom}_{\mathcal{U}_\pm^\star}(\Lambda_\pm^{2,\mathcal{U}_\pm}, \Lambda_\pm^{2,\mathcal{U}_\pm})\} \geq 1,\\
&\dim\{\mathrm{Hom}_{\mathcal{U}_\pm^\star}(S_{0,\mp}^{2,\mathcal{U}_\pm^\star}, S_{0,\mp}^{2,\mathcal{U}_\pm})\} \geq 1, \ \dim\{\mathrm{Hom}_{\mathcal{U}_\pm^\star}(\Lambda_{0,\mp}^{2,\mathcal{U}_\pm}, \Lambda_{0,\mp}^{2,\mathcal{U}_\pm})\} \geq 1.
\end{aligned}$$

Since we may identify $\mathrm{Hom}_G(\xi, \xi) = \mathrm{Hom}_G(\xi \otimes \xi, \mathbb{R})$, we may use the formalism of Definition 2.3.2 to establish reverse inequalities; this will then complete the proof. We consider strings $\mathcal{I}\{(\nu_1, \nu_2; \nu_3, \nu_4)\}$ where $\nu_1 \leq \nu_2$ and $\nu_3 \leq \nu_4$.

(1) General remarks. We stratify the invariants by the number of indices which are decorated by $J_\pm$. Suppose $J_\pm$ decorates both indices. We can exchange the $J_\pm$ decoration on one occurrence of an index with the other occurrence of the other index. Thus we could replace

$\mathcal{I}\{(J_\pm 1, 2; J_\pm 2, 1)\}$ and $\mathcal{I}\{(J_\pm 1, 2; 1, J_\pm 2)\}$ by $-\mathcal{I}\{(J_\pm 1, J_\pm 2; 2, 1)\}$ and $-\mathcal{I}\{(J_\pm, 1, J_\pm 2, ; 1, 2)\}$, respectively. Furthermore, we could then express $\mathcal{I}\{(J_\pm 1, J_\pm 2; \star, \star)\} = \pm\mathcal{I}\{(1, 2; \star, \star)\}$ so we will not consider such invariants. Thus if both indices are decorated, we need only consider $\mathcal{I}\{(1, J_\pm 1; 2, J_\pm 2)\}$.

(2) Let $\xi_\pm := S_\pm^{2,\mathcal{U}_\pm}$.

(a) There are no $J_\pm$ terms. We have

i. $\psi_1 := \mathcal{I}\{(1, 2; 1, 2)\}$.

ii. $\mathcal{I}\{(1, 1; 2, 2)\} = \mp\mathcal{I}\{(J_\pm 1, J_\pm 1; 2, 2)\} = -\mathcal{I}\{(1, 1; 2, 2)\} = 0$.

(b) There is one $J_\pm$ term. This is not possible for $\mathcal{U}_\pm^\star$.

i. $\mathcal{I}\{(1, J_\pm 1; \star, \star)\} = \mp\mathcal{I}\{(J_\pm 1, J_\pm 1 J_\pm 1; \star, \star)\}$ by changing the basis we see this equals $-\mathcal{I}\{(1, J_\pm 1; \star, \star)\}$ by equivariance – thus this invariant does not appear.

ii. $\psi_2 := \mathcal{I}\{(J_\pm 1, 2; 1, 2)\}$.

(c) There are two $J_\pm$ terms; $\mathcal{I}\{(1, J_\pm 1; 2, J_\pm 2)\} = 0$ by (2)(b)i.

This shows $\dim\{\text{Hom}_{\mathcal{U}_-}(\xi_-, \xi_-)\} \leq 2$ and $\dim\{\text{Hom}_{\mathcal{U}_\pm^\star}(\xi_\pm, \xi_\pm)\} \leq 1$.

(3) We clear the previous notation. Let $\xi_\pm := \Lambda_\pm^{2,\mathcal{U}_\pm}$.

(a) There are no $J_\pm$ terms. We have $\psi_1 := \mathcal{I}\{(1, 2; 1, 2)\}$.

(b) There is one $J_\pm$ term. This is not possible for $\mathcal{U}_\pm^\star$.

i. $\psi_2 := \mathcal{I}\{(J_\pm 1, 2; 1, 2)\}$.

ii. $\mathcal{I}\{(1, J_\pm 1; 2, 2)\}$ does not appear by (2)(b)i.

(c) There are two $J_\pm$ terms. But $\mathcal{I}\{(1, J_\pm 1; 2, J_\pm 2)\} = 0$ by (2)(b)i.

This shows $\dim\{\text{Hom}_{\mathcal{U}_-}(\xi_-, \xi_-)\} \leq 2$ and $\dim\{\text{Hom}_{\mathcal{U}_\pm^\star}(\xi_\pm, \xi_\pm)\} \leq 1$.

(4) We clear the previous notation. Let $\xi_\pm := S_{0,\mp}^{2,\mathcal{U}_\pm}$.

(a) There are no $J_\pm$ terms. We have

i. $\mathcal{I}\{(1, 1; 2, 2)\} = 0$ as $\Theta \perp \langle\cdot, \cdot\rangle$.

ii. $\psi_1 := \mathcal{I}\{(1, 2; 1, 2)\}$.

(b) There is one $J_\pm$ term. We have

i. $\mathcal{I}\{(1, J_\pm 1; 2, 2)\} = -\mathcal{I}\{(J_\pm 1, 1; 2, 2)\} = -\mathcal{I}\{(1, J_\pm 1; 2, 2)\}$ so this term vanishes. Similarly $\mathcal{I}\{(1, 1; 2, J_\pm 2)\}$ vanishes.

ii. $\mathcal{I}\{(1, J_\pm 2; 1, 2)\} = \mp\mathcal{I}\{(J_\pm 1, J_\pm J_\pm 2; 1, 2)\} = -\mathcal{I}\{(J_\pm 1, 2; 1, 2)\}$ $= -\mathcal{I}\{(2, J_\pm 1; 2, 1)\} = -\mathcal{I}\{(1, J_\pm 2; 1, 2)\}$ so this term vanishes.

(c) There are two $J_\pm$ terms. We have $\mathcal{I}\{(1, J_\pm 2; J_\pm 1, 2)\}$ $= \mp\mathcal{I}\{(1, J_\pm 2; 1, J_\pm 2)\} = \mathcal{I}\{(1, 2; 1, 2)\} = \psi_1$.

This shows $\dim\{\mathrm{Hom}_{\mathcal{U}_-}(\xi_-,\xi_-)\}\le 1$ and $\dim\{\mathrm{Hom}_{\mathcal{U}_\pm^\star}(\xi_\pm,\xi_\pm)\}\le 1$.

(5) We clear the previous notation. Let $\xi := \Lambda_{0,\mp}^{2,\mathcal{U}_\pm}$.

(a) There are no $J_\pm$ terms. This leads to one invariant $\psi_1 = \mathcal{I}\{(1,2;1,2)\}$.

(b) There is one $J_\pm$ term. This leads to one possibility $\mathcal{I}\{(J_\pm 1,2)(1,2)\} = \mp\mathcal{I}\{(J_\pm J_\pm 1, J_\pm 2;1,2)\}$ $= -\mathcal{I}\{(1,J_\pm 2;1,2)\} = -\mathcal{I}\{(J_\pm 2,1;2,1)\} = -\mathcal{I}\{(J_\pm 1,2;1,2)\}$ so this does not appear.

(c) There are two $J_\pm$ terms. This leads to one possibility $\mathcal{I}\{(1,J_\pm 1;2,J_\pm 2)\}$. This is zero as $\Theta\perp\Omega$.

This shows $\dim\{\mathrm{Hom}_{\mathcal{U}_-}(\xi_-,\xi_-)\}\le 1$ and $\dim\{\mathrm{Hom}_{\mathcal{U}_\pm^\star}(\xi_\pm,\xi_\pm)\}\le 1$.

The proof is now complete. □

2.6 Compact Lie Groups

The existence of a non-degenerate invariant inner product was central in the previous discussion. For compact groups, the metric can always be chosen to be positive definite. Since such an inner product is non-degenerate on every linear subset, the results established previously in Chapter 2 also pertain in this setting.

Lemma 2.6.1 *Let G be a compact Lie group.*

(1) There is a unique invariant smooth measure $|\,\mathrm{dvol}(g)|$ with $|\,\mathrm{dvol}(g^{-1})| = |\,\mathrm{dvol}(g)|$ on G so $\int_G|\,\mathrm{dvol}(g)| = 1$ which is both left and right invariant under the group action.

(2) Any module $\xi = (V,\sigma)$ for the group G admits a positive definite inner product which is invariant under the action of G.

Proof. Let $\{\phi_1,\dots,\phi_\ell\}$ be a frame for the cotangent bundle of G which is invariant under the left action of G. Note that $\ell = \dim(G)$. Then $\phi_1\wedge\dots\wedge\phi_\ell$ is a smooth measure $|\,\mathrm{dvol}(g)|$ which is invariant under the left action of G. We rescale if necessary to ensure that $\int_G|\,\mathrm{dvol}(g)| = 1$. Let R_h denote right multiplication by h. Then $R_h^*|\,\mathrm{dvol}(g)|$ is again a smooth measure which is invariant under the left action of G. Consequently,

$$R_h^*|\,\mathrm{dvol}(g)| = c(h)|\,\mathrm{dvol}(g)|$$

where $c(h) > 0$. Since R_h is a diffeomorphism and integration is invariant under diffeomorphism,

$$\int_G R_h^* |\operatorname{dvol}(g)| = \int_G |\operatorname{dvol}(g)|\,;$$

this implies that $c(h) = 1$ for all h in G and hence $|\operatorname{dvol}(g)|$ is invariant under both the left and right group action. Since $|\operatorname{dvol}(g^{-1})|$ is also a measure of total mass 1 which is invariant under both the left and right group action, $|\operatorname{dvol}(g^{-1})| = |\operatorname{dvol}(g)|$; Assertion (1) now follows. If f is a smooth function on G, then:

$$\begin{aligned} &\int_G f(gh)|\operatorname{dvol}(g)| = \int_G f(hg)|\operatorname{dvol}(g)| \\ &= \int_G f(g)|\operatorname{dvol}(g)| = \int_G f(g^{-1})|\operatorname{dvol}(g)|. \end{aligned}$$

Let $\langle\cdot,\cdot\rangle_0$ be any positive definite inner product on $V = V_\xi$. We average $\langle\cdot,\cdot\rangle_0$ over G to define a positive definite inner product:

$$\langle v_1, v_2\rangle := \int_G \langle\sigma(g)v_1, \sigma(g)v_2\rangle_0 |\operatorname{dvol}(g)|.$$

We show $\langle\cdot,\cdot\rangle$ is invariant under the group action and establish Assertion (2) by computing:

$$\begin{aligned} \langle\sigma(h)v_1, \sigma(h)v_2\rangle &= \int_G \langle\sigma(g)\sigma(h)v_1, \sigma(g)\sigma(h)v_2\rangle_0 |\operatorname{dvol}(g)| \\ &= \int_G \langle\sigma(gh)v_1, \sigma(gh)v_2\rangle_0 |\operatorname{dvol}(g)| = \int_G \langle\sigma(gh)v_1, \sigma(gh)v_2\rangle_0 |\operatorname{dvol}(gh)| \\ &= \int_G \langle\sigma(g_1)v_1, \sigma(g_1)v_2\rangle_0 |\operatorname{dvol}(g_1)| = \langle v_1, v_2\rangle. \end{aligned}$$

□

Let $\langle\cdot,\cdot\rangle$ be such an inner product.

Lemma 2.6.2 *Let G be a compact Lie group. Let $\xi = (V, \sigma)$ be a module for the group G. Let $V^G = \{v \in V : \sigma(g)v = v\ \forall\ g \in G\}$. Let π define orthogonal projection on V^G; π is given by:*

$$\pi := \int_G \sigma(g)|\operatorname{dvol}(g)|.$$

Proof. If $v \in V^G$ is a fixed vector, then $\sigma(g)v = v$ for all g in G. Thus

$$\pi v = \int_G \sigma(g)v|\operatorname{dvol}(g)| = \int_G v|\operatorname{dvol}(g)| = v.$$

This shows $\pi = \text{Id}$ on V^G. We show $\pi(v) \in V^G$ for any $v \in V$ by computing:

$$\begin{aligned}\sigma(h)\pi v &= \int_G \sigma(h)\sigma(g)v|\,\text{dvol}(g)| = \int_G \sigma(hg)v|\,\text{dvol}(g)| \\ &= \int_G \sigma(g)v|\,\text{dvol}(h^{-1}g)| = \int_G \sigma(g)v|\,\text{dvol}(g)| = \pi v.\end{aligned}$$

This shows $\text{Range}\,\pi = V^G$. Since π is the identity on V^G, $\pi^2 = \pi$. Since $\langle\cdot,\cdot\rangle$ is an inner product invariant under the action of G, $\sigma(g)^* = \sigma(g)^{-1}$. We see that π is self-adjoint by computing:

$$\begin{aligned}\pi^* v &= \int_G \sigma^*(g)v|\,\text{dvol}(g)| = \int_G \sigma(g)^{-1}v|\,\text{dvol}(g)| \\ &= \int_G \sigma(g^{-1})v|\,\text{dvol}(g)| = \int_G \sigma(g)v|\,\text{dvol}(g^{-1})| \\ &= \int_G \sigma(g)v|\,\text{dvol}(g)| = \pi v.\end{aligned}$$

It now follows that π is an orthogonal projection. □

We have dealt with real representations. We now turn to the complex theory. The analysis of complex modules for the group G is a bit simpler than the corresponding theory for real modules for the group G. We begin our analysis with:

Lemma 2.6.3 *Let G be a compact Lie group.*

(1) If ξ is an irreducible complex module for the group G, then

$$\dim_{\mathbb{C}}\{Hom_{\mathbb{C}}^G(\xi,\xi)\} = 1.$$

(2) Every complex module for the group G ξ is completely reducible.

(3) Expand $\xi = \sum_i n_i\xi_i$ where the ξ_i are inequivalent irreducible modules. Then $n_i = \dim\{Hom_{\mathbb{C}}^G(\xi_i,\xi)\}$.

Proof. We use the arguments given to prove Lemma 2.1.6, Lemma 2.6.1, and Lemma 2.6.2. Let T be an equivariant homomorphism from ξ to ξ. Let λ be a complex eigenvalue of T. Set

$$E_\lambda := \{v \in V : Tv = \lambda v\}.$$

If $v \in E_\lambda$ and if $g \in G$, then $T(g \cdot v) = g \cdot Tv = \lambda g \cdot v$ and thus E_λ is a module for the group G. Since $E_\lambda \neq \{0\}$ and since ξ is irreducible, $E_\lambda = V$ and thus $T = \lambda \cdot \text{Id}$. Assertion (1) now follows. We average over the group to construct an invariant positive definite Hermitian inner product on ξ and establish Assertion (2). Assertion (3) is immediate from Assertions (1) and (2). □

We can now study the orthogonality relations. Let $\xi = (V, \sigma)$ be an irreducible complex module for the group G. Choose an orthonormal basis $\{e_i\}$ for V and let $\sigma(g)e_i = \sum_j \xi_{ij}(g)e_j$ be the matrix coefficients.

Lemma 2.6.4 *Let G be a compact Lie group. Let ξ and η be irreducible complex modules for the group G.*

(1) If ξ and η are inequivalent, then $(\xi_{uv}, \eta_{ij})_{L^2(G)} = 0$.
(2) $(\xi_{uv}, \xi_{ij})_{L^2(G)} = \frac{1}{\dim(\xi)}\delta_{ui}\delta_{vj}$.

Proof. The space of complex linear maps $\mathrm{Hom}_{\mathbb{C}}(V_\eta, V_\xi)$ from V_η to V_ξ is a complex representation space for G. If $T \in \mathrm{Hom}_{\mathbb{C}}(V_\eta, V_\xi)$, then

$$\sigma_{\mathrm{Hom}_{\mathbb{C}}(V_\eta, V_\xi)}(g)T := \sigma(g)_\xi \circ T \circ \sigma_\eta(g)^{-1}.$$

Choose inner products on V_ξ and on V_η to be invariant under the group action. We then have $\sigma_\eta(g)^{-1} = \sigma_\eta(g)^*$. Let

$$\pi : \mathrm{Hom}_{\mathbb{C}}(V_\eta, V_\xi) \to \mathrm{Hom}_{\mathbb{C}}(\eta, \xi) = \mathrm{Hom}_{\mathbb{C}}(V_\eta, V_\xi)^G$$

denote orthogonal projection. We use Lemma 2.6.1 and Lemma 2.6.2 to express:

$$\begin{aligned}(\pi T)_{uj} &= \sum_{iv} \int_G \xi_{uv}(g) T_{vi} \eta^*_{ij}(g) |\,\mathrm{dvol}(g)| \\ &= \sum_{iv} T_{vi} \int_G \xi_{uv}(g)\eta^*_{ij}(g) |\,\mathrm{dvol}(g)| = \sum_{iv} T_{vi}(\xi_{uv}, \eta_{ji})_{L^2(G)}.\end{aligned}$$

The collection of numbers $(\xi_{uv}, \eta_{ji})_{L^2(G)}$ defines a linear transformation of $\mathrm{Hom}_{\mathbb{C}}(V_\eta, V_\xi)$. There are $\dim(\xi)^2$ of the first sort of indices (these comprise the $1 \le u, v \le \dim(\xi)$ indices) and $\dim(\eta)^2$ of the second sort of indices (these are the $1 \le i, j \le \dim(\eta)$ indices). So this is a huge matrix that acts on a vector space of dimension $\dim(\xi)\dim(\eta)$. If ξ and η are not equivalent, then $\pi = 0$. This means this linear transformation vanishes. Hence all of the entries in this huge matrix vanish; Assertion (1) follows.

On the other hand, if $\xi = \eta$, then $\mathrm{Hom}_{\mathbb{C}}(\xi, \xi) = \mathrm{Id} \cdot \mathbb{C}$ by Lemma 2.6.3. Let A^{uj} be the matrix whose only non-zero entry is in position (u, j). The $\{A^{uj}\}$ are an orthonormal basis for $\mathrm{Hom}_{\mathbb{C}}(V_{\xi,}, V_{\xi,})$. Since $\mathrm{Id} = \sum_u A^{uu}$,

$||\operatorname{Id}||^2 = \dim(\xi)$ and we may express:

$$\pi(T)_{uj} = \left\{\frac{1}{\dim(\xi)}\langle T, \operatorname{Id}\rangle \cdot \operatorname{Id}\right\}_{uj} = \frac{1}{\dim(\xi)} T_{kk}\delta_{uj} = (\xi_{uv}, \xi_{ji})_{L^2(G)} T_{vi}.$$

If $u \neq j$, then $\pi(T)$ has no component in A^{uj} and thus all the coefficients vanish. This shows $(\xi_{uv}, \xi_{ji})_{L^2(G)} = 0$ for $u \neq j$. Similarly, since T_{vi} plays no role for $v \neq i$, we have $(\xi_{uv}, \xi_{ji})_{L^2(G)} = 0$ for $v \neq i$. Finally, if $u = j$ and $v = i$, the coefficient is $\frac{1}{\dim(\xi)}$. Assertion (2) now follows. □

The left regular action

$$(L_g f)(h) := f(gh)$$

makes $L^2(G)$ into a representation space G which has infinite dimension. Let $\operatorname{Irr}_{\mathbb{C}}(G)$ be the set of equivalence classes of irreducible complex modules for the group G. We have shown that $\{\xi_{ij}\}_{\xi \in \operatorname{Irr}_{\mathbb{C}}(G)}$ is an orthogonal subset of $L^2(G)$. Thus in particular all these functions are linearly independent. Set

$$A_\xi^j := \operatorname{Span}_{1\leq i\leq \dim(\xi)}\{\xi_{ij}\} \subset L^2(G),$$
$$A_\xi := \operatorname{Span}_{1\leq i,j\leq \dim(\xi)}\{\xi_{ij}\} \subset L^2(G).$$

Lemma 2.6.5

(1) $\dim(A_\xi^j) = \dim(\xi)$ *and* $\dim(A_\xi) = \dim(\xi)^2$.
(2) $A_\xi^j \perp A_\xi^i$ *in* $L^2(G)$ *for* $i \neq j$.
(3) $A_\xi \perp A_\tau$ *in* $L^2(G)$ *for* $\xi \neq \tau$.
(4) $L_g A_\xi^j = A_\xi^j$ *so* A_ξ^j *is a representation space for* G *of finite dimension.*
(5) A_ξ^j *is isomorphic to* V_ξ.
(6) Let $\tilde{A}$ *be a subspace of* $L^2(G)$ *whose dimension is finite. Assume the left action of* G *preserves* $\tilde{A}$. *Also assume* $\tilde{A}$ *is abstractly isomorphic to* V_ξ *as a representation space. Then* $\tilde{A} \subset A_\xi$ *in* $L^2(G)$.

Proof. Assertions (1)–(3) follow from the orthogonality relations of Lemma 2.6.4. We now establish Assertions (4) and (5). We have that

$$\{L_g\xi_{ij}\}(h) = \xi_{ij}(gh) = \textstyle\sum_\ell \xi_{i\ell}(g)\xi_{\ell j}(h).$$

This means that as functions, we have the corresponding identity

$$L_g\xi_{ij} = \textstyle\sum_\ell \xi_{i\ell}(g)\xi_{\ell j}.$$

This shows that the space A_ξ^j is invariant under L_g and that the matrix representation is given by ξ_{ij} relative to the canonical basis. Finally, suppose $\tilde{A}$ is a subspace of $L^2(G)$ whose dimension is finite. Also assume $\tilde{A}$ is abstractly isomorphic to V_ξ as a representation space. Choose a basis f_i for $\tilde{A}$ so that

$$L_g f_\nu = \textstyle\sum_\mu \xi_{\nu\mu}(g) f_\mu .$$

Evaluating at 1 then yields

$$f_\nu(g) = (L_g f_\nu)(1) = \textstyle\sum_\mu \xi_{\nu\mu}(g) f_\mu(1).$$

Thus f_ν is a linear combination of the $\xi_{\nu\mu}$ so $\tilde{A} \subset A$. □

If $\eta \in \mathrm{Irr}_{\mathbb{C}}(G)$, let $\mu_\eta(\xi)$ be the multiplicity with which η appears in ξ. We can now establish a basic density result:

Lemma 2.6.6

(1) $\{\xi_{ij}\}_{\xi\in\mathrm{Irr}_{\mathbb{C}}(G)}$ *is a complete orthogonal basis for* $L^2(G)$.
(2) $L^2(G) = \oplus_{\xi\in\mathrm{Irr}_{\mathbb{C}}(G)} \oplus_{1\le j\le\dim(\xi)} A_\xi^j$.

Since $||\xi_{ij}||_{L^2} = \frac{1}{\dim(\xi)}$, $\{\xi_{ij}\}_{\xi\in\mathrm{Irr}_{\mathbb{C}}(G)}$ is not an orthonormal basis.

Proof. We decompose $L^2(G) = \oplus_\lambda E(\lambda)$ into the eigenspaces of the Laplacian; see, for example, [Seeley (1969)]. Each eigenspace $E(\lambda)$ is a representation space for G whose dimension is finite. We decompose each $E(\lambda)$ as the direct sum of irreducible modules. Each irreducible module is a subspace of some A_ξ by Lemma 2.6.5 and thus $E(\lambda) \subset \oplus_\xi A_\xi$. This shows $L^2(G) \subset \oplus_\xi A_\xi$; the reverse inclusion is trivial. □

The following result [Peter and Weyl (1927)] follows from the discussion above.

Theorem 2.6.1 *Let G be a compact Lie group. Then*

(1) $\{\xi_{ij}\}_{\xi\in\mathrm{Irr}_{\mathbb{C}}(G)}$ *is a complete orthogonal basis for* $L^2(G;\mathbb{C})$.

(2) $L^2(G;\mathbb{C}) = \oplus_{\xi\in\mathrm{Irr}_{\mathbb{C}}(G)} \dim(\xi)\cdot\xi$ *as a representation space for G under left multiplication.*

Chapter 3

Connections, Curvature, and Differential Geometry

In Chapter 3, we present elementary results in curvature theory. Section 3.1 deals with affine connections. These are connections on the tangent bundle whose torsion tensor vanishes. Section 3.2 deals with equiaffine connections; these are the Ricci symmetric affine connections. In Section 3.3, we consider the Levi-Civita connection defined by a pseudo-Riemannian metric of arbitrary signature. In Section 3.4, we give an introduction to complex and para-complex geometry. In Section 3.5, we turn our attention to para-Hermitian geometry and to pseudo-Hermitian geometry where we assume the (para)-complex structure $J_\pm$ is integrable and $J_\pm^* g = \mp g$. There is an additional identity the Riemann curvature tensor satisfies in this setting. This identity was found by [Gray (1976)] in positive definite signature. Rather than following Gray's original proof, we present a different combinatorial proof of this identity in arbitrary signature. A useful result is established in Section 3.6; if the (para)-Kähler form vanishes at a point, there exist (para)-holomorphic coordinates so the first derivatives of the metric vanish at this point.

3.1 Affine Connections

Let $x = (x^1, \dots, x^m)$ be a system of local coordinates on a smooth manifold M of dimension m. Set:

$$\partial_{x_i} := \frac{\partial}{\partial x_i}.$$

Let $C^\infty(TM)$ denote the space of *smooth vector fields* on M. A *connection* on TM is a partial differential operator of order one:

$$\nabla : C^\infty(TM) \to C^\infty(T^*M \otimes TM).$$

There is an associated *directional derivative.* If x and y are smooth vector fields, we set $\nabla_x y = x \cdot \nabla y$ using the natural pairing between TM and T^*M to define a smooth vector field $\nabla_x y$. For fixed x, the map $y \to \nabla_x y$ is a differential operator from $C^\infty(TM)$ to itself. We may then express

$$\nabla y = \sum_i dx^i \otimes \nabla_{\partial_{x_i}} y.$$

We expand

$$\nabla_{\partial_{x_i}} \partial_{x_j} = \Gamma_{ij}{}^k \partial_{x_k}$$

to define the *Christoffel symbols* of ∇ of the first kind. The connection is determined by the Christoffel symbols; since ∇ is a partial differential operator of order one, we have

$$\nabla_{a^i \partial_{x_i}} (b^j \partial_{x_j}) = \{a^i \partial_{x_i} b^k + a^i b^j \Gamma_{ij}{}^k\} \partial_{x_k}.$$

If x and y are vector fields, the *Lie bracket* $[\cdot, \cdot]$ is defined to be:

$$[x, y] := xy - yx. \tag{3.1.a}$$

This can be written in coordinates in the form

$$[a^i \partial_{x_i}, b^j \partial_{x_j}] = \{a^i \partial_{x_i} b_j - b^i \partial_{x_i} a^j\} \partial_{x_j}.$$

If ∇ is an arbitrary connection on TM, the *torsion tensor* $\mathcal{T}$ of ∇ is defined by setting:

$$\mathcal{T}(x, y) = \nabla_x y - \nabla_y x - [x, y].$$

It is clear that this is anti-symmetric in $\{x, y\}$. Furthermore this is a tensor since we have for any smooth function f that:

$$\mathcal{T}(fx, y) = \mathcal{T}(x, fy) = f\mathcal{T}(x, y).$$

We say that ∇ is an *affine connection* if $\mathcal{T} = 0$ or, equivalently, if ∇ is *without torsion.*

Lemma 3.1.1 *Let ∇ be a connection on TM. Fix $P \in M$. The following conditions are equivalent:*

(1) There exist local coordinates centered at P so $\Gamma_{ij}{}^k(P) = 0$.
(2) The torsion tensor $\mathcal{T}$ vanishes at P.

Proof. Let $x = (x^1, \ldots, x^m)$ be an initial system of local coordinates on M. The torsion tensor $\mathcal{T}$ vanishes at P if and only if we have the symmetry $\Gamma_{ij}{}^k(P) = \Gamma_{ji}{}^k(P)$. In particular, if there exists a coordinate system where $\Gamma(P) = 0$, then necessarily $\mathcal{T}$ vanishes at P. Thus Assertion (1) implies Assertion (2). Conversely, assume that Assertion (2) holds. Define a new system of coordinates by setting:

$$z^i = x^i + \tfrac{1}{2} \sum_{j,k} c^i{}_{jk} x^j x^k$$

where $c^i{}_{jk} = c^i{}_{kj}$ remains to be chosen. As $\partial_{x_j} = \partial_{z_j} + c^l{}_{ji} x^i \partial_{z_l}$,

$$\nabla_{\partial_{x_i}} \partial_{x_j}(P) = \nabla_{\partial_{z_i}} \partial_{z_j}(P) + c^l{}_{ji} \partial_{z_l}(P).$$

Lemma 3.1.1 now follows by setting $c^l{}_{ij} := \Gamma_{ji}{}^l(P)$; the fact that $c^l{}_{ij} = c^l{}_{ji}$ is exactly the assumption that the torsion tensor of ∇ vanishes P. □

We now show that the curvature tensor of an affine connection has the symmetries given in Equation (1.3.a) and that the covariant derivative of the curvature has the symmetries given in Equation (1.3.b):

Lemma 3.1.2 *Let ∇ be an affine connection on TM. Then:*

(1) $R_{ijk}{}^l = -R_{jik}{}^l$.
(2) $R_{ijk}{}^l + R_{jki}{}^l + R_{kij}{}^l = 0$.
(3) $R_{ijk}{}^l{}_{;n} = -R_{jik}{}^l{}_{;n}$.
(4) $R_{ijk}{}^l{}_{;n} + R_{jki}{}^l{}_{;n} + R_{kij}{}^l{}_{;n} = 0$.
(5) $R_{ijk}{}^l{}_{;n} + R_{jnk}{}^l{}_{;i} + R_{nik}{}^l{}_{;j} = 0$.

Proof. The $\mathbb{Z}_2$ symmetries of Assertions (1) and (3) are immediate. One may compute directly that

$$R_{ijk}{}^l = \partial_{x_i} \Gamma_{jk}{}^l - \partial_{x_j} \Gamma_{ik}{}^l + \Gamma_{in}{}^l \Gamma_{jk}{}^n - \Gamma_{jn}{}^l \Gamma_{ik}{}^n. \tag{3.1.b}$$

Fix a point P of M and choose local coordinates centered at P so $\Gamma(P) = 0$. Equation (3.1.b) yields

$$\begin{aligned} R_{ijk}{}^l(P) &= \{\partial_{x_i} \Gamma_{jk}{}^l - \partial_{x_j} \Gamma_{ik}{}^l\}(P) \\ R_{ijk}{}^l{}_{;n}(P) &= \{\partial_{x_n} \partial_{x_i} \Gamma_{jk}{}^l - \partial_{x_n} \partial_{x_j} \Gamma_{ik}{}^l\}(P). \end{aligned} \tag{3.1.c}$$

We use the symmetry $\Gamma_{ab}{}^c = \Gamma_{ba}{}^c$ and apply Equation (3.1.c) to compute:

$$\begin{aligned} &(R_{ijk}{}^l + R_{jki}{}^l + R_{kij}{}^l)(P) \\ &= (\partial_{x_i} \Gamma_{jk}{}^l - \partial_{x_j} \Gamma_{ik}{}^l + \partial_{x_j} \Gamma_{ki}{}^l - \partial_{x_k} \Gamma_{ji}{}^l + \partial_{x_k} \Gamma_{ij}{}^l - \partial_{x_i} \Gamma_{kj}{}^l)(P) \\ &= (\partial_{x_i} \Gamma_{jk}{}^l - \partial_{x_j} \Gamma_{ik}{}^l + \partial_{x_j} \Gamma_{ik}{}^l - \partial_{x_k} \Gamma_{ji}{}^l + \partial_{x_k} \Gamma_{ji}{}^l - \partial_{x_i} \Gamma_{jk}{}^l)(P) = 0, \end{aligned}$$

$$
\begin{aligned}
&(R_{ijk}{}^l{}_{;n} + R_{jki}{}^l{}_{;n} + R_{kij}{}^l{}_{;n})(P)\\
&\quad = \{\partial_{x_n}(\partial_{x_i}\Gamma_{jk}{}^l - \partial_{x_j}\Gamma_{ik}{}^l + \partial_{x_j}\Gamma_{ki}{}^l - \partial_{x_k}\Gamma_{ji}{}^l + \partial_{x_k}\Gamma_{ij}{}^l - \partial_{x_i}\Gamma_{kj}{}^l)\}(P)\\
&\quad = \{\partial_{x_n}(\partial_{x_i}\Gamma_{jk}{}^l - \partial_{x_j}\Gamma_{ik}{}^l + \partial_{x_j}\Gamma_{ik}{}^l - \partial_{x_k}\Gamma_{ji}{}^l + \partial_{x_k}\Gamma_{ji}{}^l - \partial_{x_i}\Gamma_{jk}{}^l)\}(P)\\
&\quad = 0,\\
&(R_{ijk}{}^l{}_{;n} + R_{jnk}{}^l{}_{;i} + R_{nik}{}^l{}_{;j})(P)\\
&\quad = (\partial_{x_n}\partial_{x_i}\Gamma_{jk}{}^l - \partial_{x_n}\partial_{x_j}\Gamma_{ik}{}^l + \partial_{x_i}\partial_{x_j}\Gamma_{nk}{}^l - \partial_{x_i}\partial_{x_n}\Gamma_{jk}{}^l + \partial_{x_j}\partial_{x_n}\Gamma_{ik}{}^l\\
&\qquad - \partial_{x_j}\partial_{x_i}\Gamma_{nk}{}^l)(P) = 0. \qquad \square
\end{aligned}
$$

Let $\Theta \in V^* \otimes S^2 \otimes V$ be a tensor with $\Theta_{ijk}{}^l = \Theta_{ikj}{}^l$. We define a corresponding affine connection by setting

$$ {}^\Theta\Gamma_{jk}{}^l := \Theta_{ijk}{}^l x^i $$

with corresponding curvature operator, specialized at the origin, given by:

$$ {}^\Theta R_{ijk}{}^l(0) = \Theta_{ijk}{}^l - \Theta_{jik}{}^l. $$

The following will be an important observation:

Lemma 3.1.3 *The map $\Theta \to {}^\Theta R(0)$ defines a* GL *equivariant surjective linear map from $V^* \otimes S^2 \otimes V$ onto $\mathfrak{A}$.*

Proof. The GL equivariance is immediate from the definition since as maps of $\otimes^3 V^* \otimes V$ we are simply anti-symmetrizing in the first two indices. The fact the map is onto follows from Equation (3.1.c) and from Theorem 1.3.4 where we show every element of $\mathfrak{A}$ is geometrically realized (the proof of Theorem 1.3.4 does not use this result). $\square$

3.2 Equiaffine Connections

Although the following result is classic [Schirokow and Schirokow (1962)], we present the proof as it is short and the constructions involved play a crucial role in our development. Let $\vec{x} = (x^1, \ldots, x^m)$ be a system of local coordinates on M. Let

$$ \omega_{\vec{x}} := \Gamma_{ij}{}^j dx^i\,; $$

$\omega_{\vec{x}}$ is not a tensor but depends on the coordinates chosen.

Theorem 3.2.1 *Let ∇ be an affine connection. The following assertions are equivalent. If any is satisfied, then ∇ is said to be an* equiaffine *or a* Ricci symmetric *connection.*

(1) $d\omega_{\vec{x}} = 0$ *for any system of local coordinates* $\vec{x}$ *on* M.
(2) $\operatorname{Tr}(\mathcal{R}) = 0$.
(3) The connection ∇ *is Ricci symmetric.*
(4) The connection ∇ *locally admits a parallel volume form.*

Remark 3.2.1 Such connections play a central role in many settings – see, for example, the discussion in [Bokan, Djorić, and Simon (2003)], [Manhart (2003)], [Mizuhara and Shima (1999)], and [Pinkall, Schwenk-Schellschmidt, and Simon (1994)].

Proof. We use the Bianchi identity of Equation (1.3.a) to see:

$$0 = \operatorname{Tr}\{z \to \mathcal{R}(x,y)z\} + \operatorname{Tr}\{z \to \mathcal{R}(y,z)x\} + \operatorname{Tr}\{z \to \mathcal{R}(z,x)y\}.$$

Consequently:

$$\operatorname{Tr}\{\mathcal{R}(x,y)\} - \rho(y,x) + \rho(x,y) = 0.$$

This shows that Assertions (2) and (3) are equivalent. We use Equation (3.1.b) to show that Assertions (1) and (2) are equivalent:

$$\begin{aligned}
&\operatorname{Tr}\{\mathcal{R}_{ij}\}dx^i \wedge dx^j \\
&\quad = \{\partial_{x_i}\Gamma_{jk}{}^k - \partial_{x_j}\Gamma_{ik}{}^k + \Gamma_{in}{}^k\Gamma_{jk}{}^n - \Gamma_{jn}{}^k\Gamma_{ik}{}^n\}dx^i \wedge dx^j \\
&\quad = \{\partial_{x_i}\Gamma_{jk}{}^k - \partial_{x_j}\Gamma_{ik}{}^k\}dx^i \wedge dx^j = 2d\{\Gamma_{jk}{}^k dx^j\}.
\end{aligned}$$

Since $\nabla_{\partial_{x_i}} dx^j = -\Gamma_{ik}{}^j dx^k$, we may compute:

$$\nabla_{\partial_{x_i}}\{e^\Phi dx^1 \wedge \cdots \wedge dx^m\} = \{\partial_{x_i}\Phi - \textstyle\sum_k \Gamma_{ik}{}^k\}\{e^\Phi dx^1 \wedge \cdots \wedge dx^m\}.$$

Thus there exists a local parallel volume form on an open subset of M if and only if $\Gamma_{ik}{}^k dx^i$ is exact on that open subset. As every closed 1-form is locally exact, Assertions (1) and (4) are equivalent. □

3.3 The Levi-Civita Connection

Let (M, g) be a pseudo-Riemannian manifold; here g is a smooth non-degenerate symmetric bilinear form of signature (p, q) on the tangent bundle TM. We say that a connection ∇ is a *Riemannian connection* if it makes g parallel; this means we have the identity:

$$xg(y,z) = g(\nabla_x y, z) + g(y, \nabla_x z)$$

for all smooth vector fields x, y, and z. There is a unique connection, called the *Levi-Civita connection*, which is both Riemannian and affine. Let $g_{ij} := g(\partial_{x_i}, \partial_{x_j})$ give the components of the metric relative to the coordinate frame. The *Christoffel symbols of the second kind* associated by the Levi-Civita connection are given by:

$$\Gamma_{ijk} := g(\nabla_{\partial_{x_i}} \partial_{x_j}, \partial_{x_k}).$$

The *Koszul formula* yields:

$$\Gamma_{ijk} = \tfrac{1}{2}\{\partial_{x_i} g_{jk} + \partial_{x_j} g_{ik} - \partial_{x_k} g_{ij}\}. \tag{3.3.a}$$

Let g^{ij} be the inverse matrix; $g^{ij} g_{jl} = \delta^i_l$ where δ is the *Kronecker symbol.* We can raise indices to see the *Christoffel symbols of the first kind*, which determine ∇, are given by:

$$\Gamma_{ij}{}^k = g^{kl} \Gamma_{ijl} = \tfrac{1}{2} g^{kl}\{\partial_{x_i} g_{jl} + \partial_{x_j} g_{il} - \partial_{x_l} g_{ij}\}.$$

Lemma 3.3.1 generalizes Lemma 3.1.1 from the affine setting to the pseudo-Riemannian setting; it is a classic result. In Lemma 3.6.1 we will establish a similar result in the pseudo-Hermitian and in the para-Hermitian setting if the derivative of the (para)-Kähler form vanishes at the point in question.

Lemma 3.3.1 *Let P be a point of a pseudo-Riemannian manifold (M, g). There exist local coordinates on M centered at P so that $\Gamma_{ij}{}^k(P) = 0$ and so that $g = g(P) + O(x^2)$.*

Proof. Since the Levi-Civita connection is an affine connection, we may apply Lemma 3.1.1 to make a quadratic change of coordinates to ensure that $\Gamma_{ij}{}^k(P) = 0$. We lower indices to conclude that $\Gamma_{ijk}(P) = 0$. Lemma 3.3.1 now follows from Equation (3.3.a) since one may express the first derivatives of the metric in terms of the Christoffel symbols:

$$\partial_{x_i} g_{jk} = \Gamma_{ijk} + \Gamma_{ikj}. \qquad \square$$

We use the metric to lower indices and to define the curvature tensor $R \in \otimes^4 V^*$ by setting:

$$R(x, y, z, w) := g(\mathcal{R}(x, y)z, w).$$

Lemma 3.3.2 *Let P be a point of a pseudo-Riemannian manifold (M,g). Apply Lemma 3.3.1 to choose local coordinates centered at P so that*

$$g = g(P) + O(|x|^2)$$

and so that

$$R_{ijkl} = \tfrac{1}{2}\{\partial_{x_i}\partial_{x_k}g_{jl} + \partial_{x_j}\partial_{x_l}g_{ik} - \partial_{x_i}\partial_{x_l}g_{jk} - \partial_{x_j}\partial_{x_k}g_{il}\} + O(|x|^2).$$

Proof. Since the first derivatives of the metric vanish at P, the Christoffel symbols vanish at P. Consequently we may lower an index in Equation (3.1.b) to compute:

$$\begin{aligned}
R_{ijkl} &= g_{ln}(\partial_{x_i}\Gamma_{jk}{}^n + \Gamma_{ip}{}^n\Gamma_{jk}{}^p - \partial_{x_j}\Gamma_{ik}{}^n - \Gamma_{jp}{}^n\Gamma_{ik}{}^p)\\
&= g_{ln}(\partial_{x_i}\Gamma_{jk}{}^n - \partial_{x_j}\Gamma_{ik}{}^n) + O(|x|^2)\\
&= \partial_{x_i}(g_{ln}\Gamma_{jk}{}^n) - \partial_{x_j}(g_{ln}\Gamma_{ik}{}^n) + O(|x|^2)\\
&= (\partial_{x_i}\Gamma_{jkl} - \partial_{x_j}\Gamma_{ikl}) + O(|x|^2)\\
&= \tfrac{1}{2}\{\partial_{x_i}\partial_{x_k}g_{jl} + \partial_{x_i}\partial_{x_j}g_{kl} - \partial_{x_i}\partial_{x_l}g_{jk}\}\\
&\quad + \tfrac{1}{2}\{-\partial_{x_j}\partial_{x_k}g_{il} - \partial_{x_j}\partial_{x_i}g_{kl} + \partial_{x_j}\partial_{x_l}g_{ik}\} + O(|x|^2).
\end{aligned}$$

We cancel the term $\partial_{x_i}\partial_{x_j}g_{kl}$ to prove the result. □

We now establish the basic symmetries of R given in Equation (1.6.a):

Lemma 3.3.3 *Let (M,g) be a pseudo-Riemannian manifold. Then*

(1) $R_{ijkl} = -R_{jikl} = R_{klij}$.
(2) $R_{ijkl} + R_{jkil} + R_{kijl} = 0$.

Proof. Fix $P \in M$ and apply Lemma 3.3.1 to normalize the coordinates so $\Gamma(P) = 0$. It is then immediate from Lemma 3.3.2 that $R_{ijkl} = R_{klij}$. The remaining symmetries follow from Lemma 3.1.2. □

Let $\mathrm{ext}(e)$ be exterior multiplication by a covector e in T^*M; if θ is a differential form, then $\mathrm{ext}(e)\theta := e \wedge \theta$. Let $\mathrm{int}(e)$ be the dual, interior multiplication;

$$g(\mathrm{ext}(e)\theta, \psi) = g(\theta, \mathrm{int}(e)\psi).$$

Let d be the exterior derivative and let δ be the interior co-derivative:

$$\begin{aligned}
d &: C^\infty(\Lambda^p M) \to C^\infty(\Lambda^{p+1}M),\\
\delta &: C^\infty(\Lambda^p M) \to C^\infty(\Lambda^{p-1}M).
\end{aligned}$$

Let $\{e_i\}$ be a local frame for TM and let $\{e^i\}$ be the dual frame for T^*M.

Lemma 3.3.4

(1) If ∇ is an affine connection, then $d = \operatorname{ext}(e^i)\nabla_{e_i}$.
(2) If ∇ is the Levi-Civita connection, then $\delta = -\operatorname{int}(e_i)\nabla_{e_i}$.

Proof. Suppose that ∇ is an affine connection. Fix a point P of M. If $x = (x^1, ..., x^m)$ is any system of local coordinates on M, then we have by definition that $d = \operatorname{ext}(dx^i)\partial_{x_i}$. Consequently

$$d - \operatorname{ext}(e^i)\nabla_{e_i} = d - \operatorname{ext}(dx^i)\nabla_{\partial_{x_i}}$$

is linear in the Christoffel symbols. This difference is invariantly defined. Furthermore, we can always choose coordinates centered at P so $\Gamma(P) = 0$ by Lemma 3.1.1. Consequently, this difference vanishes at P. Since P was arbitrary, Assertion (1) follows. Taking the adjoint and applying Assertion (1) shows that $\delta + \operatorname{int}(dx^i)\nabla_{\partial_{x_i}}$ is linear in the first order jets of the metric. We apply Lemma 3.3.1 to normalize the coordinate system so the first order jets of the metric vanish and complete the proof for the special case that ∇ is the Levi-Civita connection (one needs to have $\nabla g = 0$). □

The following technical result will be useful in the proof of Lemma 7.4.1. It will also be useful in the proof of Lemma 7.7.7.

Lemma 3.3.5 *Fix a point P of a pseudo-Riemannian manifold (M, g). Let $V := T_PM$ and let $\langle\cdot,\cdot\rangle := g_P$. Choose local coordinates to identify M with V near P. There exists the germ of a smooth map from V to $\mathrm{GL}(V)$ with $\kappa(0) = \mathrm{Id}$ so that $g_Q(x, y) = \langle\kappa(Q)x, \kappa(Q)y\rangle$ for Q near P. If g is real analytic, then κ can be chosen to be real analytic.*

Proof. Consider the squaring map $\kappa \to \kappa^2$ for $\kappa \in \mathrm{GL}$. Since the Jacobian of this map at the identity is multiplication by 2, the squaring map is invertible in a neighborhood of Id; we shall denote the inverse map by $\kappa \to \kappa^{1/2}$. Express $g(x, y) = \langle x, \kappa y\rangle$ where κ is the germ of a map from V to GL with $\kappa(0) = \mathrm{Id}$. Since g is symmetric, $\kappa = \kappa^*$, or equivalently that $\langle x, \kappa y\rangle = \langle\kappa x, y\rangle$. Let $\{e_i\}$ be an orthonormal basis for $(V, \langle\cdot,\cdot\rangle)$ and let $\{x^i\}$ be the dual system of coordinates; identify $\partial_{x_i} = e_i$. We show that κ is a smooth map by computing:

$$g_{ij} = g(\partial_{x_i}, \partial_{x_j}) = \langle\partial_{x_i}, \kappa\partial_{x_j}\rangle = \varepsilon_{ik}\kappa_j^k \quad \text{so} \quad \kappa_i^j = g_{ik}\varepsilon^{jk}.$$

Set $\vartheta = \kappa^{1/2}$. Since $\kappa = \kappa^*$, $\vartheta = \vartheta^*$. We show $g = g^\vartheta$ by computing:

$$g(x, y) = \langle x, \kappa y\rangle = \langle x, \vartheta^2 y\rangle = \langle\vartheta^* x, \vartheta y\rangle = \langle\vartheta x, \vartheta y\rangle = g^\vartheta(x, y).$$

If g is real analytic, then κ is real analytic and, consequently, $\kappa^{1/2}$ is real analytic as well. □

3.4 Complex Geometry

Let M be a smooth manifold of dimension m. Let $\mathcal{D}$ be a smooth subbundle of the tangent bundle TM of dimension r. We say that $\mathcal{D}$ is an *integrable distribution* if given any smooth sections x and y to $\mathcal{D}$, we also have that the Lie bracket $[x,y]$ is a smooth section to $\mathcal{D}$. We have the classical result of [Frobenius (1877)]:

Theorem 3.4.1 *Let $\mathcal{D}$ be an integrable subbundle of TM. Fix a point P of M. There exist local coordinates $(x^1,\dots,x^m)$ on M centered at P so $\mathcal{D} = \operatorname{Span}\{\partial_{x_1},\dots,\partial_{x_r}\}$. Furthermore, the functions $(x^{r+1},\dots,x^m)$ are constant on the leaves of $\mathcal{D}$ and the differentials $\{dx^{r+1},\dots,dx^m\}$ are linearly independent.*

Let $\sqrt{-1}$ be an indeterminate with $\sqrt{-1}^2 = -1$. Let $\mathbb{C} = \mathbb{R} + \sqrt{-1}\mathbb{R}$ be the associated field of *complex numbers*. Let M be a smooth manifold of even dimension $m = 2m$. Assume there exists an endomorphism J_- of the tangent bundle TM satisfying $J_-^2 = -\operatorname{Id}$. Then (M, J_-) is said to be an *almost complex manifold.* This reduces the structure group of the tangent bundle from the real general linear group GL to the complex general linear group GL_- defined in Equation (1.2.a).

We complexify and let $T_{\mathbb{C}}M := TM \otimes_{\mathbb{R}} \mathbb{C}$. We extend J_- to be complex linear on $T_{\mathbb{C}}M$ and we extend the Lie bracket of Equation (3.1.a) to be complex bilinear as well. Since $J_-^2 = -\operatorname{Id}$, J_- has two eigenvalues $\pm\sqrt{-1}$. We let $T'(M)$ and $T''(M)$ be the $\sqrt{-1}$ and $-\sqrt{-1}$ eigenspaces, respectively. They are given by:

$$T'(M) := \{\xi \in T_{\mathbb{C}}M : J_-\xi = \sqrt{-1}\xi\},$$
$$T''(M) := \{\xi \in T_{\mathbb{C}}M : J_-\xi = -\sqrt{-1}\xi\}.$$

We say that $T'(M)$ is a *complex integrable distribution* if given any two smooth sections ξ_1 and ξ_2 to $T'(M)$, then $[\xi_1, \xi_2]$ also is a smooth section to $T'(M)$. Let N_- be the Nijenhuis tensor defined in Equation (1.1.f). One has the following result [Newlander and Nirenberg (1957)]. It can be regarded as a complex analogue of Theorem 3.4.1:

Theorem 3.4.2 *Let (M, J_-) be an almost complex manifold. The following assertions are equivalent. If any is satisfied, then (M, J_-) is said to be a* complex manifold *and J_- is said to be an* integrable almost complex structure*:*

(1) The distribution $T'(M)$ is complex integrable.
(2) $N_- = 0$.
(3) There exist adapted local coordinates $(x^1, \ldots, x^{\bar{m}}, y^1, \ldots, y^{\bar{m}})$ centered at any point of the manifold so that $J_-\partial_{x_i} = \partial_{y_i}$ and $J_-\partial_{y_i} = -\partial_{x_i}$.

Proof. If $\xi = u + \sqrt{-1}v$ is a section to $T'(M)$, then the conjugate $\bar{\xi} := u - \sqrt{-1}v$ is a section to $T''(M)$. Let $\{\xi_1, \ldots, \xi_{\bar{m}}\}$ be a $\mathbb{C}$ basis for $T'(M)$; $J_-\xi_i = \sqrt{-1}\xi_i$. If Assertion (1) holds, then additionally $J_-[\xi_i, \xi_j] = \sqrt{-1}[\xi_i, \xi_j]$. We show that Assertion (1) implies Assertion (2) by computing:

$$\begin{aligned}
&N_-(\xi_i, \xi_j)\\
&\quad = [\xi_i, \xi_j] + \sqrt{-1}\left[\sqrt{-1}\xi_i, \xi_j\right] + \sqrt{-1}\left[\xi_i, \sqrt{-1}\xi_j\right] - \left[\sqrt{-1}\xi_i, \sqrt{-1}\xi_j\right]\\
&\quad = (1 - 1 - 1 + 1)\left[\xi_i, \xi_j\right] = 0,\\
&N_-(\xi_i, \bar{\xi}_j)\\
&\quad = \left[\xi_i, \bar{\xi}_j\right] + J_-\left[\sqrt{-1}\xi_i, \bar{\xi}_j\right] + J_-\left[\xi_i, -\sqrt{-1}\bar{\xi}_j\right] - \left[\sqrt{-1}\xi_i, -\sqrt{-1}\bar{\xi}_j\right]\\
&\quad = (1 - 1)[\xi_i, \bar{\xi}_j] + (1 - 1)\sqrt{-1}J_-[\xi_i, \bar{\xi}_j] = 0,\\
&N_-(\bar{\xi}_i, \bar{\xi}_j) = \bar{N}_-(\xi_i, \xi_j) = 0.
\end{aligned}$$

Conversely, assume Assertion (2) holds. Then:

$$\begin{aligned}
0 &= N_-(\xi_i, \xi_j)\\
&= [\xi_i, \xi_j] + J_-\left[\sqrt{-1}\xi_i, \xi_j\right] + J_-\left[\xi_i, \sqrt{-1}\xi_j\right] - \left[\sqrt{-1}\xi_i, \sqrt{-1}\xi_j\right]\\
&= (2 + 2J_-\sqrt{-1})[\xi_i, \xi_j].
\end{aligned}$$

Consequently, $J_-[\xi_i, \xi_j] = \sqrt{-1}[\xi_i, \xi_j]$ and $[\xi_i, \xi_j]$ belongs to $T'(M)$. Thus Assertions (1) and (2) are equivalent.

If Assertion (3) holds, we choose local adapted coordinates and set

$$\begin{aligned}
&z^i := x^i + \sqrt{-1}y^i, \quad dz^i := dx^i + \sqrt{-1}dy^i,\\
&\partial_{z_i} := \tfrac{1}{2}(\partial_{x_i} - \sqrt{-1}\partial_{y_i}).
\end{aligned}$$

We then have $J_-\partial_{z_i} = \frac{1}{2}\left(\partial_{y_i} + \sqrt{-1}\partial_{x_i}\right) = \sqrt{-1}\partial_{z_i}$ so $\{\partial_{z_i}\}$ span $T'(M)$. Since $[\partial_{z_i}, \partial_{z_j}] = 0$, $T'(M)$ is complex integrable.

The proof that Assertion (1) implies Assertion (3) is a deep result in the theory of partial differential equations. As it is beyond the scope of this book, we shall refer to [Newlander and Nirenberg (1957)] for the proof. □

We now present a brief introduction to the para-complex setting; for further details, the reader should consult [Cortés et al. (2005)]; see also [Cruceanu, Fortuny, and Gadea (1996)]. Let (M, g, J_+) be an *almost para-complex manifold* of dimension $m = 2\bar{m}$; this means that J_+ is an endomorphism of the tangent bundle with $J_+^2 = \mathrm{Id}$ and $\mathrm{Tr}(J_+) = 0$. Let $T^\pm(M)$ be the ± 1 eigenbundles of J_+. This reduces the structure group from GL to $\mathrm{GL}_+ = \mathrm{GL}(\bar{m}, \mathbb{R}) \oplus \mathrm{GL}(\bar{m}, \mathbb{R})$. Let N_+ be the para-Nijenhuis tensor of Equation (1.1.g). Theorem 3.4.2 generalizes to this setting to become:

Theorem 3.4.3 *Let (M, J_+) be an almost para-complex manifold. The following assertions are equivalent. If any is satisfied, then (M, J_+) is said to be a* para-complex manifold *and J_+ is said to be an* integrable almost para-complex structure*:*

(1) $T^+(M)$ and $T^-(M)$ are integrable distributions.
(2) $N_+ = 0$.
(3) There exist adapted local coordinates $(x^1, \ldots, x^{\bar{m}}, y^1, \ldots, y^{\bar{m}})$ centered at any point of the manifold so $J_+\partial_{x_i} = \partial_{y_i}$ and so $J_+\partial_{y_i} = \partial_{x_i}$.

Proof. Let $\{\xi_1^\pm, \ldots, \xi_{\bar{m}}^\pm\}$ be a real local eigenframe for $T^\pm(M)$ such that $J_+\xi_i^\pm = \pm\xi_i^\pm$. If Assertion (1) holds, then $J_+[\xi_i^\pm, \xi_j^\pm] = \pm[\xi_i^\pm, \xi_j^\pm]$. We prove Assertion (1) implies Assertion (2) by computing:

$$\begin{aligned} N_+(\xi_i^+, \xi_j^+) &= [\xi_i^+, \xi_j^+] - [\xi_i^+, \xi_j^+] - [\xi_i^+, \xi_j^+] + [\xi_i^+, \xi_j^+] \\ &= (1 - 1 - 1 + 1)\,[\xi_i^+, \xi_j^+] = 0, \\ N_+(\xi_i^-, \xi_j^-) &= [\xi_i^-, \xi_j^-] + [-\xi_i^-, \xi_j^-] + [\xi_i^-, -\xi_j^-] + [-\xi_i^-, -\xi_j^-] \\ &= (1 - 1 - 1 + 1)\,[\xi_i^-, \xi_j^-] = 0, \\ N_+(\xi_i^+, \xi_j^-) &= [\xi_i^+, \xi_j^-] - J_+\,[\xi_i^+, \xi_j^-] - J_+\,[\xi_i^+, -\xi_j^-] + [\xi_i^+, -\xi_j^-] \\ &= (1 - 1)[\xi_i^+, \xi_j^-] + (1 - 1)J_+[\xi_i^+, \xi_j^-] = 0. \end{aligned}$$

Conversely, assume Assertion (2) holds. Then:

$$\begin{aligned} 0 = N_+(\xi_i^+, \xi_j^+) &= [\xi_i^+, \xi_j^+] - J_+\,[\xi_i^+, \xi_j^+] - J_+\,[\xi_i^+, \xi_j^+] + [\xi_i^+, \xi_j^+] \\ &= (2 - 2J_+)[\xi_i^+, \xi_j^+], \\ 0 = N_+(\xi_i^-, \xi_j^-) &= [\xi_i^-, \xi_j^-] - J_+\,[-\xi_i^-, \xi_j^-] - J_+\,[\xi_i^-, -\xi_j^-] + [\xi_i^-, \xi_j^-] \\ &= (2 + 2J_+)[\xi_i^-, \xi_j^-]. \end{aligned}$$

Consequently

$$J_+[\xi_i^+,\xi_j^+] = [\xi_i^+,\xi_j^+] \text{ and } J_+[\xi_i^-,\xi_j^-] = -[\xi_i^-,\xi_i^-].$$

For this reason $T^+(M)$ and $T^-(M)$ are integrable distributions. Thus Assertions (1) and (2) are equivalent. If we can choose local coordinates

$$\{x^1,\ldots,x^{\bar{m}},y^1,\ldots,y^{\bar{m}}\}$$

so that $J_+\partial_{x_i} = \partial_{y_i}$ and $J_+\partial_{y_i} = \partial_{x_i}$, then $\xi_i^\pm := \partial_{x_i} \pm \partial_{y_i}$ provide local frames showing $T^\pm M$ are integrable distributions. Thus Assertion (3) implies Assertion (1).

Finally, suppose Assertion (1) holds. Fix $P \in M$. Since the distributions $T^\pm(M)$ are integrable, we can apply Theorem 3.4.1 to find smooth functions

$$\{u^{1,\pm},\ldots,u^{\bar{m},\pm}\}$$

defined near P so these functions are constant on the leaves of $T^\mp(M)$ and so the differentials $\{du^{1,\pm},\ldots,du^{\bar{m},\pm}\}$ are linearly independent. Since the distributions $T^+(M)$ and $T^-(M)$ are transversal, the functions

$$(u^{1,+},\ldots,u^{\bar{m},+},u^{1,-},\ldots,u^{\bar{m},-})$$

define a system of local coordinates. The functions $u^{i,\pm}$ are constant on the leaves of $T^\mp(M)$. Consequently, $du^{i,\pm}$ vanishes on $T^\mp(M)$ and hence $du^{i,\pm}$ belongs to $(T^*M)^\pm$. Thus, dually, $J_+\partial_{u^{i,\pm}} = \pm\partial_{u^{i,\pm}}$. Set

$$x^i = u^{i,+} + u^{i,-} \quad \text{and} \quad y^i = u^{i,+} - u^{i,-}.$$

This yields the desired coordinate system. □

Let ∇ be an affine connection on an almost (para)-complex manifold $(M, J_\pm)$. If $\nabla(J_\pm) = 0$, then ∇ is called a *(para)-Kähler affine connection.* The following is a useful observation:

Lemma 3.4.1 *If an almost (para)-complex manifold $(M, J_\pm)$ admits a (para)-Kähler affine connection, then $J_\pm$ is an integrable (para)-complex structure.*

Proof. Let ∇ be an affine connection or, equivalently, a connection with vanishing torsion tensor. Suppose that J_- is an almost complex structure on M. We extend ∇ to be complex bilinear on $T_{\mathbb{C}}M$. Assume $\nabla(J_-) = 0$; this means that $J_-\nabla_x y = \nabla_x J_- y$. If $x, y \in T'(M)$, we have:

$$\begin{aligned} J_-[x,y] &= J_-\nabla_x y - J_-\nabla_y x = \nabla_x J_- y - \nabla_y J_- x \\ &= \nabla_x\sqrt{-1}y - \nabla_y\sqrt{-1}x = \sqrt{-1}[x,y]. \end{aligned}$$

This shows $T'(M)$ is complex integrable and hence J_- is an integrable complex structure.

The argument is similar in the para-complex setting. Suppose that J_+ is an almost para-complex structure on M. If $x, y \in T^\varepsilon(M)$ for $\varepsilon = \pm$, we have:

$$\begin{aligned} J_+[x,y] &= J_+\nabla_x y - J_+\nabla_y x = \nabla_x J_+ y - \nabla_y J_+ x \\ &= \nabla_x \varepsilon y - \nabla_y \varepsilon x = \varepsilon[x,y]. \end{aligned}$$

This shows the distributions $T^\pm(M)$ are integrable distributions and hence the almost para-complex structure J_+ is integrable. □

3.5 The Gray Identity

In Section 3.5, we prove Theorem 1.9.1. Let $(M, g, J_\pm)$ be a para-Hermitian manifold $(+)$ or a pseudo-Hermitian manifold $(-)$. Let $\mathcal{G}_\pm$ be the *Gray symmetrizer* defined in Equation (1.8.b). Let "/" denote ordinary partial differentiation. We lower indices in Equation (3.1.b) to see:

$$\begin{aligned} R_{abcd} &= g_{df}\partial_{u_a}\Gamma_{bc}{}^f - g_{df}\partial_{u_b}\Gamma_{ac}{}^f + g_{df}\Gamma_{ae}{}^f\Gamma_{bc}{}^e - g_{df}\Gamma_{be}{}^f\Gamma_{ac}{}^e \\ &= \Gamma_{bcd/a} - g_{df/a}\Gamma_{bc}{}^f - \Gamma_{acd/b} + g_{df/b}\Gamma_{ac}{}^f \\ &\quad + g^{el}\Gamma_{aed}\Gamma_{bcl} - g^{el}\Gamma_{bed}\Gamma_{acl} \\ &= \Gamma_{bcd/a} - g^{fl}g_{df/a}\Gamma_{bcl} - \Gamma_{acd/b} + g^{fl}g_{df/b}\Gamma_{acl} \\ &\quad + g^{el}\Gamma_{aed}\Gamma_{bcl} - g^{el}\Gamma_{bed}\Gamma_{acl}. \end{aligned} \tag{3.5.a}$$

Let (M, g, J_-) be a pseudo-Hermitian manifold. We wish to show $\mathcal{G}_-(R) = 0$. As J_- is an integrable complex structure, we may introduce local coordinates $(u^1, \ldots, u^{2\bar{m}})$ so

$$\begin{aligned} J_-\partial_{u_1} &= \partial_{u_{\bar{m}+1}}, \ \ldots\ , J_-\partial_{u_{\bar{m}}} = \partial_{u_{2\bar{m}}}, \\ J_-\partial_{u_{\bar{m}+1}} &= -\partial_{u_1}, \ \ldots\ , J_-\partial_{u_{2\bar{m}}} = -\partial_{u_{\bar{m}}}. \end{aligned}$$

We let Greek indices $\{\alpha, \beta, \gamma, \delta, \varepsilon, \theta\}$ and corresponding Roman indices $\{a, b, c, d, e, f\}$ range from 1 to $2\bar{m}$; set $\partial_\alpha := J_-\partial_{u_a}$. Thus

$$\begin{aligned} g_{ab} &:= g(\partial_{u_a}, \partial_{u_b}), & g_{\alpha\beta} &:= g(J_-\partial_{u_a}, J_-\partial_{u_b}), \\ g_{a\beta} &:= g(\partial_{u_a}, J_-\partial_{u_b}), & g_{\alpha b} &:= g(J_-\partial_{u_a}, \partial_{u_b}). \end{aligned}$$

We have $g_{ab} = g_{\alpha\beta}$ and $g_{a\beta} = -g_{\alpha b}$. Let g^{ab} be the inverse matrix.

We first study the linear terms in the second derivatives of the metric:

$$\Gamma_{bcd/a} - \Gamma_{acd/b} = \tfrac{1}{2}\{g_{bd/ac} + g_{ac/bd} - g_{bc/ad} - g_{ad/bc}\}. \tag{3.5.b}$$

We shall examine the role $T^1_{abcd} := g_{bd/ac}$ plays in the Gray identity; the remaining three terms play similar roles and the argument is similar after permuting the indices appropriately and applying Equation (1.8.c). We use the fact that $J_-^* g = g$, that $J_-^2 = -\operatorname{Id}$, and apply $\mathcal{G}_-$ to compute:

$$\begin{aligned}
\mathcal{G}_-(T^1)_{abcd} &= g_{bd/ac} + g_{\beta\delta/\alpha\gamma} - g_{\beta d/\alpha c} - g_{bd/\alpha\gamma} \\
&\qquad - g_{b\delta/\alpha c} - g_{\beta d/a\gamma} - g_{\beta\delta/ac} - g_{b\delta/a\gamma} \\
&= g_{bd/ac} + g_{bd/\alpha\gamma} + g_{b\delta/\alpha c} - g_{bd/\alpha\gamma} \\
&\qquad - g_{b\delta/\alpha c} + g_{b\delta/a\gamma} - g_{bd/ac} - g_{b\delta/a\gamma} \\
&= 0.
\end{aligned}$$

We examine the quadratic terms:

$$\begin{aligned}
&T^2_{abcd} := g^{fe} g_{ad/f} g_{bc/e}, \qquad T^3_{abcd} := g^{fe} g_{af/d} g_{bc/e}, \\
&T^4_{abcd} := g^{fe} g_{af/d} g_{be/c}.
\end{aligned} \tag{3.5.c}$$

By Equation (1.8.c), the Gray symmetrizer is invariant under permuting the factors. Since the other possible quadratic terms arise by permuting the roles of $\{a, b, c, d\}$ in these expressions, it suffices to study T^2, T^3, and T^4. We have:

$$\begin{aligned}
\mathcal{G}_-(T^2)_{abcd} &= g^{fe}\{g_{ad/f}g_{bc/e} + g_{\alpha\delta/f}g_{\beta\gamma/e} - g_{\alpha d/f}g_{\beta c/e} - g_{\alpha d/f}g_{b\gamma/e} \\
&\qquad - g_{\alpha\delta/f}g_{bc/e} - g_{ad/f}g_{\beta\gamma/e} - g_{a\delta/f}g_{\beta c/e} - g_{a\delta/f}g_{b\gamma/e}\} \\
&= g^{fe}\{g_{ad/f}g_{bc/e} + g_{ad/f}g_{bc/e} - g_{a\delta/f}g_{b\gamma/e} + g_{a\delta/f}g_{b\gamma/e} \\
&\qquad - g_{ad/f}g_{bc/e} - g_{ad/f}g_{bc/e} + g_{a\delta/f}g_{b\gamma/e} - g_{a\delta/f}g_{b\gamma/e}\} \\
&= 0,
\end{aligned}$$

$$\begin{aligned}
\mathcal{G}_-(T^3)_{abcd} &= g^{fe}\{g_{af/d}g_{bc/e} + g_{\alpha f/\delta}g_{\beta\gamma/e} - g_{\alpha f/d}g_{\beta c/e} - g_{\alpha f/d}g_{b\gamma/e} \\
&\qquad - g_{\alpha f/\delta}g_{bc/e} - g_{af/d}g_{\beta\gamma/e} - g_{af/\delta}g_{\beta c/e} - g_{af/\delta}g_{b\gamma/e}\} \\
&= g^{fe}\{g_{af/d}g_{bc/e} + g_{\alpha f/\delta}g_{bc/e} + g_{\alpha f/d}g_{b\gamma/e} - g_{\alpha f/d}g_{b\gamma/e} \\
&\qquad - g_{\alpha f/\delta}g_{bc/e} - g_{af/d}g_{bc/e} + g_{af/\delta}g_{b\gamma/e} - g_{af/\delta}g_{b\gamma/e}\} \\
&= 0,
\end{aligned}$$

$$\begin{aligned}
\mathcal{G}_-(T^4)_{abcd} &= g^{fe}\{g_{af/d}g_{be/c} + g_{\alpha f/\delta}g_{\beta e/\gamma} - g_{\alpha f/d}g_{\beta e/c} - g_{\alpha f/d}g_{be/\gamma} \\
&\qquad - g_{\alpha f/\delta}g_{be/c} - g_{af/d}g_{\beta e/\gamma} - g_{af/\delta}g_{\beta e/c} - g_{af/\delta}g_{be/\gamma}\}.
\end{aligned}$$

We continue the analysis of $\mathcal{G}_-(T^4)_{abcd}$ by computing:

$$\begin{aligned}
&g^{fe}g_{af/d}g_{be/c} - g^{fe}g_{\alpha f/d}g_{\beta e/c} = g^{fe}g_{af/d}g_{be/c} - g^{\theta\varepsilon}g_{a\theta/d}g_{b\varepsilon/c} = 0,\\
&g^{fe}g_{\alpha f/\delta}g_{\beta e/\gamma} - g^{fe}g_{af/\delta}g_{be/\gamma} = g^{\theta\varepsilon}g_{a\theta/\delta}g_{b\varepsilon/\gamma} - g^{fe}g_{af/\delta}g_{be/\gamma} = 0,\\
&-g^{fe}g_{\alpha f/d}g_{be/\gamma} - g^{fe}g_{af/d}g_{\beta e/\gamma} = g^{\theta\varepsilon}g_{a\theta/d}g_{\beta\varepsilon/\gamma} - g^{fe}g_{af/d}g_{\beta e/\gamma} = 0,\\
&-g^{fe}g_{\alpha f/\delta}g_{be/c} - g^{fe}g_{af/\delta}g_{\beta e/c} = g^{\theta\varepsilon}g_{a\theta/\delta}g_{\beta\varepsilon/c} - g^{fe}g_{af/\delta}g_{\beta e/c} = 0.
\end{aligned}$$

The desired result now follows.

We now turn to the para-Hermitian setting. Let (M, g, J_+) be a para-Hermitian manifold. We must show $\mathcal{G}_+(R) = 0$. The situation is similar to that given in the pseudo-Hermitian setting modulo the occasional change of sign. Introduce coordinates $(u^1, \dots, u^{2\bar{m}})$ on M satisfying the relations:

$$\begin{aligned}
J_+\partial_{u_1} &= \partial_{u_{\bar{m}+1}}, \ \dots\ , J_+\partial_{u_{\bar{m}}} = \partial_{u_{2\bar{m}}},\\
J_+\partial_{u_{\bar{m}+1}} &= \partial_{u_1}, \ \dots\ , J_+\partial_{u_{2\bar{m}}} = \partial_{u_{\bar{m}}}.
\end{aligned}$$

We let $\partial_\alpha := J_+\partial_{u_a}$. Thus

$$\begin{aligned}
g_{ab} &:= g(\partial_{u_a}, \partial_{u_b}), & g_{\alpha\beta} &:= g(J_+\partial_{u_a}, J_+\partial_{u_b}),\\
g_{a\beta} &:= g(\partial_{u_a}, J_+\partial_{u_b}), & g_{\alpha b} &:= g(J_+\partial_{u_a}, \partial_{u_b}).
\end{aligned}$$

We have $g_{ab} = -g_{\alpha\beta}$ and $g_{a\beta} = -g_{\alpha b}$. We use Equation (3.5.a). We first study the linear terms in the second derivatives of the metric given in Equation (3.5.b). We examine the role $T^1_{abcd} := g_{bd/ac}$ plays in the Gray identity in the para-Hermitian context; the remaining three terms play similar roles and the argument is similar after permuting the indices appropriately. We use the fact that $J_+^*g = -g$ and apply $\mathcal{G}_+$ to compute:

$$\begin{aligned}
\mathcal{G}_+(T^1)_{abcd} &= g_{bd/ac} + g_{\beta\delta/\alpha\gamma} + g_{\beta d/\alpha c} + g_{bd/\alpha\gamma}\\
&\quad + g_{b\delta/\alpha c} + g_{\beta d/a\gamma} + g_{\beta\delta/ac} + g_{b\delta/a\gamma}\\
&= g_{bd/ac} - g_{bd/\alpha\gamma} - g_{b\delta/\alpha c} + g_{bd/\alpha\gamma}\\
&\quad + g_{b\delta/\alpha c} - g_{b\delta/a\gamma} - g_{bd/ac} + g_{b\delta/a\gamma} = 0.
\end{aligned}$$

Next we examine the quadratic terms in the first derivatives of the metric and adopt the notation of Equation (3.5.c).

$$\begin{aligned}
\mathcal{G}_+(T^2)_{abcd} &= g^{fe}\{g_{ad/f}g_{bc/e} + g_{\alpha\delta/f}g_{\beta\gamma/e} + g_{\alpha d/f}g_{\beta c/e} + g_{\alpha d/f}g_{b\gamma/e}\\
&\quad + g_{\alpha\delta/f}g_{bc/e} + g_{ad/f}g_{\beta\gamma/e} + g_{a\delta/f}g_{\beta c/e} + g_{a\delta/f}g_{b\gamma/e}\}\\
&= g^{fe}\{g_{ad/f}g_{bc/e} + g_{ad/f}g_{bc/e} + g_{a\delta/f}g_{b\gamma/e} - g_{a\delta/f}g_{b\gamma/e}\\
&\quad - g_{ad/f}g_{bc/e} - g_{ad/f}g_{bc/e} - g_{a\delta/f}g_{b\gamma/e} + g_{a\delta/f}g_{b\gamma/e}\}\\
&= 0,
\end{aligned}$$

$$\begin{aligned}
\mathcal{G}_+(T^3)_{abcd} &= g^{fe}\{g_{af/d}g_{bc/e} + g_{\alpha f/\delta}g_{\beta\gamma/e} + g_{\alpha f/d}g_{\beta c/e} + g_{\alpha f/d}g_{b\gamma/e} \\
&\quad + g_{\alpha f/\delta}g_{bc/e} + g_{af/d}g_{\beta\gamma/e} + g_{af/\delta}g_{\beta c/e} + g_{af/\delta}g_{b\gamma/e}\} \\
&= g^{fe}\{g_{af/d}g_{bc/e} - g_{\alpha f/\delta}g_{bc/e} - g_{\alpha f/d}g_{b\gamma/e} + g_{\alpha f/d}g_{b\gamma/e} \\
&\quad + g_{\alpha f/\delta}g_{bc/e} - g_{af/d}g_{bc/e} - g_{af/\delta}g_{b\gamma/e} + g_{af/\delta}g_{b\gamma/e}\} = 0, \\
\mathcal{G}_+(T^4)_{abcd} &= g^{fe}\{g_{af/d}g_{be/c} + g_{\alpha f/\delta}g_{\beta e/\gamma} + g_{\alpha f/d}g_{\beta e/c} + g_{\alpha f/d}g_{be/\gamma} \\
&\quad + g_{\alpha f/\delta}g_{be/c} + g_{af/d}g_{\beta e/\gamma} + g_{af/\delta}g_{\beta e/c} + g_{af/\delta}g_{be/\gamma}\}.
\end{aligned}$$

We continue our study of $\mathcal{G}_+(T^4)_{abcd}$ by computing:

$$\begin{aligned}
g^{fe}g_{af/d}g_{be/c} + g^{fe}g_{\alpha f/d}g_{\beta e/c} &= g^{fe}g_{af/d}g_{be/c} - g^{\theta\varepsilon}g_{a\theta/d}g_{b\varepsilon/c} = 0, \\
g^{fe}g_{\alpha f/\delta}g_{\beta e/\gamma} + g^{fe}g_{af/\delta}g_{be/\gamma} &= -g^{\theta\varepsilon}g_{a\theta/\delta}g_{b\varepsilon/\gamma} + g^{fe}g_{af/\delta}g_{be/\gamma} = 0, \\
g^{fe}g_{\alpha f/d}g_{be/\gamma} + g^{fe}g_{af/d}g_{\beta e/\gamma} &= -g^{\theta\varepsilon}g_{a\theta/d}g_{\beta\varepsilon/\gamma} + g^{fe}g_{af/d}g_{\beta e/\gamma} = 0, \\
g^{fe}g_{\alpha f/\delta}g_{be/c} + g^{fe}g_{af/\delta}g_{\beta e/c} &= -g^{\theta\varepsilon}g_{a\theta/\delta}g_{\beta\varepsilon/c} + g^{fe}g_{af/\delta}g_{\beta e/c} = 0.
\end{aligned}$$

This completes the proof of Theorem 1.9.1. We have gone through the computations in some detail as the various sign changes are a bit tricky and care must be taken. The original discussion of Gray was in the Hermitian (positive definite) setting and was quite different in flavor. □

3.6 Kähler Geometry in the Riemannian Setting II

Let ι be an indeterminate with $\iota^2 = 1$. Let $\tilde{\mathbb{C}} := \mathbb{R} + \iota\mathbb{R}$ with the associated commutative ring structure. Although $\tilde{\mathbb{C}}$ is not a field, we can still attempt to model the holomorphic setting. Let $(M, J_\pm)$ be a (para)-complex manifold. A system of local coordinates $(x^1, ..., x^{\bar{m}}, y^1, ..., y^{\bar{m}})$ on M is said to be a system of *(para)-holomorphic coordinates* if $J_\pm\partial_{x_i} = \partial_{y_i}$ and $J_\pm\partial_{y_i} = \pm\partial_{x_i}$. In the complex setting, we let $z_i := x_i + \sqrt{-1}y_i$ and in the para-complex setting, we let $z_i := x_i + \iota y_i$ for $1 \leq i \leq \bar{m}$.

Lemma 3.3.1 generalizes to this context to become:

Lemma 3.6.1 *Let P be a point of a para-Hermitian manifold (M, g, J_+) or of a pseudo-Hermitian manifold (M, g, J_-). Then $d\Omega_\pm(P) = 0$ if and only if there exist (para)-holomorphic coordinates $(w^1, \dots, w^{\bar{m}})$ for M centered at P so that $g = g(P) + O(|w|^2)$.*

Proof. We follow [Gilkey (1973)] to study the complex case. Let

$$T_{\mathbb{C}}(M) := T(M) \otimes_{\mathbb{R}} \mathbb{C}$$

be the complex tangent bundle of a pseudo-Hermitian manifold. Extend the metric g to $T_{\mathbb{C}}(M)$ to be complex linear in each factor. Let $w^a = x^a + \sqrt{-1}y^a$ be a system of local holomorphic coordinates on M for $1 \le a \le \bar{m}$. We have

$$\partial_{w_a} := \tfrac{1}{2}(\partial_{x_a} - \sqrt{-1}\partial_{y_a}) \quad\text{and}\quad \partial_{\bar{w}_a} := \tfrac{1}{2}(\partial_{x_a} + \sqrt{-1}\partial_{y_a}).$$

Since J_- is compatible with g, $g(\partial_{w_a}, \partial_{w_b}) = 0$. We set

$$g^w_{a\bar{b}} := g(\partial_{w_a}, \partial_{\bar{w}_b}) = \tfrac{1}{2}\{g(\partial_{x_a}, \partial_{x_b}) - \sqrt{-1}g(\partial_{x_a}, \partial_{y_b})\} = \bar{g}^w_{b\bar{a}}.$$

The Kähler form is given by $\Omega(\xi_1, \xi_2) = g(\xi_1, J_-\xi_2)$. Then

$$\begin{aligned}
\Omega_- &= \tfrac{1}{2\sqrt{-1}} g^w_{b\bar{d}} dw^b \wedge d\bar{w}^d, \\
d\Omega_- &= \tfrac{1}{2\sqrt{-1}} \textstyle\sum_{b<c,d} (g^w_{c\bar{d}/b} - g^w_{b\bar{d}/c}) dw^b \wedge dw^c \wedge d\bar{w}^d \\
&\quad - \tfrac{1}{2\sqrt{-1}} \textstyle\sum_{b,c<d} (g^w_{b\bar{d}/\bar{c}} - g^w_{b\bar{c}/\bar{d}}) dw^b \wedge d\bar{w}^c \wedge d\bar{w}^d.
\end{aligned}$$

Thus the condition $d\Omega_-(P) = 0$ is equivalent to the symmetry:

$$g^w_{b\bar{d}/c}(P) = g^w_{c\bar{d}/b}(P). \tag{3.6.a}$$

Clearly if we can choose holomorphic coordinates so all the first order jets of the metric vanish at P, then Equation (3.6.a) is satisfied so $d\Omega_-(P) = 0$. Conversely, suppose Equation (3.6.a) is satisfied. Let $\varepsilon_{a\bar{b}} := g^w_{a\bar{b}}(0)$. As in the proof of Lemma 3.1.1, make a quadratic change of coordinates to set:

$$z^a := w^a + \xi^a{}_{bc} w^b w^c;$$

the complex constants $\xi^a{}_{bc} = \xi^a{}_{cb}$ remain to be determined. Express

$$\begin{aligned}
&\partial_{w_c} = \partial_{z_c} + 2\xi^a{}_{bc} w^b \partial_{z_a}, \\
&g^w_{c\bar{d}} = g^z_{c\bar{d}} + 2\varepsilon_{a\bar{d}} \xi^a{}_{bc} w^b + 2\varepsilon_{c\bar{a}} \bar{\xi}^a{}_{bd} \bar{w}^b + O(|w|^2), \\
&g^w_{c\bar{d}/b} = g^z_{c\bar{d}/b} + 2\varepsilon_{a\bar{d}} \xi^a{}_{bc} + O(|w|).
\end{aligned}$$

We set $\xi^a{}_{bc} := \frac{1}{2}\varepsilon^{a\bar{d}} g^w_{c\bar{d}/b}$; this is symmetric in $\{b, c\}$ by Equation (3.6.a) and thus defines an admissible change of coordinates with $g^z_{c\bar{d}/b}(0) = 0$. Taking the complex conjugate yields $g^z_{d\bar{c}/\bar{b}}(0) = 0$ as well and thus $dg^z(0) = 0$. This completes the proof in the pseudo-Hermitian setting.

We now turn to the para-Hermitian setting. We assume J_+ is an integrable para-complex structure. If there exist para-holomorphic coordinates with $dg(P) = 0$, then clearly $d\Omega_+(P) = 0$. We assume $d\Omega_+(P) = 0$ and attempt to construct a suitable coordinate system. Let $\{x^i, y^i\}$ be para-holomorphic coordinates. By making a linear change of coordinates, we can assume the metric g is hyperbolic at the point in question. Consequently:

$$\begin{aligned} &g(\partial_{x_i}, \partial_{x_j})(P) = \delta_{ij}, \quad g(\partial_{x_i}, \partial_{y_j})(P) = 0, \\ &g(\partial_{y_i}, \partial_{y_j})(P) = -\delta_{ij}. \end{aligned} \tag{3.6.b}$$

We set

$$z^i := x^i + \iota y^i \in \tilde{\mathbb{C}}.$$

We consider $TM \otimes \tilde{\mathbb{C}}$ and $T^*M \otimes \tilde{\mathbb{C}}$ and define:

$$\begin{aligned} &\partial_{z_i} := \tfrac{1}{2}(\partial_{x_i} + \iota\partial_{y_i}), \quad \partial_{\bar{z}_i} := \tfrac{1}{2}(\partial_{x_i} - \iota\partial_{y_i}), \\ &dz^i := dx^i + \iota dy^i, \qquad d\bar{z}^i := dx^i - \iota dy^i. \end{aligned}$$

We then have the standard relations

$$\begin{aligned} &\partial_{z_i} \cdot dz^i = 1, \qquad \partial_{z_i} \cdot d\bar{z}^i = 0, \\ &\partial_{\bar{z}_i} \cdot dz^i = 0, \qquad \partial_{\bar{z}_i} \cdot \bar{d}z^i = 1. \end{aligned} \tag{3.6.c}$$

Extend g to be $\tilde{\mathbb{C}}$ bilinear.

$$\begin{aligned} g(\partial_{z_i}, \partial_{z_j}) &= \tfrac{1}{4}\{g(\partial_{x_i}, \partial_{x_j}) + g(\partial_{y_i}, \partial_{y_j}) + \iota g(\partial_{x_i}, \partial_{y_j}) + \iota g(\partial_{y_i}, \partial_{x_j})\} = 0, \\ g(\partial_{\bar{z}_i}, \partial_{\bar{z}_j}) &= \tfrac{1}{4}\{g(\partial_{x_i}, \partial_{x_j}) + g(\partial_{y_i}, \partial_{y_j}) - \iota g(\partial_{x_i}, \partial_{y_j}) - \iota g(\partial_{y_i}, \partial_{x_j})\} = 0, \\ g(\partial_{z_i}, \partial_{\bar{z}_j}) &= \tfrac{1}{4}\{g(\partial_{x_i}, \partial_{x_j}) - g(\partial_{y_i}, \partial_{y_j}) - \iota g(\partial_{x_i}, \partial_{y_j}) + \iota g(\partial_{y_i}, \partial_{x_j})\} \\ &= \tfrac{1}{2}\{g(\partial_{x_i}, \partial_{x_j}) - \iota g(\partial_{x_i}, \partial_{y_j})\} = \bar{g}(\partial_{\bar{z}_i}, \partial_{z_j}). \end{aligned}$$

We extend the Kähler form to be $\tilde{\mathbb{C}}$ bilinear and set $g^z_{i\bar{j}} := g(\partial_{z_i}, \partial_{\bar{z}_j})$. Then:

$$\begin{aligned} &\Omega_+(\partial_{z_i}, \partial_{z_j}) = g(\partial_{z_i}, J_+\partial_{z_j}) = \iota g(\partial_{z_i}, \partial_{z_j}) = 0, \\ &\Omega_+(\partial_{\bar{z}_i}, \partial_{\bar{z}_j}) = g(\partial_{\bar{z}_i}, J_+\partial_{\bar{z}_j}) = -\iota g(\partial_{\bar{z}_i}, \partial_{\bar{z}_j}) = 0, \\ &\Omega_+(\partial_{z_i}, \partial_{\bar{z}_j}) = g(\partial_{z_i}, J_+\partial_{\bar{z}_j}) = -\iota g(\partial_{z_i}, \partial_{\bar{z}_j}), \\ &\Omega_+ = -\tfrac{1}{2\iota} g^z_{i\bar{j}} dz^i \wedge d\bar{z}^j, \\ &d\Omega_+ = -\tfrac{1}{2\iota}\left\{\partial_{z_i} g^z_{j\bar{k}} dz^i \wedge dz^j \wedge d\bar{z}^k - \partial_{\bar{z}_i} g^z_{j\bar{k}} dz^j \wedge d\bar{z}^i \wedge d\bar{z}^k\right\}. \end{aligned}$$

Consequently, as before, the Kähler relations become

$$d\Omega_+(P) = 0 \quad \Leftrightarrow \quad \partial_{z_i}\{g^z_{j\bar{k}}\}(P) = \partial_{z_j}\{g^z_{i\bar{k}}\}(P) \tag{3.6.d}$$

since the relation $\partial_{\bar{z}_j}\{g^z_{i\bar{k}}\}(P) = \partial_{\bar{k}}\{g^z_{i\bar{j}}\}(P)$ follows by conjugation. Let $c_{ij} \in \tilde{\mathbb{C}}$ and consider a coordinate system formally given by:

$$w^i = z^i + c^i{}_{jk} z^j z^k.$$

If we expand $w^i = u^i + \iota v^i$ and $c^i{}_{jk} = a^i{}_{jk} + \iota b^i{}_{jk}$, we have:

$$u^i + \iota v^i = x^i + \iota y^i + (a^i{}_{jk} + \iota b^i{}_{jk})(x^j + \iota y^j)(x^k + \iota y^k),$$

and consequently we have in terms of real coordinates that:

$$\begin{aligned} u^i &= x^i + a^i{}_{jk}(x^j x^k + y^j y^k) + b^i{}_{jk}(x^j y^k + y^j x^k),\\ v^i &= y^i + a^i{}_{jk}(x^j y^k + y^k x^j) + b^i{}_{jk}(x^j x^k + y^j y^k). \end{aligned}$$

We take this as our definition of a change of coordinates near the origin and compute:

$$\begin{aligned} \partial_{x_l} &= \partial_{u_l} + 2(a^i{}_{lk}x^k + b^i{}_{lk}y^k)\partial_{u_i} + 2(a^i{}_{lk}y^k + b^i{}_{lk}x^k)\partial_{v_i},\\ \partial_{y_l} &= \partial_{v_l} + 2(a^i{}_{lk}y^k + b^i{}_{lk}x^k)\partial_{u_i} + 2(a^i{}_{lk}x^k + b^i{}_{lk}y^k)\partial_{v_i},\\ du^i &= dx^i + 2(a^i{}_{jk}x^k + b^i{}_{jk}y^k)dx^j + 2(a^i{}_{jk}y^k + b^i{}_{jk}x^k)dy^j,\\ dv^i &= dy^i + 2(a^i{}_{jk}x^k + b^i{}_{jk}y^k)dy^j + 2(a^i{}_{jk}y^k + b^i{}_{jk}x^k)dx^j. \end{aligned} \tag{3.6.e}$$

As $J_+\partial_{x_i} = \partial_{y_i}$ and $J_+\partial_{y_i} = \partial_{x_i}$, $J_+dx^i = dy^i$ and $J_+dy^i = dx^i$. Thus Equation (3.6.e) implies $J_+du^i = dv^i$ and $J_+dv^i = du^i$ and dually $J_+\partial_{u_i} = \partial_{v_i}$ and $J_+\partial_{v_i} = \partial_{u_i}$. Thus $w = u + \iota v$ also are para-holomorphic coordinates. We compute:

$$\begin{aligned} &\partial_{w_l} + 2c^i{}_{lk}z^k\partial_{w_i}\\ &= \tfrac{1}{2}\left\{\partial_{u_l} + \iota\partial_{v_l} + (a^i{}_{lk} + \iota b^i{}_{lk})(x^k + \iota y^k)(\partial_{u_i} + \iota\partial_{v_i})\right\}\\ &= \tfrac{1}{2}\left\{\partial_{u_l} + 2(a^i{}_{lk}x^k + b^i{}_{lk}y^k)\partial_{u_i} + 2(a^i{}_{lk}y^k + b^i{}_{lk}x^k)\partial_{v_i}\right\}\\ &\quad + \tfrac{1}{2}\iota\left\{\partial_{v_l} + 2(a^i{}_{lk}y^k + b^i{}_{lk}x^k)\partial_{u_i} + 2(a^i{}_{lk}x^k + b^i{}_{lk}y^k)\partial_{v_i}\right\}. \end{aligned}$$

Combining this equation with Equation (3.6.e) then shows that:

$$\partial_{z_l} = \partial_{w_l} + 2c^i{}_{lk}z^k\partial_{w_i}.$$

The remainder of the argument is the same as in the complex setting. One chooses $c^i{}_{jk}$ to be a suitable multiple of $\partial_{z_j}g^{z}_{k\bar{i}}$ and uses Equation (3.6.d) to see this is symmetric in j and k. □

Proof of Theorem 1.10.1. Since $\nabla g = 0$, $\nabla\Omega_\pm(\cdot,\cdot) = g(\cdot,\nabla(J_\pm)\cdot)$. As g is non-degenerate, the condition $\nabla(J_\pm) = 0$ is equivalent to the condition $\nabla\Omega_\pm = 0$. Thus Assertion (1a) is equivalent to Assertion (1c) in Theorem 1.10.1.

Suppose that $\nabla(J_\pm) = 0$ or equivalently that $\nabla\Omega_\pm = 0$. We apply Lemma 3.3.4 to show that $d\Omega_\pm = 0$ by expressing:

$$d\Omega_\pm = e^i \wedge \nabla_{e_i}\Omega_\pm = 0.$$

Thus Assertion (1a) implies Assertion (1b) in Theorem 1.10.1. Conversely, suppose that Assertion (1b) of Theorem 1.10.1 holds. By Lemma 3.6.1, there are local (para-)holomorphic coordinates so $g = g(P) + O(|z|^2)$. Since the matrix of $J_\pm$ is constant, $\nabla J_\pm - J_\pm \nabla$ is linear in the first order jets of the metric. As the first order jets of the metric vanish at P, $\nabla J_\pm - J_\pm \nabla$ vanishes at P. Since P was arbitrary, $\nabla(J_\pm) = 0$. This establishes Assertion (1).

We use Lemma 3.3.4 to see that $\nabla\Omega_\pm = 0$ implies $d\Omega_\pm = 0$ and $\delta\Omega_\pm = 0$. Since $\nabla J_\pm = 0$, $\nabla_x J_\pm = J_\pm \nabla_x$ for any tangent vector x. Consequently $\mathcal{R}(x,y)J_\pm = J_\pm \mathcal{R}(x,y)$. Assertion (2) of Theorem 1.10.1 follows from:

$$\begin{aligned} R(J_\pm x, J_\pm y, z, w) &= R(z, w, J_\pm x, J_\pm y) = \langle \mathcal{R}(z,w) J_\pm x, J_\pm y\rangle \\ &= \langle J_\pm \mathcal{R}(z,w)x, J_\pm y\rangle = \mp\langle \mathcal{R}(z,w)x, y\rangle = \mp R(z,w,x,y) \\ &= \mp R(x,y,z,w). \end{aligned}$$

□

Remark 3.6.1 The Kähler condition is a much stronger condition than complex integrability. Let (M, g, J_-) be a compact Kähler manifold with positive definite g. As $\nabla\Omega_\pm = 0$, $\Omega_\pm$ is a harmonic form of degree two. Since $\Omega_\pm^{\bar{m}}$ is a multiple of the volume form, results of Hodge and of de Rham show the associated element $x = [\Omega_\pm]$ in de Rham cohomology satisfies $x^{\bar{m}} \neq 0$. This has topological implications and shows that not every manifold M admits a Kähler metric. For example, we may identify the product

$$S^1 \times S^3 = \left(\mathbb{C}^2 - \{0\}\right)/\mathbb{Z}$$

by introducing the equivalence relation $(z_1, z_2) \sim \lambda^k(z_1, z_2)$ where $\lambda > 1$ generates a cyclic subgroup $\{\lambda^k\}_{k\in\mathbb{N}}$ of $\mathbb{R}^+$. This description shows that $S^1 \times S^3$ admits an integrable complex structure. On the other hand, as the second cohomology group $H^2(S^1 \times S^3) = 0$, $S^1 \times S^3$ admits no Kähler metric.

Chapter 4

Real Affine Geometry

In Chapter 4, we study questions related to real affine differential geometry. The structure group in Riemannian geometry is the orthogonal group $\mathcal{O}$. The structure group in affine differential geometry is the affine group. This is the group of linear transformations and translations of a vector space; one can also study unimodular geometry where a given volume form is preserved (equivalently, if the linear transformations have determinant 1) and centro-affine geometry (where there is a distinguished origin). This leads to the study of non-degenerate hypersurfaces in affine space (equivalently, a vector space without a distinguished origin) with various structures inherited from an arbitrary transversal vector field where certain non-degeneracy conditions are imposed. The Gauss equations and the Weingarten equations are the equations of structure in this setting. For a non-degenerate hypersurface, there is a bijective correspondence between the class of conormal fields and a conformal class of pseudo-Riemannian metrics; it is remarkable that these two quite different structures can be linked. We refer to [Simon, Schwenk-Schellschmidt, and Viesel (1991)] for further details and additional bibliographic references as the field is a vast one; additionally consult [Blaschke (1985)], [Li, Simon, and Zhao (1993)], [Nomizu and Sasaki (1993)], and [Schirokow and Schirokow (1962)].

Passing from the context of flat space to more general topologies, this leads naturally to the consideration of affine connections on the tangent bundle of an arbitrary smooth manifold. While one can consider more general connections with torsion, this is a natural context in which to work. In the pseudo-Riemannian setting, the metric permits one to raise and lower indices and thereby to identify the tangent and the cotangent bundles. This identification is not available in the affine setting without the structure of an auxiliary inner product, as will be the case in Weyl geometry. Consequently, it is important to keep in mind the covariance (in other words, the distinction between the tangent and cotangent bundles) as there is no natural way in this context to raise and lower indices. The curvature operator $\mathcal{R}$ of an affine connection satisfies the identities given in Equation (1.3.a). We pass to the algebraic setting by considering the space $\mathfrak{A}$ of affine curvature operators. These are the $(1,3)$ tensors satisfying the symmetries of Equation (1.3.a), namely:

$$\mathcal{A}(x,y) = -\mathcal{A}(y,x) \quad \text{and} \quad \mathcal{A}(x,y)z + \mathcal{A}(y,z)x + \mathcal{A}(z,x)y = 0.$$

The first identity is a $\mathbb{Z}_2$ symmetry; the second identity is the *Bianchi identity.* In Section 4.1, the structure of an auxiliary non-degenerate symmetric bilinear form $\langle\cdot,\cdot\rangle$ on V will be given. We shall discuss the corresponding decomposition of [Bokan (1990)] of $\mathfrak{A}$ as an orthogonal module presented previously in Theorem 1.4.1; we obtain as a consequence the structure of $\mathfrak{R}$ as an orthogonal module given in [Singer and Thorpe (1969)]. Although the results of [Bokan (1990)] require the imposition of an inner product, in fact only the conformal class of the inner product is needed and this structure often appears naturally in the context of hypersurface geometry given the additional structure of an affine conormal. Such structures also play an important role in Weyl geometry as we shall see subsequently. In Section 4.2, we continue our study of the orthogonal module structure of $\mathfrak{A}$ and examine the modules $\mathbb{R}$, S_0^2, and Λ^2. These modules appear in the decomposition of $\mathfrak{A}$ given in Theorem 1.4.1. We identify every possible orthogonal submodule of $\mathfrak{A}$ abstractly isomorphic either to S_0^2 or to Λ^2. In Section 4.3, we perform a similar examination of the modules $W_6^{\mathcal{O}}$, $W_7^{\mathcal{O}}$, and $W_8^{\mathcal{O}}$ and exhibit the orthogonal projectors from $\ker(\rho) \cap \ker(\rho_{13}) \cap \mathfrak{A}$ to these spaces.

In Section 4.4, we establish Theorem 1.3.1 giving the decomposition of $\mathfrak{A}$ as a general linear module [Strichartz (1988)]; it is convenient to postpone our discussion of the general linear module structure until after we have discussed the orthogonal module structure although the general linear module structure is in a certain sense more fundamental. We will take a similar approach in Section 5.6 when we discuss the structure of the space of Kähler affine curvature operators as modules over the groups GL_- and $\mathrm{GL}_\pm^\star$ given in Theorem 1.5.1. We shall use results in the para-Hermitian and, similarly, in the pseudo-Hermitian settings. The existence of a suitable metric is a very convenient structure in many settings.

In Section 4.5, we report on results contained in [Gilkey, Nikčević, and Westerman (2009)], and [Gilkey, Nikčević, and Westerman (2009a)] to prove Theorem 1.3.3. This result discusses the eight natural geometric realization questions related to the decomposition of $\mathfrak{A}$ into three components as a general linear module given in Theorem 1.3.1. We show any affine curvature operator and any covariant affine curvature operator is geometrically realizable in Theorem 1.3.4. In Theorem 4.5.1, we study realization questions related to the Ricci tensor. In Theorem 4.5.2 and in Theorem 4.5.3, we study realization questions related to the projective curvature tensor. Theorem 1.3.3 then follows from these results; similarly, Theorem 1.4.3 will follow from Theorem 4.5.1. We conclude our discussion in Section 4.6 by presenting the two decompositions in [Bokan (1990)] of $\mathfrak{A}$ as an orthogonal module.

4.1 Decomposition of $\mathfrak{A}$ and $\mathfrak{R}$ as Orthogonal Modules

Fix a non-degenerate inner product $\langle \cdot, \cdot \rangle$ of signature (p, q) on V. We shall follow the discussion in [Bokan (1990)]; see also related results contained in [Blažić et al. (2006)]. If $\mathcal{A} \in \mathfrak{A}$, we may lower indices to define an associated tensor $A \in \otimes^4 V^*$ by setting

$$A(x, y, z, w) := \langle \mathcal{A}(x, y)z, w \rangle.$$

Under this isomorphism, A belongs to $\mathfrak{A}$ if and only if we have the two symmetries listed below for all x, y, z, and w in V:

$$A(x, y, z, w) = -A(y, x, z, w), \tag{4.1.a}$$

$$A(x, y, z, w) + A(y, z, x, w) + A(z, x, y, w) = 0. \tag{4.1.b}$$

The space $\mathfrak{R}$ of Riemann curvature tensors is the subspace of $\mathfrak{A}$ consisting of all tensors with the same symmetries as the curvature tensor defined by the Levi-Civita connection of a pseudo-Riemannian metric; A belongs to $\mathfrak{R}$ if and only if $A \in \mathfrak{A}$ and we have the additional two symmetries listed below for all x, y, z, and w in V:

$$A(x, y, z, w) = A(z, w, x, y), \tag{4.1.c}$$

$$A(x, y, z, w) = -A(x, y, w, z). \tag{4.1.d}$$

The following is a useful, if elementary, observation relating Equation (4.1.c), $\mathfrak{R}$, and Equation (4.1.d):

Lemma 4.1.1 *Let $A \in \mathfrak{A}$. The following assertions are equivalent:*

(1) $A \in \mathfrak{R}$.
(2) We have $A(x,y,z,w) = A(z,w,x,y)$ for all x, y, z, and w in V.
(3) We have $A(x,y,z,w) = -A(x,y,w,z)$ for all x, y, z, and w in V.

We note that the relations of Assertion (2) are simply those of Equation (4.1.c) and that the relations of Assertion (3) are simply those of Equation (4.1.d).

Proof. It is immediate that Equation (4.1.a) and Equation (4.1.c) imply Equation (4.1.d). Conversely, suppose that Equation (4.1.a), Equation (4.1.b), and Equation (4.1.d) hold. We must show that Equation (4.1.c) holds. Let $x, y, z, w \in V$. We compute:

$$\begin{aligned}
0 &= A(x,y,z,w) + A(y,z,x,w) + A(z,x,y,w) \\
&= A(x,y,z,w) - A(y,z,w,x) - A(z,x,w,y) \\
&= A(x,y,z,w) + A(z,w,y,x) + A(w,y,z,x) \\
&\quad + A(x,w,z,y) + A(w,z,x,y) \\
&= A(x,y,z,w) - 2A(z,w,x,y) - A(w,y,x,z) - A(x,w,y,z) \\
&= 2A(x,y,z,w) - 2A(z,w,x,y).
\end{aligned}$$ □

We have defined several tensors in Equation (1.1.e) that are analogous to the Ricci tensor. As the following is a simple algebraic consequence of the symmetries defining $\mathfrak{A}$, we shall omit the proof; it permits us to restrict our attention to $\rho = \rho_{14}$ and to ρ_{13}.

Lemma 4.1.2

(1) $\rho_{12,jk}(A) := \varepsilon^{il} A_{iljk} = 0$.
(2) $\rho_{13,jk}(A) := \varepsilon^{il} A_{ijlk}$.
(3) $\rho_{14,jk}(A) = \rho_{jk}(A) := \varepsilon^{il} A_{ijkl}$.
(4) $\rho_{23,jk}(A) := \varepsilon^{il} A_{jilk} = -\rho_{13,jk}$.
(5) $\rho_{24,jk}(A) := \varepsilon^{il} A_{jikl} = -\rho_{jk}$.
(6) $\rho_{34,jk}(A) := \varepsilon^{il} A_{jkil} = -\rho_{jk}(A) + \rho_{kj}(A)$.

Definition 4.1.1 Let $\langle\cdot,\cdot\rangle$ be a non-degenerate symmetric inner product on V. Let $c \in \mathbb{R}$, let $\phi \in S^2$, and let $\psi \in \Lambda^2$. Set

(1) $\sigma_1(c)(x,y,z,w) := c\{\langle x,w\rangle\langle y,z\rangle - \langle x,z\rangle\langle y,w\rangle\}$.
(2) $\sigma_2(\phi)(x,y,z,w) := \phi(x,w)\langle y,z\rangle - \phi(y,w)\langle x,z\rangle$.
(3) $\sigma_3(\phi)(x,y,z,w) := \langle x,w\rangle\phi(y,z) - \langle y,w\rangle\phi(x,z)$.

(4) $\sigma_4(\psi)(x, y, z, w) := 2\psi(x, y)\langle z, w\rangle + \psi(x, z)\langle y, w\rangle - \psi(y, z)\langle x, w\rangle$.
(5) $\sigma_5(\psi)(x, y, z, w) := \psi(x, w)\langle y, z\rangle - \psi(y, w)\langle x, z\rangle$.

Lemma 4.1.3

(1) $\sigma_1 : \mathbb{R} \to \mathfrak{A}$.
(2) $\sigma_2 : S^2 \to \mathfrak{A}$ *and* $\sigma_3 : S^2 \to \mathfrak{A}$.
(3) $\sigma_4 : \Lambda^2 \to \mathfrak{A}$ *and* $\sigma_5 : \Lambda^2 \to \mathfrak{A}$.

Proof. Suppose ϕ_1 and ϕ_2 are symmetric tensors of rank two. We define

$$A(x, y, z, w) = \phi_1(x, w)\phi_2(y, z) - \phi_1(y, w)\phi_2(x, z).$$

Clearly A satisfies the $\mathbb{Z}_2$ symmetry of Equation (4.1.a). We verify that $A \in \mathfrak{A}$ by checking that the Bianchi identity is satisfied:

$$\begin{aligned}
&A(x, y, z, w) + A(y, z, x, w) + A(z, x, y, w) \\
&= \ \ \phi_1(x, w)\phi_2(y, z) - \phi_1(y, w)\phi_2(x, z) \\
&\quad +\phi_1(y, w)\phi_2(z, x) - \phi_1(z, w)\phi_2(y, x) \\
&\quad +\phi_1(z, w)\phi_2(x, y) - \phi_1(x, w)\phi_2(z, y) \\
&= 0.
\end{aligned}$$

Taking $\phi_1 = \phi_2 = \langle\cdot,\cdot\rangle$ defines σ_1, taking $\phi_1 = \phi$ and $\phi_2 = \langle\cdot,\cdot\rangle$ defines σ_2, and taking $\phi_1 = \langle\cdot,\cdot\rangle$ and $\phi_2 = \phi$ defines σ_3. Assertion (1) and Assertion (2) now follow. Let ψ be an anti-symmetric tensor of rank two. Again, the $\mathbb{Z}_2$ symmetry of Equation (4.1.a) is immediate so all we need do is check the Bianchi identity. We compute:

$$\begin{aligned}
&(\sigma_4\psi)(x, y, z, w) + (\sigma_4\psi)(y, z, x, w) + (\sigma_4\psi)(z, x, y, w) \\
&= \ \ 2\psi(x, y)\langle z, w\rangle + \psi(x, z)\langle y, w\rangle - \psi(y, z)\langle x, w\rangle \\
&\quad +2\psi(y, z)\langle x, w\rangle + \psi(y, x)\langle z, w\rangle - \psi(z, x)\langle y, w\rangle \\
&\quad +2\psi(z, x)\langle y, w\rangle + \psi(z, y)\langle x, w\rangle - \psi(x, y)\langle z, w\rangle \\
&= 0, \\
&(\sigma_5\psi)(x, y, z, w) + (\sigma_5\psi)(y, z, x, w) + (\sigma_5\psi)(z, x, y, w) \\
&= \psi(x, w)\langle y, z\rangle - \psi(y, w)\langle x, z\rangle + \psi(y, w)\langle z, x\rangle \\
&\quad -\psi(z, w)\langle y, x\rangle + \psi(z, w)\langle x, y\rangle - \psi(x, w)\langle z, y\rangle \\
&= 0.
\end{aligned}$$

□

We now study the space of quadratic invariants for $\mathfrak{A}$.

Lemma 4.1.4 $\dim\{\mathcal{I}_2^{\mathcal{O}}(\mathfrak{A})\} \le 10$.

Proof. We adopt the notation of Definition 2.3.1 to construct invariants. We consider the invariant $\mathcal{I}\{(\nu_1,\nu_2,\nu_3,\nu_4)(\mu_1,\mu_2,\mu_3,\mu_4)\}$ where $\{\nu_1,\nu_2,\nu_3,\nu_4,\mu_1,\mu_2,\mu_3,\mu_4\}$ is a rearrangement of $\{1,1,2,2,3,3,4,4\}$. We can permute the indices, we can use the $\mathbb{Z}_2$ symmetry $A_{ijkl} = -A_{jikl}$, and we can use the Bianchi identity $0 = A_{ijkl} + A_{jkil} + A_{kijl}$ to reduce the number of strings to be considered. We establish Lemma 4.1.4 by considering the various possible cases:

(1) The indices decouple in the two variables. This means that $\nu_i \in \{1,2\}$ and $\mu_i \in \{3,4\}$. Given the $\mathbb{Z}_2$ symmetry, we may assume $\nu_1 < \nu_2$ and $\mu_1 < \mu_2$ so the invariant under consideration is given by:

$$\mathcal{I}\{(1,2,\nu_3,\nu_4)(3,4,\mu_3,\mu_4)\}.$$

Since $\mathcal{I}\{(1,2,1,2)\} = -\mathcal{I}\{(2,1,1,2)\} = -\mathcal{I}\{(1,2,2,1)\}$, we may assume the invariant is

$$\psi_1 = \mathcal{I}\{(1,2,2,1)(3,4,4,3)\}.$$

(2) Next suppose that two indices appear with multiplicity one in each variable; this means that $(\nu_1,\nu_2,\nu_3,\nu_4)$ is a rearrangement of $(1,2,2,3)$ and $(\mu_1,\mu_2,\mu_3,\mu_4)$ is a rearrangement of $(1,4,4,3)$. We use the Bianchi identity to express

$$\begin{aligned}&\mathcal{I}\{(1,3,2,2)(*,*,*,*)\}\\ &= -\mathcal{I}\{(2,1,3,2)(*,*,*,*)\} - \mathcal{I}\{(3,2,1,2)(*,*,*,*)\}.\end{aligned}$$

Thus we may assume that $(\nu_3,\nu_4) \ne (2,2)$ and that $(\mu_3,\mu_4) \ne (4,4)$. And by interchanging the roles of the indices "1" and "3", we can assume $1 \notin \{\nu_3,\nu_4\}$. This leads to two classes of strings:

(a) The invariant $\mathcal{I}\{(1,2,2,3)(*,*,*,*)\}$ gives rise to four possibilities:

i. $\psi_2 := \mathcal{I}\{(1,2,2,3)(1,4,4,3)\}$.
ii. $\psi_3 := \mathcal{I}\{(1,2,2,3)(1,4,3,4)\}$.
iii. $\psi_4 := \mathcal{I}\{(1,2,2,3)(3,4,4,1)\}$.
iv. $\psi_5 := \mathcal{I}\{(1,2,2,3)(3,4,1,4)\}$.

(b) The invariant $\mathcal{I}\{(1,2,3,2)(*,*,*,*)\}$ also gives rise to four possibilities, only two of which give rise to new invariants:

i. $\psi_6 := \mathcal{I}\{(1,2,3,2)(1,4,3,4)\}$.

ii. Interchange the indices $2 \leftrightarrow 4$ to see:
$\mathcal{I}\{(1,2,3,2)(1,4,4,3)\} = \mathcal{I}\{(1,4,3,4)(1,2,2,3)\} = \psi_3$.
iii. $\psi_7 := \mathcal{I}\{(1,2,3,2)(3,4,1,4)\}$.
iv. Interchange the indices $2 \leftrightarrow 4$ and $1 \leftrightarrow 3$ to see:
$\mathcal{I}\{(1,2,3,2)(3,4,4,1)\} = \mathcal{I}\{(3,4,1,4)(1,2,2,3)\} = \psi_5$.

(3) Finally, we suppose that $\{\nu_1, \nu_2, \nu_3, \nu_4\}$ and $\{\mu_1, \mu_2, \mu_3, \mu_4\}$ are rearrangements of $\{1,2,3,4\}$; this means that that every index appears in each monomial. By permuting the indices, we may suppose

$$S = (1,2,3,4)(\mu_1, \mu_2, \mu_3, \mu_4).$$

By applying the Bianchi identity, we may assume that $(\mu_3, \mu_4) \neq (3,4)$, that $(\mu_3, \mu_4) \neq (4,3)$, that $(\mu_3, \mu_4) \neq (1,2)$, and that $(\mu_3, \mu_4) \neq (2,1)$. We can also interchange the roles of the indices "1" and "2" to assume that $1 \notin \{\mu_3, \mu_4\}$. This gives rise to four possibilities; only three yield new invariants:

(a) $\psi_8 := \mathcal{I}\{(1,2,3,4)(1,3,2,4)\}$.
(b) $\psi_9 := \mathcal{I}\{(1,2,3,4)(1,3,4,2)\}$.
(c) Permute the indices $2 \to 3 \to 4 \to 2$ to see:
$\mathcal{I}\{(1,2,3,4)(1,4,2,3)\} = \mathcal{I}\{(1,3,4,2)(1,2,3,4)\} = \psi_9$.
(d) $\psi_{10} := \mathcal{I}\{(1,2,3,4)(1,4,3,2)\}$.

Lemma 4.1.1 now follows as we have exhausted all the possibilities. □

Remark 4.1.1 We will perform a similar analysis in Lemma 5.1.1 to examine the space of unitary quadratic invariants in the Kähler affine setting and in Lemma 7.1.1 to examine the space of unitary quadratic invariants in the para-Hermitian setting or in the pseudo-Hermitian setting. This examination is considerably more delicate owing to the presence of the complex structure in the invariants; Lemma 4.1.1 is a crucial input to the analysis there.

We adopt the notation of Definition 1.4.1 defining $\{W_6^{\mathcal{O}}, W_7^{\mathcal{O}}, W_8^{\mathcal{O}}\}$. Theorem 1.4.1 and Theorem 1.6.2 will follow from the following result of [Bokan (1990)], [Singer and Thorpe (1969)], and [Strichartz (1988)].

Theorem 4.1.1 *Let $m \geq 4$.*

(1) There are orthogonal module decompositions:

(a) $\mathfrak{R} \approx \mathbb{R} \oplus S_0^2 \oplus W_6^{\mathcal{O}}$.
(b) $\mathfrak{A} \approx \mathbb{R} \oplus 2 \cdot S_0^2 \oplus 2 \cdot \Lambda^2 \oplus W_6^{\mathcal{O}} \oplus W_7^{\mathcal{O}} \oplus W_8^{\mathcal{O}}$.

(2) The orthogonal modules $\{\mathbb{R}, S_0^2, \Lambda^2, W_6^{\mathcal{O}}, W_7^{\mathcal{O}}, W_8^{\mathcal{O}}\}$ *are inequivalent and irreducible.*

Remark 4.1.2 Theorem 4.1.1 is a result concerning the decomposition of $\mathfrak{R}$ and $\mathfrak{A}$ into irreducible orthogonal modules. It does not actually identify the submodules corresponding to the various factors. The submodules of $\mathfrak{A}$ isomorphic to $S_0^2 \oplus S_0^2$ or to $\Lambda^2 \oplus \Lambda^2$ or to $\mathbb{R}$ are identified explicitly in Lemma 4.2.1. We have that $W_8^{\mathcal{O}}$ is a submodule of $\Lambda^2 \otimes S^2$; it is **not** a submodule of $\mathfrak{A}$ although there is a submodule of $\mathfrak{A}$ abstractly isomorphic to $W_8^{\mathcal{O}}$; this module is identified as well in Lemma 4.3.1.

Proof. If $A \in \mathfrak{R}$, we compute the components of the Ricci tensor to be:

$$\rho_{jk} = \varepsilon^{il} R_{ijkl} = \varepsilon^{li} R_{klij} = \varepsilon^{li} R_{lkji} = \rho_{kj}.$$

Thus the Ricci tensor is symmetric. Since $W_6^{\mathcal{O}} = \mathfrak{R} \cap \ker(\rho)$, ρ defines an $\mathcal{O}$ equivariant map

$$0 \longrightarrow W_6^{\mathcal{O}} \longrightarrow \mathfrak{R} \xrightarrow{\rho} S^2.$$

We wish to show that ρ is surjective and that $W_6^{\mathcal{O}}$ is non-trivial. Since $m \geq 4$, we can consider the metric:

$$g := dx^i \otimes dx^i + 2x^2x^3(dx^1 \otimes dx^4 + dx^4 \otimes dx^1).$$

Let $A = R(0) \in \mathfrak{R}$ be the curvature at the origin of the Levi-Civita connection. We apply Lemma 3.3.2 to compute A and to see that the non-zero components of A up to the $\mathbb{Z}_2$ symmetries given in Equation (4.1.a), in Equation (4.1.c), and in Equation (4.1.d) are determined by:

$$A_{1234} = A_{1324} = -1.$$

Thus, in particular, $A_{ijkl} = 0$ unless the indices $\{i, j, k, l\}$ are a permutation of the indices $\{1, 2, 3, 4\}$. This shows that $\rho(A) = 0$ and demonstrates that

$$W_6^{\mathcal{O}} \neq \{0\}.$$

Next we show that ρ is surjective. If $\theta \in S^2$, set $\tau = \tau(\theta) := \varepsilon^{ij}\theta_{ij}$. We adopt the notation of Definition 4.1.1 and set

$$\sigma(\theta) = \tfrac{1}{m-2}(\sigma_2(\theta) + \sigma_3(\theta)) - \tfrac{1}{(m-1)(m-2)}\sigma_1(\tau(\theta)).$$

More explicitly

$$\begin{aligned}\sigma(\theta)(x,y,z,w):=\ & \tfrac{1}{m-2}\{\theta(x,w)\langle y,z\rangle+\langle x,w\rangle\theta(y,z)\}\\ & -\tfrac{1}{m-2}\{\theta(x,z)\langle y,w\rangle+\langle x,z\rangle\theta(y,w)\}\\ & -\tfrac{\tau}{(m-1)(m-2)}\{\langle x,w\rangle\langle y,z\rangle-\langle x,z\rangle\langle y,w\rangle\}.\end{aligned}\tag{4.1.e}$$

We will examine σ in more detail in Section 6.2. By Lemma 4.1.3, σ takes values in $\mathfrak{A}$. Since $\sigma(\theta)(x,y,z,w)=\sigma(\theta)(z,w,x,y)$, σ takes values in $\mathfrak{R}$ by Lemma 4.1.1. We use Equation (1.1.d) to verify that σ splits the map defined by ρ by computing:

$$\begin{aligned}\rho(\sigma(\theta))(y,z)=\ & \varepsilon^{ij}\tfrac{1}{m-2}\{\theta(e_i,e_j)\langle y,z\rangle+\langle e_i,e_j\rangle\theta(y,z)\}\\ & -\varepsilon^{ij}\tfrac{1}{m-2}\{\theta(e_i,z)\langle y,e_j\rangle+\langle e_i,z\rangle\theta(y,e_j)\}\\ & -\varepsilon^{ij}\tfrac{\tau}{(m-1)(m-2)}\{\langle e_i,e_j\rangle\langle y,z\rangle-\langle e_i,z\rangle\langle y,e_j\rangle\}\\ =\ & \tfrac{1}{m-2}\{\langle y,z\rangle\tau(\theta)+m\theta(y,z)\}-\tfrac{1}{m-2}\{\theta(y,z)+\theta(y,z)\}\\ & -\tfrac{\tau}{(m-1)(m-2)}\{m\langle y,z\rangle-\langle y,z\rangle\}\\ =\ & \theta(y,z).\end{aligned}$$

This shows that $\text{Range}(\rho)=S^2$. Thus $\mathfrak{R}\approx S^2\oplus W_6^{\mathcal{O}}$. By Theorem 2.4.1, $S^2=\mathbb{R}\cdot\langle\cdot,\cdot\rangle\oplus S_0^2$. Assertion (1a) now follows:

$$\mathfrak{R}\approx S^2\oplus W_6^{\mathcal{O}}\approx\mathbb{R}\oplus S_0^2\oplus W_6^{\mathcal{O}}.$$

If $A\in\mathfrak{A}$, symmetrize the last two indices to define:

$$\pi_{\Lambda\otimes S}(A)(x,y,z,w):=\tfrac{1}{2}\{A(x,y,z,w)+A(x,y,w,z)\}.$$

If $A\in\ker(\pi_{\Lambda\otimes S})$, then A is anti-symmetric in the last two indices. Consequently $A\in\mathfrak{R}$ by Lemma 4.1.1. Thus we have a short exact sequence

$$0\longrightarrow\mathfrak{R}\longrightarrow\mathfrak{A}\stackrel{\pi_{\Lambda\otimes S}}{\longrightarrow}\Lambda^2\otimes S^2.\tag{4.1.f}$$

We wish to show that this map is surjective. If $\Theta\in\Lambda^2\otimes S^2$, set

$$(\sigma_{\Lambda\otimes S}\Theta)_{ijkl}:=\Theta_{ijkl}+\tfrac{1}{2}\{\Theta_{kjil}+\Theta_{ikjl}-\Theta_{ljik}-\Theta_{iljk}\}.\tag{4.1.g}$$

Let $A=\sigma_{\Lambda\otimes S}\Theta$. It is then immediate that $A_{ijkl}=-A_{jikl}$. We verify the Bianchi identity is satisfied, and thus $A\in\mathfrak{A}$, by computing:

$$\begin{aligned}& A_{ijkl}+A_{jkil}+A_{kijl}\\ =\ & \Theta_{ijkl}+\tfrac{1}{2}\{\Theta_{kjil}+\Theta_{ikjl}-\Theta_{ljik}-\Theta_{iljk}\}\\ & +\Theta_{jkil}+\tfrac{1}{2}\{\Theta_{ikjl}+\Theta_{jikl}-\Theta_{lkji}-\Theta_{jlki}\}\\ & +\Theta_{kijl}+\tfrac{1}{2}\{\Theta_{jikl}+\Theta_{kjil}-\Theta_{likj}-\Theta_{klij}\}\end{aligned}$$

$$\begin{aligned}&= \quad \Theta_{ijkl}(1-\tfrac{1}{2}-\tfrac{1}{2})+\Theta_{jkil}(1-\tfrac{1}{2}-\tfrac{1}{2})+\Theta_{kijl}(1-\tfrac{1}{2}-\tfrac{1}{2})\\ &\quad +\Theta_{ljik}(-\tfrac{1}{2}+\tfrac{1}{2})+\Theta_{iljk}(-\tfrac{1}{2}+\tfrac{1}{2})+\theta_{klji}(-\tfrac{1}{2}+\tfrac{1}{2})\\ &= 0.\end{aligned}$$

Let $\Theta \in \Lambda^2 \otimes S^2$. To show that $\sigma_{\Lambda\otimes S}$ is a splitting, we use the $\mathbb{Z}_2$ symmetries to compute:

$$\begin{aligned}&(\pi_{\Lambda\otimes S}\sigma_{\Lambda\otimes S}\Theta)_{ijkl}\\ &\quad = \tfrac{1}{2}\Theta_{ijkl}+\tfrac{1}{4}\{\Theta_{kjil}+\Theta_{ikjl}-\Theta_{ljik}-\Theta_{iljk}\}\\ &\qquad +\tfrac{1}{2}\Theta_{ijlk}+\tfrac{1}{4}\{\Theta_{ljik}+\Theta_{iljk}-\Theta_{kjil}-\Theta_{ikjl}\}\\ &\quad = \tfrac{1}{2}\{\Theta_{ijkl}+\Theta_{ijlk}\}=\Theta_{ijkl}.\end{aligned}$$

We can now establish Assertion (1b). We apply Theorem 2.4.1 to decompose $\Lambda^2 \otimes S^2$, we apply Assertion (1a) to decompose $\mathfrak{R}$, and we use the fact that $\sigma_{\Lambda^2\otimes S^2}$ is surjective to see that there are orthogonal module isomorphisms:

$$\begin{aligned}&\Lambda^2\otimes S^2 \approx S_0^2\oplus 2\cdot\Lambda^2\oplus W_7^{\mathcal{O}}\oplus W_8^{\mathcal{O}},\\ &\mathfrak{R}\approx \mathbb{R}\oplus S_0^2\oplus W_6^{\mathcal{O}},\\ &\mathfrak{A}\approx \mathfrak{R}\oplus\Lambda^2\otimes S^2\approx \mathbb{R}\oplus 2\cdot S_0^2\oplus 2\cdot\Lambda^2\oplus W_6^{\mathcal{O}}\oplus W_7^{\mathcal{O}}\oplus W_8^{\mathcal{O}}.\end{aligned}$$

By Lemma 4.1.4, $\dim\{\mathcal{I}_2^{\mathcal{O}}(\mathfrak{A})\}\le 10$. We have shown that the modules appearing in Assertion (1b) are all non-trivial for $m \ge 4$. Thus we may apply the estimate of Lemma 2.2.2 to see that

$$\dim\{\mathcal{I}_2^{\mathcal{O}}(\mathfrak{A})\}\ge 1+3+3+1+1+1=10.$$

Since we have equality in these estimates, Assertion (2) also follows from Lemma 2.2.2. □

The following is an immediate corollary of the proof of Theorem 4.1.1 involving the map σ of Equation (4.1.e):

Corollary 4.1.1 *We have a short exact sequence*

$$0\longrightarrow W_6^{\mathcal{O}}\longrightarrow\mathfrak{R}\xrightarrow{\rho}S^2\longrightarrow 0$$

which is equivariantly split by the map σ. The complementary orthogonal projection from $\mathfrak{R}$ to $W_6^{\mathcal{O}}$ is given by $\mathrm{Id}-\sigma(\rho)$.

We can draw another consequence from the argument given to prove Theorem 4.1.1:

Corollary 4.1.2

(1) $\dim\{\mathfrak{R}\} = \frac{1}{12}(m^4 - m^2)$.

(2) $\dim\{\mathfrak{A}\} = \frac{1}{3}m^2(m^2 - 1)$.

(3) $\dim\{W_6^{\mathcal{O}}\} = \frac{m(m+1)(m-3)(m+2)}{12}$.

We refer to Theorem 2.4.1 where $\dim\{S_0^2\}$, $\dim\{\Lambda^2\}$, $\dim\{W_7^{\mathcal{O}}\}$, and $\dim\{W_8^{\mathcal{O}}\}$ are computed.

Proof. Assertion (1) follows from Theorem 1.6.1; this result will be proved independently in Section 6.1. From Equation (4.1.f) and Assertion (1), we may establish Assertion (2) by computing:

$$\begin{aligned}\dim\{\mathfrak{A}\} &= \dim\{\mathfrak{R}\} + \dim\{\Lambda^2\} \cdot \dim\{S^2\}\\ &= \tfrac{1}{12}\{m^4 - m^2\} + \tfrac{1}{4}\{m(m-1)m(m+1)\}\\ &= \tfrac{1}{12}m^2\{m^2 - 1 + 3m^2 - 3\} = \tfrac{1}{3}m^2(m^2 - 1).\end{aligned}$$

Assertion (3) follows from Theorem 2.4.1 and Theorem 4.1.1. □

We can draw a final consequence of the proof we have given of Theorem 4.1.1. Set:

$$\begin{aligned}&\psi_1^{\mathcal{O}} := \mathcal{I}\{(1,2,2,1)(3,4,4,3)\},\ \psi_2^{\mathcal{O}} := \mathcal{I}\{(1,2,2,3)(1,4,4,3)\},\\ &\psi_3^{\mathcal{O}} := \mathcal{I}\{(1,2,2,3)(3,4,4,1)\},\ \psi_4^{\mathcal{O}} := \mathcal{I}\{(1,2,2,3)(1,4,3,4)\},\\ &\psi_5^{\mathcal{O}} := \mathcal{I}\{(1,2,2,3)(3,4,1,4)\},\ \psi_6^{\mathcal{O}} := \mathcal{I}\{(1,2,3,2)(1,4,3,4)\},\\ &\psi_7^{\mathcal{O}} := \mathcal{I}\{(1,2,3,2)(3,4,1,4)\},\ \psi_8^{\mathcal{O}} := \mathcal{I}\{(1,2,3,4)(1,3,2,4)\},\\ &\psi_9^{\mathcal{O}} := \mathcal{I}\{(1,2,3,4)(1,3,4,2)\},\ \psi_{10}^{\mathcal{O}} := \mathcal{I}\{(1,2,3,4)(1,4,3,2)\}.\end{aligned} \tag{4.1.h}$$

Corollary 4.1.3 *Let $m \geq 4$. The invariants $\{\psi_1^{\mathcal{O}}, \dots, \psi_{10}^{\mathcal{O}}\}$ given above are a basis for $\mathcal{I}_2^{\mathcal{O}}(\mathfrak{A})$.*

4.2 The Modules $\mathbb{R}$, S_0^2, and Λ^2 in $\mathfrak{A}$

In Section 4.2, we continue our discussion of the $\mathcal{O}$ decomposition of $\mathfrak{A}$ given in Theorem 1.4.1.

Let ρ and ρ_{13} be as defined in Equation (1.1.e). Let ρ_a and $\rho_{13,a}$ be the alternating part of these tensors and let $\rho_{0,s}$ and $\rho_{13,0,s}$ be the symmetric part of these tensors which has zero trace. We adopt the notation of Definition 4.1.1 and study the factors $\mathbb{R} \oplus 2 \cdot S_0^2 \oplus 2 \cdot \Lambda^2$ in $\mathfrak{A}$:

Lemma 4.2.1 *Let $m \geq 3$.*

(1) σ_1 is an $\mathcal{O}$ equivariant map from $\mathbb{R}$ to $\mathfrak{A}$ with $\tau \circ \sigma_1 = m(m-1)\,\mathrm{Id}$. Thus $\frac{1}{m(m-1)}\sigma_1$ splits the projection $\mathfrak{A}\xrightarrow{\tau}\mathbb{R}\longrightarrow 0$ and $\mathrm{Range}(\sigma_1)$ is the submodule of $\mathfrak{A}$ abstractly isomorphic to $\mathbb{R}$; these are the tensors of constant sectional curvature. This module is detected by the scalar curvature τ.

(2) σ_2 and σ_3 are $\mathcal{O}$ equivariant maps from S_0^2 to $\mathfrak{A}$ such that

$$\begin{pmatrix} \rho_s \circ \sigma_2 & \rho_{13,s} \circ \sigma_2 \\ \rho_s \circ \sigma_3 & \rho_{13,s} \circ \sigma_3 \end{pmatrix} = \begin{pmatrix} -\,\mathrm{Id} & (1-m)\,\mathrm{Id} \\ (m-1)\,\mathrm{Id} & \mathrm{Id} \end{pmatrix}.$$

Thus σ_2 and σ_3 can be used to split the map $\mathfrak{A}\xrightarrow{\rho_{0,s}\oplus\rho_{13,0,s}} S_0^2 \oplus S_0^2 \longrightarrow 0$ and $\mathrm{Range}\{\sigma_2 \oplus \sigma_3\}$ is the submodule of $\mathfrak{A}$ abstractly isomorphic to $S_0^2 \oplus S_0^2$. This module is detected by $\rho_{0,s} \oplus \rho_{13,0,s}$.

(3) σ_4 and σ_5 are $\mathcal{O}$ equivariant maps from Λ^2 to $\mathfrak{A}$ such that

$$\begin{pmatrix} \rho_a \circ \sigma_4 & \rho_{13,a} \circ \sigma_4 \\ \rho_a \circ \sigma_5 & \rho_{13,a} \circ \sigma_5 \end{pmatrix} = \begin{pmatrix} (-1-m)\,\mathrm{Id} & -3\,\mathrm{Id} \\ -\,\mathrm{Id} & (1-m)\,\mathrm{Id} \end{pmatrix}.$$

Thus σ_4 and σ_5 can be used to split the map $\mathfrak{A}\xrightarrow{\rho_a\oplus\rho_{13,a}}\Lambda^2 \oplus \Lambda^2 \to 0$ and $\mathrm{Range}\{\sigma_4 \oplus \sigma_5\}$ is the submodule of $\mathfrak{A}$ abstractly isomorphic to $\Lambda^2 \oplus \Lambda^2$. This module is detected by $\rho_a \oplus \rho_{13,a}$.

Remark 4.2.1 We caution the reader that the maps of Assertion (2) and of Assertion (3) do not preserve the natural inner products involved; they are not isometries – we postpone until Lemma 4.2.2 the question of finding all isometric embeddings of $S_0^2 \oplus S_0^2$ and of $\Lambda^2 \oplus \Lambda^2$ into $\mathfrak{A}$. The two decompositions of [Bokan (1990)] arise by choosing different decompositions of the modules given in Assertion (2) and in Assertion (3).

Proof. We use Equation (1.1.d) to establish Assertion (1) by computing:

$$\begin{aligned} \tau(\sigma_1 c) &= c\varepsilon^{il}\varepsilon^{jk}\{\langle e_i, e_l\rangle\langle e_j, e_k\rangle - \langle e_i, e_k\rangle\langle e_j, e_l\rangle\} \\ &= c\varepsilon^{il}\varepsilon^{jk}\{\varepsilon_{il}\varepsilon_{jk} - \varepsilon_{ik}\varepsilon_{jl}\} = c\{\delta_i^i\delta_j^j - \delta_k^l\delta_l^k\} \\ &= m(m-1)c. \end{aligned}$$

Let $\phi \in S_0^2$ and let $\psi \in \Lambda^2$. Since the modules S_0^2 and Λ^2 are inequivalent irreducible modules, necessarily $\rho(\sigma_k)$ belongs to S_0^2 for $k = 2, 3$ and belongs to Λ^2 for $k = 4, 5$. Recall that $\tau(\phi) := \varepsilon^{ij}\phi_{ij}$; this vanishes for $\phi \in S_0^2$. We establish the matrix identities of Assertions (2) and (3) by computing:

$$\begin{aligned}
\rho(\sigma_2(\phi))_{uv} &:= \varepsilon^{ij}\{\phi(e_i,e_j)\langle e_u,e_v\rangle - \phi(e_u,e_j)\langle e_i,e_v\rangle\}\\
&= \tau(\phi)\varepsilon_{uv} - \phi(e_u,e_v) = -\phi_{uv},\\
\rho(\sigma_3(\phi))_{uv} &:= \varepsilon^{ij}\{\langle e_i,e_j\rangle\phi(e_u,e_v) - \langle e_u,e_j\rangle\phi(e_i,e_v)\}\\
&= m\phi(e_u,e_v) - \phi(e_u,e_v) = (m-1)\phi_{uv},\\
\rho(\sigma_4(\psi))_{uv} &:= \varepsilon^{ij}\{2\psi(e_i,e_u)\langle e_v,e_j\rangle + \psi(e_i,e_v)\langle e_u,e_j\rangle\\
&\qquad - \psi(e_u,e_v)\langle e_i,e_j\rangle\}\\
&= 2\psi(e_v,e_u) + \psi(e_u,e_v) - m\psi(e_u,e_v) = (-1-m)\psi_{uv},\\
\rho(\sigma_5(\psi))_{uv} &:= \varepsilon^{ij}\{\psi(e_i,e_j)\langle e_u,e_v\rangle - \psi(e_u,e_j)\langle e_i,e_v\rangle\}\\
&= 0 - \psi(e_u,e_v) = -\psi_{uv},
\end{aligned}$$

and that

$$\begin{aligned}
\rho_{13}(\sigma_2(\phi))_{uv} &:= \varepsilon^{ij}\{\phi(e_i,e_v)\langle e_u,e_j\rangle - \phi(e_u,e_v)\langle e_i,e_j\rangle\}\\
&= \phi(e_u,e_v) - m\phi(e_u,e_v) = (1-m)\phi_{uv},\\
\rho_{13}(\sigma_3(\phi))_{uv} &:= \varepsilon^{ij}\{\langle e_i,e_v\rangle\phi(e_u,e_j) - \langle e_u,e_v\rangle\phi(e_i,e_j)\}\\
&= \phi(e_u,e_v) - \varepsilon_{uv}\tau(\phi) = \phi_{uv},\\
\rho_{13}(\sigma_4(\psi))_{uv} &:= \varepsilon^{ij}\{2\psi(e_i,e_u)\langle e_j,e_v\rangle + \psi(e_i,e_j)\langle e_u,e_v\rangle\\
&\qquad -\psi(e_u,e_j)\langle e_i,e_v\rangle\}\\
&= 2\psi(e_v,e_u) + 0 - \psi(e_u,e_v) = -3\psi_{uv},\\
\rho_{13}(\sigma_5(\psi))_{uv} &:= \varepsilon^{ij}\{\psi(e_i,e_v)\langle e_u,e_j\rangle - \psi(e_u,e_v)\langle e_i,e_j\rangle\}\\
&= \psi(e_u,e_v) - m\psi(e_u,e_v) = (1-m)\psi_{uv}.
\end{aligned}$$

We compute:

$$\begin{aligned}
\det\begin{pmatrix} -\operatorname{Id} & (1-m)\operatorname{Id}\\ (m-1)\operatorname{Id} & \operatorname{Id}\end{pmatrix} &= (m-1)^2 - 1,\\
\det\begin{pmatrix} (-1-m)\operatorname{Id} & -3\operatorname{Id}\\ -\operatorname{Id} & (1-m)\operatorname{Id}\end{pmatrix} &= m^2 - 4.
\end{aligned}$$

Since $m \geq 3$, these determinants are non-zero. This shows that $\sigma_2 \oplus \sigma_3$ and $\sigma_4 \oplus \sigma_5$ are injective maps. These maps are detected by $\rho_{0,s} \oplus \rho_{13,0,s}$ and $\rho_a \oplus \rho_{13,a}$. The remaining conclusions of Assertions (2) and (3) now follow from the fact that S_0^2 and Λ^2 are inequivalent irreducible modules appearing in $\mathfrak{A}$ with multiplicity two. $\square$

We use σ_1 to identify the orthogonal module isomorphic to $\mathbb{R}$ and set:

$$W_1^{\mathcal{O}} := \operatorname{Range}\{\sigma_1\}.$$

The components $W_1^{\mathcal{O}}$, $W_6^{\mathcal{O}}$, $W_7^{\mathcal{O}}$, and $W_8^{\mathcal{O}}$ all appear with multiplicity one and hence are unique in any decomposition of $\mathfrak{A}$ as an orthogonal module. However, the factors S_0^2 and Λ^2 appear with multiplicity two. Thus many decompositions are possible. We discuss these decompositions as follows using the notation of Definition 4.1.1. If $(a_2, a_3) \neq (0,0)$ and $(a_4, a_5) \neq (0,0)$, define:

$$\begin{aligned} S^{\mathfrak{A}}(a_2, a_3) &= \{a_2\sigma_2(\phi) + a_3\sigma_3(\phi)\}_{\phi \in S_0^2}, \\ \Lambda^{\mathfrak{A}}(a_4, a_5) &= \{a_4\sigma_4(\psi) + a_5\sigma_5(\psi)\}_{\psi \in \Lambda^2}. \end{aligned} \tag{4.2.a}$$

These are $\mathcal{O}$ invariant subspaces of $\mathfrak{A}$; let $\xi_S^{\mathfrak{A}}(a_2, a_3)$ and $\xi_\Lambda^{\mathfrak{A}}(a_4, a_5)$ denote the associated orthogonal submodule of $\mathfrak{A}$. We note that if $\lambda \neq 0$, then

$$\xi_S^{\mathfrak{A}}(a_2, a_3) = \xi_S^{\mathfrak{A}}(\lambda a_2, \lambda a_3) \quad \text{and} \quad \xi_\Lambda^{\mathfrak{A}}(a_4, a_5) = \xi_\Lambda^{\mathfrak{A}}(\lambda a_4, \lambda a_5).$$

Let $\mathbb{RP}^1 := \{\mathbb{R}^2 - \{0\}\}/\{\mathbb{R} - \{0\}\}$ denote the real projective space of dimension one; we can regard (a_2, a_3) and (a_4, a_5) as elements of $\mathbb{RP}^1$ in defining $\xi_S^{\mathfrak{A}}$ and $\xi_\Lambda^{\mathfrak{A}}$.

Lemma 4.2.2

(1) Any orthogonal submodule of $\mathfrak{A}$ isomorphic to S_0^2 is equal to $\xi_S^{\mathfrak{A}}(a_2, a_3)$ where $(a_2, a_3) \neq (0,0)$.

(2) Any orthogonal submodule of $\mathfrak{A}$ isomorphic to Λ^2 is equal to $\xi_\Lambda^{\mathfrak{A}}(a_4, a_5)$ where $(a_4, a_5) \neq (0,0)$.

(3) If $\tilde{\xi}$ is the submodule of $\mathfrak{A}$ isomorphic to S_0^2 and perpendicular to $\xi_S^{\mathfrak{A}}(a_2, a_3)$, then $\tilde{\xi} = \xi_S^{\mathfrak{A}}((m-1)a_3 - a_2, a_3 - (m-1)a_2)$.

(4) If $\tilde{\xi}$ is the submodule of $\mathfrak{A}$ isomorphic to Λ^2 and perpendicular to $\xi_\Lambda^{\mathfrak{A}}(a_4, a_5)$, then $\tilde{\xi} = \xi_\Lambda^{\mathfrak{A}}((m-1)a_5 + 3a_4, -(3m+3)a_4 - 3a_5)$.

(5) Any orthogonal decomposition of $\mathfrak{A}$ into irreducible submodules is given (up to rearranging the factors) by the following, for suitably chosen $(a_2, a_3) \neq (0,0)$ and $(a_4, a_5) \neq (0,0)$.

$$\begin{aligned} \mathfrak{A} = \ & W_1^{\mathcal{O}} \oplus W_6^{\mathcal{O}} \oplus W_7^{\mathcal{O}} \oplus W_8^{\mathcal{O}} \oplus \xi_S^{\mathfrak{A}}(a_2, a_3) \oplus \xi_\Lambda^{\mathfrak{A}}(a_3, a_4) \\ & \oplus \xi_S^{\mathfrak{A}}((m-1)a_3 - a_2, a_3 - (m-1)a_2) \\ & \oplus \xi_\Lambda^{\mathfrak{A}}((m-1)a_5 + 3a_4, -(3m+3)a_4 - 3a_5). \end{aligned}$$

(6) We have $\mathfrak{R} = W_1^{\mathcal{O}} \oplus \xi_S^{\mathfrak{A}}(1,1) \oplus W_6^{\mathcal{O}}$.

Proof. In the proof of Theorem 4.1.1, we showed $\dim\{\mathcal{I}_2^{\mathcal{O}}(\mathfrak{A})\} = 10$. It then follows that

$$\dim\{\mathcal{I}_2^{\mathcal{O}}(S_0^2 \oplus S_0^2)\} = \dim\{\mathcal{I}_2^{\mathcal{O}}(\Lambda^2 \oplus \Lambda^2)\} = 3.$$

We have $\sigma_2 \oplus \sigma_3$ embeds $S_0^2 \oplus S_0^2$ in $\mathfrak{A}$ and similarly $\sigma_4 \oplus \sigma_5$ embeds $\Lambda^2 \oplus \Lambda^2$ in $\mathfrak{A}$. Assertions (1) and (2) now follow from Lemma 2.2.2.

Let $A := (a_2\sigma_2 + a_3\sigma_3)\phi$ where $\phi := (e^1 \otimes e^2 + e^2 \otimes e^1)$. Then the (possibly) non-zero components of A are:

$$\begin{aligned}
&A_{1222} = -A_{2122} = (a_2 - a_3)\varepsilon_{22},\\
&A_{2111} = -A_{1211} = (a_2 - a_3)\varepsilon_{11},\\
&A_{1ii2} = -A_{i1i2} = A_{2ii1} = -A_{i2i1} = a_2\varepsilon_{ii} \quad \text{for} \quad 3 \le i \le m,\\
&A_{i12i} = -A_{1i2i} = A_{i21i} = -A_{2i1i} = a_3\varepsilon_{ii} \quad \text{for} \quad 3 \le i \le m.
\end{aligned} \tag{4.2.b}$$

This implies that:

$$\langle\sigma_2\phi, \sigma_2\phi\rangle = \langle\sigma_3\phi, \sigma_3\phi\rangle = 4(m-1)\varepsilon_{11}\varepsilon_{22}, \quad \langle\sigma_2\phi, \sigma_3\phi\rangle = -4\varepsilon_{11}\varepsilon_{22}.$$

Choose (b_2, b_3) so that $\xi_S^{\mathfrak{A}}(b_2, b_3) \perp \xi_S^{\mathfrak{A}}(a_2, a_3)$. Then we must have:

$$\begin{aligned}
0 &= \langle(a_2\sigma_2 + a_3\sigma_3)\phi, (b_2\sigma_2 + b_3\sigma_3)\phi\rangle\\
&= \{4(m-1)(a_2b_2 + a_3b_3) - 4(a_2b_3 + a_3b_2)\}\varepsilon_{11}\varepsilon_{22}\\
&= \{(4(m-1)a_2 - 4a_3)b_2 + (4(m-1)a_3 - 4a_2)b_3\}\varepsilon_{11}\varepsilon_{22}.
\end{aligned}$$

We solve this relation and establish Assertion (3) by setting:

$$b_2 = (m-1)a_3 - a_2 \quad \text{and} \quad b_3 = a_3 - (m-1)a_2.$$

Let $A := (a_4\sigma_4 + a_5\sigma_5)\psi$ where $\psi := (e^1 \otimes e^2 - e^2 \otimes e^1)$. We now perform the same analysis for ξ_Λ. The (possibly) non-zero components of A are:

$$\begin{aligned}
&A_{1222} = -A_{2122} = (3a_4 + a_5)\varepsilon_{22},\\
&A_{1211} = -A_{2111} = (3a_4 + a_5)\varepsilon_{11},\\
&A_{12ii} = -A_{21ii} = 2a_4\varepsilon_{ii} \quad \text{for} \quad 3 \le i \le m,\\
&A_{1i2i} = -A_{2i1i} = -A_{i12i} = A_{i21i} = a_4\varepsilon_{ii} \quad \text{for} \quad 3 \le i \le m,\\
&A_{1ii2} = -A_{2ii1} = -A_{i1i2} = A_{i2i1} = a_5\varepsilon_{ii} \quad \text{for} \quad 3 \le i \le m.
\end{aligned} \tag{4.2.c}$$

We may now compute

$$\begin{aligned}
&\langle\sigma_4\psi, \sigma_4\psi\rangle = (12m + 12)\varepsilon_{11}\varepsilon_{22}, \quad \langle\sigma_4\psi, \sigma_5\psi\rangle = 12\varepsilon_{11}\varepsilon_{22},\\
&\langle\sigma_5\psi, \sigma_5\psi\rangle = 4(m-1)\varepsilon_{11}\varepsilon_{22}.
\end{aligned}$$

Choose (b_4, b_5) so $\xi_\Lambda^{\mathfrak{A}}(a_4, a_5) \perp \xi_\Lambda^{\mathfrak{A}}(b_4, b_5)$. We then have

$$\begin{aligned}
0 &= (12m + 12)a_4b_4 + 4(m-1)a_5b_5 + 12(a_4b_5 + a_5b_4)\\
&= \{(12m + 12)a_4 + 12a_5\}b_4 + \{4(m-1)a_5 + 12a_4\}b_5.
\end{aligned}$$

We solve this relation and complete the proof of Assertion (4) by setting:

$$b_4 = (m-1)a_5 + 3a_4 \quad \text{and} \quad b_5 = -(3m+3)a_4 - 3a_5.$$

Assertion (5) follows from the previous Assertions and from Theorem 4.1.1.

To prove Assertion (6), we must identify the summand of $\mathfrak{R}$ isomorphic to S_0^2. Because $(a_2\sigma_2 + a_3\sigma_3)\phi$ is a Riemann curvature tensor, this tensor is anti-symmetric in $\{k, l\}$; from this we see that $a_2 = a_3$. □

4.3 The Modules $W_6^{\mathcal{O}}$, $W_7^{\mathcal{O}}$, and $W_8^{\mathcal{O}}$ in $\mathfrak{A}$

In Section 4.3, we conclude our discussion of decomposition of $\mathfrak{A}$ as an orthogonal module given in Theorem 1.4.1 by describing the spaces $W_6^{\mathcal{O}}$, $W_7^{\mathcal{O}}$, and $W_8^{\mathcal{O}}$ in a bit more detail. The space $\ker(\rho) \cap \ker(\rho_{13}) \cap \mathfrak{A}$ is the submodule of $\mathfrak{A}$ abstractly isomorphic to $W_6^{\mathcal{O}} \oplus W_7^{\mathcal{O}} \oplus W_8^{\mathcal{O}}$.

Definition 4.3.1 If $A \in \mathfrak{A}$, define $\mu(A) \in \otimes^4 V^*$ and $\psi(A) \in \otimes^4 V^*$ by setting:

$$\begin{aligned}
\psi(A)_{ijkl} &= \tfrac{1}{4}\{A_{ijkl} + A_{jilk} + A_{klij} + A_{lkji}\},\\
\mu(A)_{ijkl} &= \tfrac{1}{8}\{3A_{ijkl} + 3A_{ijlk} + A_{ilkj} + A_{iklj} + A_{ljki} + A_{kjli}\}.
\end{aligned}$$

Lemma 4.3.1

(1) $W_6^{\mathcal{O}} = \ker(\rho) \cap \mathfrak{A} \cap \Lambda^2 \otimes \Lambda^2$.
(2) $W_7^{\mathcal{O}} = \ker(\rho) \cap \mathfrak{A} \cap \Lambda^2 \otimes S^2$.
(3) ψ is orthogonal projection from $\ker(\rho) \cap \ker(\rho_{13}) \cap \mathfrak{A}$ *to* $W_6^{\mathcal{O}}$.
(4) μ is orthogonal projection from $\ker(\rho) \cap \ker(\rho_{13}) \cap \mathfrak{A}$ *to* $W_7^{\mathcal{O}}$.
(5) The complementary orthogonal projection from $\ker(\rho) \cap \ker(\rho_{13}) \cap \mathfrak{A}$ *to the submodule of* $\mathfrak{A}$ *abstractly isomorphic to* $W_8^{\mathcal{O}}$ *is given by* $\mathrm{Id} - \psi - \mu$.

Remark 4.3.1 The associated projection from all of $\mathfrak{A}$ to $W_6^{\mathcal{O}}$, to $W_7^{\mathcal{O}}$, and to $W_8^{\mathcal{O}}$ may be obtained by subtracting the corresponding projection on the factor isomorphic to $\mathbb{R} \oplus S_0^2 \oplus S_0^2 \oplus \Lambda^2 \oplus \Lambda^2$; this may be determined from Lemma 4.2.1.

Proof. Assertions (1) and (2) are immediate; they are simply restatements of Definition 1.4.1. We now establish Assertion (3). Let $A \in \mathfrak{A}$. To show that $\psi(A) \in \mathfrak{R}$, it suffices, by Lemma 4.1.1, to check the relations of Equation (4.1.a), Equation (4.1.b), and Equation (4.1.d) are satisfied:

$$\begin{aligned}
&\psi(A)_{ijkl} + \psi(A)_{jkil} + \psi(A)_{kijl} \\
&\quad = \tfrac{1}{4}\{A_{ijkl} + A_{jilk} + A_{klij} + A_{lkji}\} \\
&\qquad + \tfrac{1}{4}\{A_{jkil} + A_{kjli} + A_{iljk} + A_{likj}\} \\
&\qquad + \tfrac{1}{4}\{A_{kijl} + A_{iklj} + A_{jlki} + A_{ljik}\} = 0, \\
&\psi(A)_{jikl} = \tfrac{1}{4}\{A_{jikl} + A_{ijlk} + A_{klji} + A_{lkij}\} = -\psi(A)_{ijkl}, \\
&\psi(A)_{klij} = \tfrac{1}{4}\{A_{klij} + A_{lkji} + A_{ijkl} + A_{jilk}\} = \psi(A)_{ijkl}.
\end{aligned}$$

Furthermore, if $A \in \ker(\rho) \cap \ker(\rho_{13})$, then:

$$\begin{aligned}
\rho(\psi(A))_{jk} &= \tfrac{1}{4}\varepsilon^{il}\{A_{ijkl} + A_{jilk} + A_{klij} + A_{lkji}\} \\
&= \tfrac{1}{4}\{\rho(A)_{jk} - \rho_{13}(A)_{jk} - \rho_{13}(A)_{kj} + \rho(A)_{kj}\} = 0.
\end{aligned}$$

This shows that $\psi(A) \in \mathfrak{R} \cap \ker(\rho) = W_6^{\mathcal{O}}$. Conversely, if $A \in W_6^{\mathcal{O}}$, then the symmetries given in Equation (4.1.a), given in Equation (4.1.c), and given in Equation (4.1.d) show that

$$\psi(A)_{ijkl} = \tfrac{1}{4}\{A_{ijkl} + A_{jilk} + A_{klij} + A_{lkji}\} = A_{ijkl}.$$

Thus ψ is an equivariant projection from $W_6^{\mathcal{O}} \oplus W_7^{\mathcal{O}} \oplus W_8^{\mathcal{O}}$ to $W_6^{\mathcal{O}}$ with $\psi^2 = \psi$.

Let $A \in \mathfrak{A}$. We show that $\mu(A) \in \mathfrak{A}$ by checking that Equation (4.1.a) and Equation (4.1.b) are satisfied:

$$\begin{aligned}
&\mu(A)_{jikl} = \tfrac{1}{8}\{3A_{jikl} + 3A_{jilk} + A_{jlki} + A_{jkli} + A_{likj} + A_{kilj}\} \\
&\quad = -\mu(A)_{ijkl}, \\
&\mu(A)_{ijkl} + \mu(A)_{jkil} + \mu(A)_{kijl} \\
&\quad = \tfrac{1}{8}\{3A_{ijkl} + 3A_{ijlk} + A_{ilkj} + A_{iklj} + A_{ljki} + A_{kjli}\} \\
&\qquad + \tfrac{1}{8}\{3A_{jkil} + 3A_{jkli} + A_{jlik} + A_{jilk} + A_{lkij} + A_{iklj}\} \\
&\qquad + \tfrac{1}{8}\{3A_{kijl} + 3A_{kilj} + A_{klji} + A_{kjli} + A_{lijk} + A_{jilk}\}.
\end{aligned}$$

We use the Bianchi identity to see $3A_{ijkl} + 3A_{jkil} + 3A_{kijl} = 0$. We reorder the remaining terms and group all the terms with last index k, i, or j on a line and continue the calculation:

$$\begin{aligned}
&= \tfrac{1}{8}\{3A_{ijlk} + A_{jilk} + A_{jilk} + A_{jlik} + A_{lijk}\} \\
&\quad + \tfrac{1}{8}\{3A_{jkli} + A_{kjli} + A_{kjli} + A_{ljki} + A_{klji}\} \\
&\quad + \tfrac{1}{8}\{3A_{kilj} + A_{iklj} + A_{iklj} + A_{ilkj} + A_{lkij}\}.
\end{aligned}$$

We use the $\mathbb{Z}_2$ symmetry in the first two indices and the Bianchi identity once again to continue the calculation:

$$\begin{aligned}
&= \tfrac{1}{8}\{3A_{ijlk} - A_{ijlk} - A_{ijlk} - A_{ijlk}\} \\
&\quad + \tfrac{1}{8}\{3A_{jkli} - A_{jkli} - A_{jkli} - A_{jkli}\} \\
&\quad + \tfrac{1}{8}\{3A_{kilj} - A_{kilj} - A_{kilj} - A_{kilj}\} = 0.
\end{aligned}$$

Let $A \in \mathfrak{A} \cap \ker(\rho) \cap \ker(\rho_{13})$; by Lemma 4.1.2, $\rho_{34}(A) = 0$ as well. We verify that $\mu(A) \in \ker(\rho) \cap \ker(\rho_{13})$ by checking:

$$\begin{aligned}\rho(\mu(A))_{jk} &= \tfrac{1}{8}\varepsilon^{il}\{3A_{ijkl} + 3A_{ijlk} + A_{ilkj} + A_{iklj} + A_{ljki} + A_{kjli}\}\\ &= \tfrac{1}{8}\{3\rho_{jk} + 3\rho_{13,jk} + 0 + \rho_{13,kj} + \rho_{jk} + \rho_{34,kj}\} = 0,\end{aligned}$$

and

$$\begin{aligned}\rho_{13}(\mu(A))_{jl} &= \tfrac{1}{8}\varepsilon^{ik}\{3A_{ijkl} + 3A_{ijlk} + A_{ilkj} + A_{iklj} + A_{ljki} + A_{kjli}\}\\ &= \tfrac{1}{8}\{3\rho_{13,jl} + 3\rho_{jl} + \rho_{13,lj} + 0 + \rho_{34,lj} + \rho_{jl}\} = 0.\end{aligned}$$

This establishes that $\mu(A) \in \ker(\rho \oplus \rho_{13})$. Clearly $\mu(A)_{ijkl} = \mu(A)_{ijlk}$ is symmetric in the last two indices. On the other hand, if $A \in \mathfrak{A}$ is symmetric in the last two indices, then we use the Bianchi identity to compute:

$$\begin{aligned}\mu(A)_{ijkl} &= \tfrac{1}{8}\{6A_{ijkl} + A_{iljk} + A_{ikjl} + A_{ljik} + A_{kjil}\}\\ &= \tfrac{1}{8}\{6A_{ijkl} - A_{jilk} - A_{jikl}\} = A_{ijkl}.\end{aligned}$$

Thus $\mu^2 = \mu$ so μ is a projection with

$$\text{Range}(\mu) = \mathfrak{A} \cap \ker(\rho) \cap \ker(\rho_{13}) \cap \Lambda^2 \otimes S^2 = W_7^{\mathcal{O}}.$$

This establishes Assertion (2). Since $W_6^{\mathcal{O}}$, $W_7^{\mathcal{O}}$, and $W_8^{\mathcal{O}}$ are inequivalent irreducible modules, it follows that ψ and μ are orthogonal projections. Furthermore, the complementary projection $\text{Id} - \psi - \mu$ is defined by projection on the remaining component abstractly isomorphic to $W_8^{\mathcal{O}}$. □

4.4 Decomposition of $\mathfrak{A}$ as a General Linear Module

We regard $\mathfrak{A}$ as a space of curvature operators with $\mathfrak{A} \subset \otimes^3 V^* \otimes V$; let $A_{ijk}{}^l$ be the components of such a tensor relative to a frame $\{e_1, \ldots, e_m\}$ for V and the corresponding dual frame $\{e^1, \ldots, e^m\}$ for V^*. The general linear group GL acts on $\mathfrak{A}$ in a natural fashion; if T belongs to GL and if $\mathcal{A} \in \mathfrak{A}$, we set:

$$(T^*\mathcal{A})(x, y)z := T^{-1}\{\mathcal{A}(Tx, Ty)Tz\}.$$

This action is not irreducible, but decomposes as the direct sum of irreducible modules; Theorem 1.3.1 [Strichartz (1988)] will follow from the discussion in Section 4.4. Although our structure group is GL, we introduce an auxiliary positive definite inner product $\langle\cdot,\cdot\rangle$ to enable us to make use of the previous results of Chapter 4 and to raise and lower indices as convenient; we shall choose the frame for V so that $\varepsilon_{ij} = \delta_{ij}$.

The Ricci tensor is defined in this framework by setting:

$$\rho(y,z) := \mathrm{Tr}(x \to \mathcal{A}(x,y)z).$$

This definition agrees with the previous definition in the presence of a symmetric non-degenerate bilinear form $\langle\cdot,\cdot\rangle$ but is a GL equivariant map from $\mathfrak{A}$ to $\otimes^2 V^*$. The contraction ρ_{13} is not GL equivariant and plays no role in our decomposition. We may further decompose $\rho = \rho_a \oplus \rho_s$ into the alternating and symmetric Ricci tensors; these are GL equivariant as well.

Definition 4.4.1 If $\psi \in \Lambda^2$ and if $\phi \in S^2$, define $\sigma_a(\psi)$ and $\sigma_s(\phi)$ by setting:

$$(\sigma_a(\psi))(x,y)z = -\tfrac{1}{1+m}\{2\psi(x,y)z + \psi(x,z)y - \psi(y,z)x\},$$
$$(\sigma_s(\phi))(x,y)z = \tfrac{1}{1-m}\{\phi(x,z)y - \phi(y,z)x\}.$$

Lemma 4.4.1 *Let $m \geq 3$. The Ricci tensor $\rho = \rho_a \oplus \rho_s$ defines a* GL *equivariant short exact sequence*

$$0 \longrightarrow \ker(\rho) \longrightarrow \mathfrak{A} \xrightarrow{\rho_a \oplus \rho_s} \Lambda^2 \oplus S^2 \longrightarrow 0.$$

This sequence is equivariantly split by the map $\sigma_a \oplus \sigma_s$. Thus $\mathfrak{A}$ decomposes as a direct sum as a general linear module in the form:

$$\mathfrak{A} = (\ker(\rho) \cap \mathfrak{A}) \oplus \Lambda^2 \oplus S^2.$$

Remark 4.4.1 By Corollary 4.1.2, the dimensions of these modules are:

$$\begin{aligned}
&\dim\{S^2\} = \tfrac{1}{2}m(m+1), \qquad \dim\{\Lambda^2\} = \tfrac{1}{2}m(m-1),\\
&\dim\{\ker(\rho) \cap \mathfrak{A}\} = \dim\mathfrak{A} - \dim\{S^2\} - \dim\{\Lambda^2\}\\
&\quad = \tfrac{1}{3}m^2(m^2-1) - \tfrac{1}{2}m(m+1) - \tfrac{1}{2}m(m-1)\\
&\quad = \tfrac{m^2(m^2-4)}{3}.
\end{aligned}$$

Thus these general linear modules are inequivalent on dimensional grounds.

Proof. Adopt the notation of Definition 4.1.1. We may then express:

$$\langle(\sigma_a(\psi))(x,y)z, w\rangle = -\tfrac{1}{1+m}\sigma_4(\psi),$$
$$\langle(\sigma_s(\phi))(x,y)z, w\rangle = \tfrac{1}{m-1}\sigma_3(\phi).$$

Thus σ_a and σ_s arise from σ_4 and σ_3, respectively, by raising indices modulo appropriate scaling factors. It now follows from Lemma 4.1.3 that σ_a and

σ_s take values in $\mathfrak{A}$. We have by Lemma 4.2.1 that $\rho(\sigma_3(\phi)) = (m-1)\phi$ and $\rho(\sigma_4(\psi)) = -(m+1)\psi$. The result now follows. □

By Lemma 4.4.1, we have $\mathfrak{A} \approx \ker(\rho) \oplus \Lambda^2 \oplus S^2$. By Theorem 2.4.2, Λ^2 and S^2 are irreducible general linear modules. Thus Theorem 1.3.1 will follow from:

Lemma 4.4.2 $\ker(\rho) \cap \mathfrak{A}$ *is an irreducible general linear module.*

Proof. Let $W_4^{\mathcal{O}}$ be the submodule of $\ker(\rho)\cap\mathfrak{A}$ isomorphic to Λ^2. Similarly, let $W_5^{\mathcal{O}}$ be submodule of $\ker(\rho)\cap\mathfrak{A}$ isomorphic to S_0^2. There is then an orthogonal module orthogonal direct sum decomposition into inequivalent irreducible modules:

$$\ker(\rho)\cap\mathfrak{A} = W_4^{\mathcal{O}} \oplus W_5^{\mathcal{O}} \oplus W_6^{\mathcal{O}} \oplus W_7^{\mathcal{O}} \oplus W_8^{\mathcal{O}}.$$

Let π_i denote orthogonal projection onto $W_i^{\mathcal{O}}$ for $4 \le i \le 8$. If W is a GL invariant subspace of $\ker(\rho)$, then we may decompose

$$W = \bigoplus_{i=4}^{8} \pi_i W.$$

Note that $\pi_i W \neq \{0\}$ if and only if $W_i^{\mathcal{O}} \subset W$.

If $g : V \to V$, we extend the action to a map from V^* to V^* so that $v \cdot v^* = gv \cdot gv^*$. Then

$$(g^*A)(x,y,z,w^*) := A(gx,gy,gz,gw^*).$$

For example, let $g = g_\varepsilon$ be defined by

$$g(e_i) := \left\{\begin{matrix} e_i & \text{if } i \neq 3 \\ e_3 + \varepsilon e_4 & \text{if } i = 3 \end{matrix}\right\}, \quad g(e^i) := \left\{\begin{matrix} e^i & \text{if } i \neq 4 \\ e^4 - \varepsilon e^3 & \text{if } i = 4 \end{matrix}\right\}.$$

We then have that:

$$\{\partial_\varepsilon|_{\varepsilon=0} g^*A\}_{ijk}{}^l = \delta_{i3}A_{4jk}{}^l + \delta_{3j}A_{i4k}{}^l + \delta_{3k}A_{ij4}{}^l - \delta_{4l}A_{ijk}{}^3.$$

The lack of symmetry in the final index is crucial.

We adopt the notation of Lemma 4.2.2. In the decomposition of Lemma 4.4.1, the summand S_0^2 takes the form

$$S^2 = \operatorname{Range}(\sigma_s) = \xi_S^{\mathfrak{A}}(0,1).$$

Thus $W_5^{\mathcal{O}} = \xi_S^{\mathfrak{A}}(a_2,a_3)$ where $a_2 \neq 0$. Let

$$A := (a_2\sigma_2 + a_3\sigma_3)(e^1 \otimes e^2 + e^2 \otimes e^1) \in W_5^{\mathcal{O}}.$$

We raise indices in Equation (4.2.b). The (possibly) non-zero components of A are given by:

$$\begin{aligned}
&A_{122}{}^2 = -A_{212}{}^2 = (a_2 - a_3),\\
&A_{211}{}^1 = -A_{121}{}^1 = (a_2 - a_3),\\
&A_{1ii}{}^2 = -A_{i1i}{}^2 = A_{2ii}{}^1 = -A_{i2i}{}^1 = a_2 \quad \text{for} \quad 3 \le i \le m,\\
&A_{i12}{}^i = -A_{1i2}{}^i = A_{i21}{}^i = -A_{2i1}{}^i = a_3 \quad \text{for} \quad 3 \le i \le m.
\end{aligned}$$

Let $g_1 = g_{1,\varepsilon}$ act on V and on V^* by

$$g_1(e_i) := \left\{\begin{matrix} e_i & \text{if } i \ne 1 \\ \varepsilon e_1 & \text{if } i = 1 \end{matrix}\right\}, \qquad g_1(e^i) := \left\{\begin{matrix} e^i & \text{if } i \ne 1 \\ \varepsilon^{-1} e^1 & \text{if } i = 1 \end{matrix}\right\}. \tag{4.4.a}$$

Suppose that $\pi_5 W \ne \{0\}$ so that $W_5^{\mathcal{O}} \subset W$. This implies $A \in W$. We set $B := \lim_{\varepsilon\to 0} \varepsilon g_1^* A$. Since W is a closed subset of V, $B \in W$. The non-zero components of B are then

$$B_{2ii}{}^1 = -B_{i2i}{}^1 = a_2 \quad \text{for} \quad 3 \le i \le m. \tag{4.4.b}$$

We have $\rho(B) = 0$; the only non-zero components of $\rho_{13}(B)$ are given by:

$$\rho_{13}(B)_{21} = a_2(2 - m).$$

This is neither symmetric nor anti-symmetric. Thus we conclude

$$W_4^{\mathcal{O}} \subset W.$$

Next let $g_2 = g_{2,\varepsilon}$ act on V and on V^* by:

$$g_2(e_i) := \left\{\begin{matrix} e_i & \text{if } i \ne 3 \\ e_3 + \varepsilon e_4 & \text{if } i = 3 \end{matrix}\right\}, \qquad g_2(e^i) := \left\{\begin{matrix} e_i & \text{if } i \ne 4 \\ e^4 - \varepsilon e^3 & \text{if } i = 4 \end{matrix}\right\}.$$

Let $B \in W$ be defined by Equation (4.4.b). We set $C := \partial_\varepsilon(g_2^*)B \in W$; the non-zero components of C are given by:

$$C_{234}{}^1 = C_{243}{}^1 = -C_{324}{}^1 = -C_{423}{}^1 = a_2.$$

Thus, in particular, $C \in \ker(\rho) \cap \ker(\rho_{13})$. We compute:

$$\begin{aligned}
&\psi(C)_{234}{}^1 = \tfrac14\{C_{234}{}^1 + C_{321}{}^4 + C_{412}{}^3 + C_{143}{}^2\} = \tfrac14 a_2\\
&\mu(C)_{234}{}^1 = \tfrac18\{3C_{234}{}^1 + 3C_{231}{}^4 + C_{214}{}^3 + C_{241}{}^3 + C_{134}{}^2 + C_{431}{}^2\} = \tfrac38 a_2,\\
&(\mathrm{Id} - \psi - \mu)(C)_{234}{}^1 = \tfrac38 a_2.
\end{aligned}$$

This shows $\psi(C) \ne 0$, $\mu(C) \ne 0$, and $(\mathrm{Id} - \psi - \mu)(C) \ne 0$. Thus C has non-zero components in $W_6^{\mathcal{O}}$, $W_7^{\mathcal{O}}$, and $W_8^{\mathcal{O}}$ so $W_i^{\mathcal{O}} \subset W$ for $i = 6$, $i = 7$, and $i = 8$. We summarize our conclusions so far:

$$\pi_5 W \neq \{0\} \quad \Rightarrow \quad \ker(\rho) \cap \mathfrak{A} = W. \tag{4.4.c}$$

In the decomposition of Lemma 4.4.1, the summand Λ^2 takes the form

$$\Lambda^2 = \text{Range}(\sigma_a) = \xi_\Lambda^{\mathfrak{A}}(1,0).$$

Thus $W_4^{\mathcal{O}} = \xi_\Lambda^{\mathfrak{A}}(a_4, a_5)$ where $a_5 \neq 0$. Suppose that $\pi_4 W \neq \{0\}$. We clear the previous notation and set

$$A := (a_4\sigma_4 + a_5\sigma_5)(e^1 \otimes e^2 - e^2 \otimes e^1) \in W.$$

We raise indices in Equation (4.2.c). The (possibly) non-zero components of A are given by:

$$\begin{array}{ll}
A_{122}{}^2 = -A_{212}{}^2 = (3a_4 + a_5), & \\
A_{121}{}^1 = -A_{211}{}^1 = (3a_4 + a_5), & \\
A_{12i}{}^i = -A_{21i}{}^i = 2a_4 & \text{for} \quad 3 \leq i \leq m, \\
A_{1i2}{}^i = -A_{2i1}{}^i = -A_{i12}{}^i = A_{i21}{}^i = a_4 & \text{for} \quad 3 \leq i \leq m, \\
A_{1ii}{}^2 = -A_{i1i}{}^2 = -A_{2ii}{}^1 = A_{i2i}{}^1 = a_5 & \text{for} \quad 3 \leq i \leq m.
\end{array}$$

We argue as above. Adopt the notation of Equation (4.4.a) to define g_1. We set $B := \lim_{\varepsilon \to 0} \varepsilon g_1^* A \in W$. The non-zero components of B are then

$$B_{i2i}{}^1 = -B_{2ii}{}^1 = a_5 \quad \text{for} \quad 3 \leq i \leq m.$$

We have $B \in \ker(\rho)$. The only non-zero component of ρ_{13} is

$$\rho_{13}(B)_{21} = a_5(m-2).$$

As the tensor ρ_{13} is neither symmetric nor anti-symmetric, we conclude $W_5^{\mathcal{O}} \subset W$. Equation (4.4.c) now implies:

$$\pi_4 W \neq \{0\} \quad \Rightarrow \quad \ker(\rho) \subset W. \tag{4.4.d}$$

Let $A \in \mathfrak{A}$ have non-zero components

$$A_{123}{}^4 = A_{132}{}^4 = -A_{213}{}^4 = -A_{312}{}^4 = 1.$$

It is immediate that $A \in \ker(\rho) \cap \ker(\rho_{13})$. We then have that

$$\begin{array}{l}
\psi(A)_{1234} = \frac{1}{4}\{A_{1234} + A_{2143} + A_{3412} + A_{4321}\} = \frac{1}{4}, \\
\mu(A)_{1234} = \frac{1}{8}\{3A_{1234} + 3A_{1243} + A_{1432} + A_{1342} + A_{4231} + A_{3241}\} = \frac{3}{8}, \\
(A - \psi(A) - \mu(A))_{1234} = \frac{3}{8}.
\end{array}$$

Suppose that $\pi_i W \neq \{0\}$ for some i with $i = 6$, $i = 7$, or $i = 8$. We set

$$B := \left\{ \begin{array}{ll} 4\psi(A) & \text{if } \pi_6 W \neq \{0\} \\ \frac{8}{3}\mu(A) & \text{if } \pi_7 W \neq \{0\} \\ \frac{8}{3}(1-\psi(A)-\mu(A)) & \text{if } \pi_8 W \neq \{0\} \end{array} \right\} \in W.$$

Then $B_{1234} = 1$ so $B_{123}{}^4 = 1$. Define $g_3 = g_{3,\varepsilon} \in \text{GL}$ so

$$g_3(e_i) := \left\{ \begin{array}{ll} e_i & \text{if } i \neq 4 \\ \varepsilon e_4 & \text{if } i = 4 \end{array} \right\}, \qquad g_3(e^i) := \left\{ \begin{array}{ll} e^i & \text{if } i \neq 4 \\ \varepsilon^{-1} e^4 & \text{if } i = 4 \end{array} \right\}.$$

Let $C := \lim_{\varepsilon \to 0} \varepsilon g_3^* B \in W$. Then $C_{ijk}{}^l = 0$ unless $l = 4$ and $\{i, j, k\}$ is a permutation of $\{1, 2, 3\}$. Furthermore, $C_{123}{}^4 = -C_{213}{}^4 = 1$. Since the roles of the indices "2" and "3" in A are symmetric, they are symmetric in C as well. Thus $C_{132}{}^4 = -C_{312}{}^4 = 1$. The Bianchi identity now shows $C_{231}{}^4 = C_{321}{}^4 = 0$. Consequently $C = A$ so $A \in W$. We consider the transformation $g = g_\varepsilon$ given by:

$$g e_i := \left\{ \begin{array}{ll} e_i & i \neq 1 \\ e_1 + \varepsilon e_2 & i = 1 \end{array} \right\}, \qquad g e^i := \left\{ \begin{array}{ll} e^i & i \neq 2 \\ e^2 - \varepsilon e^1 & i = 2 \end{array} \right\}.$$

Let $D := \partial_\varepsilon|_{\varepsilon=0} g^* A \in W$;

$$D_{131}{}^4 = -D_{311}{}^4 = 1.$$

Thus $\rho(D) = 0$. Since $\rho_{13} D \neq 0$, either $\pi_4 D \neq \{0\}$ or $\pi_5 D \neq \{0\}$. In either event, this implies $W = \ker(\rho)$. Thus

$$\pi_i(W) \neq 0 \quad \text{for} \quad i \in \{6, 7, 8\} \quad \Rightarrow \quad \ker(\rho) \subset W. \qquad (4.4.e)$$

Lemma 4.4.2 now follows from the computations performed above. □

4.5 Geometric Realizability of Affine Curvature Operators

In Theorem 1.3.1, we decomposed $\mathfrak{A}$ as the direct sum of three irreducible general linear modules:

$$\mathfrak{A} = \Lambda^2 \oplus S^2 \oplus \ker(\rho).$$

By studying the possible components of $\mathfrak{A}$, this gives rise to seven different non-trivial geometric realizability questions. Note that if all the components of $\mathcal{A}$ vanish, then $\mathcal{A} = 0$ and $\mathcal{A}$ is realized by a flat connection. In Theorem 1.3.4, we imposed no restrictions on the components of $\mathcal{A}$. We showed that an arbitrary element $\mathcal{A} \in \mathfrak{A}$ is geometrically realizable. In

Theorem 4.5.1, we focus on ρ and show that a Ricci symmetric, a Ricci anti-symmetric, or a Ricci flat curvature operator $\mathcal{A}$ is geometrically realizable by a Ricci symmetric, by a Ricci anti-symmetric, or by a Ricci flat affine connection, respectively. In Theorem 4.5.2, we focus on the Weyl projective curvature tensor. We will show a projectively flat or a projectively flat and Ricci symmetric curvature operator $\mathcal{A}$ is geometrically realizable by a projectively flat or by a projectively flat and Ricci symmetric affine connection, respectively. We complete our study in Theorem 4.5.3 with a negative result dealing with the sole remaining case. We show that if $0 \neq \mathcal{A}$ is projectively flat and Ricci anti-symmetric, then $\mathcal{A}$ is not geometrically realizable by a projectively flat Ricci anti-symmetric affine connection. Theorem 1.3.3 then follows from this discussion.

Fix a basis $\{e_1, \ldots, e_m\}$ for V and let $x = (x^1, \ldots, x^m)$ be the dual system of coordinates on V so that any element of V may be expanded in the form $v = x^1 e_1 + \cdots + x^m e_m$. This identifies V with $\mathbb{R}^m$. If $\mathcal{A} \in \mathfrak{A}$, we define the components of $\mathcal{A}$ by expanding

$$\mathcal{A}(e_i, e_j)e_k = A_{ijk}{}^l e_l.$$

Let ∇ be an affine connection on the tangent bundle TV of V. For such a connection, let $\mathcal{R}_0$ be the associated curvature operator at the origin. Let $\mathfrak{A}^1 \subset \otimes^4 V^* \otimes V$ be the space of all tensors satisfying the identities of Lemma 3.1.2 (3)–(5); these are the curvature symmetries of the covariant derivative of the curvature operator. We begin our study of geometrical realization questions by establishing Theorem 1.3.4

Proof of Theorem 1.3.4. Let $\mathcal{A} \in \mathfrak{A}$ and let $\mathcal{A}^1 \in \mathfrak{A}^1$. Define a connection ∇ on TV by setting

$$\begin{aligned}\Gamma_{ij}{}^l &:= \tfrac{1}{3}(A_{kij}{}^l + A_{kji}{}^l)x^k \\ &\quad + a_1(A^1_{kij}{}^l{}_{;n} + A^1_{kji}{}^l{}_{;n})x^k x^n + a_2(A^1_{kin}{}^l{}_{;j} + A^1_{kjn}{}^l{}_{;i})x^k x^n\end{aligned}$$

where $a_1 = \frac{5}{24}$ and $a_2 = \frac{1}{24}$. Clearly $\Gamma_{ij}{}^l = \Gamma_{ji}{}^l$ so ∇ is an affine connection. As Γ vanishes at the origin, we may use Equation (3.1.c) and the curvature symmetries to see:

$$\begin{aligned}\mathcal{R}_{ijk}{}^l(0) &= \left\{\partial_{x_i}\Gamma_{jk}{}^l - \partial_{x_j}\Gamma_{ik}{}^l\right\}(0) \\ &= \tfrac{1}{3}\left\{A_{ijk}{}^l + A_{ikj}{}^l - A_{jik}{}^l - A_{jki}{}^l\right\} \\ &= \tfrac{1}{3}\left\{A_{ijk}{}^l - A_{kij}{}^l + A_{ijk}{}^l - A_{jki}{}^l\right\} \\ &= A_{ijk}{}^l \partial_{x_l}.\end{aligned}$$

Similarly, we compute:

$$\begin{aligned}
\mathcal{R}_{ijk}{}^{l}{}_{;n}(0) &= \left\{\partial_n\partial_i\Gamma_{jk}{}^{l} - \partial_n\partial_j\Gamma_{ik}{}^{l}\right\}(0)\\
&= a_1(A^1_{ijk}{}^{l}{}_{;n} + A^1_{ikj}{}^{l}{}_{;n} + A^1_{njk}{}^{l}{}_{;i} + A^1_{nkj}{}^{l}{}_{;i})\\
&\quad -a_1(A^1_{jik}{}^{l}{}_{;n} + A^1_{jki}{}^{l}{}_{;n} + A^1_{nik}{}^{l}{}_{;j} + A^1_{nki}{}^{l}{}_{;j})\\
&\quad +a_2(A^1_{ijn}{}^{l}{}_{;k} + A^1_{ikn}{}^{l}{}_{;j} + A^1_{nji}{}^{l}{}_{;k} + A^1_{nki}{}^{l}{}_{;j})\\
&\quad -a_2(A^1_{jin}{}^{l}{}_{;k} + A^1_{jkn}{}^{l}{}_{;i} + A^1_{nij}{}^{l}{}_{;k} + A^1_{nkj}{}^{l}{}_{;i})\\
&= 2a_1A^1_{ijk}{}^{l}{}_{;n} + a_1(A^1_{ikj}{}^{l}{}_{;n} + A^1_{ijk}{}^{l}{}_{;n} + A^1_{kij}{}^{l}{}_{;n})\\
&\quad +a_1(A^1_{njk}{}^{l}{}_{;i} + A^1_{nkj}{}^{l}{}_{;i} - A^1_{nik}{}^{l}{}_{;j} - A^1_{nki}{}^{l}{}_{;j})\\
&\quad +2a_2A^1_{ijn}{}^{l}{}_{;k} + a_2(A^1_{nji}{}^{l}{}_{;k} + A^1_{ijn}{}^{l}{}_{;k} + A^1_{jni}{}^{l}{}_{;k})\\
&\quad +a_2(-A^1_{jkn}{}^{l}{}_{;i} - A^1_{nkj}{}^{l}{}_{;i} + A^1_{ikn}{}^{l}{}_{;j} + A^1_{nki}{}^{l}{}_{;j})\\
&= 3a_1A^1_{ijk}{}^{l}{}_{;n} + 3a_2A^1_{ijn}{}^{l}{}_{;k}\\
&\quad +a_1(A^1_{njk}{}^{l}{}_{;i} - A^1_{jnk}{}^{l}{}_{;i} - A^1_{kjn}{}^{l}{}_{;i} - A^1_{nik}{}^{l}{}_{;j} + A^1_{ink}{}^{l}{}_{;j} + A^1_{kin}{}^{l}{}_{;j})\\
&\quad +a_2(-A^{1\ l}_{jkn;i} + A^1_{kjn}{}^{l}{}_{;i} + A^1_{jnk}{}^{l}{}_{;i} + A^1_{ikn}{}^{l}{}_{;j} - A^1_{ink}{}^{l}{}_{;j} - A^1_{kin}{}^{l}{}_{;j})\\
&= 3a_1A^1_{ijk}{}^{l}{}_{;n} + 3a_2A^1_{ijn}{}^{l}{}_{;k}\\
&\quad +(2a_1 - a_2)A^1_{njk}{}^{l}{}_{;i} + (a_1 - 2a_2)A^1_{jkn}{}^{l}{}_{;i}\\
&\quad +(2a_1 - a_2)A^1_{ink}{}^{l}{}_{;j} + (2a_2 - a_1)A^1_{ikn}{}^{l}{}_{;j}\\
&= 3a_1A^1_{ijk}{}^{l}{}_{;n} + 3a_2A^1_{ijn}{}^{l}{}_{;k}\\
&\quad +(a_2 - 2a_1)(A^1_{ink}{}^{l}{}_{;j} + A^1_{jik}{}^{l}{}_{;n}) + (2a_2 - a_1)(A^1_{ijn}{}^{l}{}_{;k} + A^1_{kin}{}^{l}{}_{;j})\\
&\quad +(2a_1 - a_2)A^1_{ink}{}^{l}{}_{;j} + (2a_2 - a_1)A^1_{ikn}{}^{l}{}_{;j}\\
&= (5a_1 - a_2)A^1_{ijk}{}^{l}{}_{;n} + (5a_2 - a_1)A^1_{ijn}{}^{l}{}_{;k}\\
&= (\tfrac{25}{24} - \tfrac{1}{24})A^1_{ijk}{}^{l}{}_{;n} + (\tfrac{5}{24} - \tfrac{5}{24})A^1_{ijn}{}^{l}{}_{;k} = A^1_{ijk}{}^{l}{}_{;n}. \qquad \square
\end{aligned}$$

The proof of Theorem 1.3.4 is purely algebraic in nature; the Christoffel symbols are exhibited explicitly. By contrast, the proof of the next theorem is analytic in nature. We will construct a sequence of connections defined by polynomial Christoffel symbols Γ_ν vanishing to higher and higher order at the origin and then set

$$\Gamma = \lim_{\nu\to\infty}\{\Gamma_1 + \cdots + \Gamma_\nu\}$$

to define a real analytic connection with the desired properties. Defining the correct sequence of norms to ensure convergence will, of course, be crucial. In what follows, we will let $C = C(\cdot)$ be a universal constant depending on the indicated parameters. Thus if $C = C(m)$, the constant depends only on the dimension m but not on any other variables.

Rather than working with real analytic functions, it is convenient to complexify and work in the holomorphic setting. Let $(z_1,\ldots,z_m)$ be the usual coordinates on $\mathbb{C}^m$, let

$$|z| := (|z_1|^2 + \cdots + |z_m|^2)^{1/2}$$

be the usual Euclidean norm, and let

$$B_\delta := \{z \in \mathbb{C}^m : |z| < \delta\}$$

be the open ball of radius $\delta > 0$ about the origin. Let W be an auxiliary real vector space. We shall take $W = S^2 \otimes V$ to study Christoffel symbols, we shall take $W = S^2$ to study symmetric Ricci tensors, and we shall take $W = \mathfrak{A}$ to study curvature operators. Let $\mathcal{H}_\delta(W)$ be the complex vector space of all holomorphic functions q from B_δ to $W \otimes_{\mathbb{R}} \mathbb{C}$ such that $q(x)$ is real for $x \in \mathbb{R}^m \cap B_\delta$. Fix a norm $|_W$ on W; any two norms on a real or complex vector space whose dimension is finite are equivalent so the precise norm involved plays an inessential role. For $\nu = 0, 1, 2, \ldots$ and for $q \in \mathcal{H}_\delta(W)$, set

$$||q||_{\delta,\nu,W} := \sup_{0<|z|<\delta} |q(z)|_W \cdot |z|^{-\nu}$$

where, of course, $||q||_{\delta,\nu,W} = \infty$ is possible. We set:

$$\mathcal{H}_{\delta,\nu}(W) := \{q \in \mathcal{H}_\delta(W) : ||q||_{\delta,\nu,W} < \infty\}.$$

Clearly if $q \in \mathcal{H}_{\delta,\nu}(W)$ and if $|z| < \delta$, then we may estimate:

$$|q(z)|_W \leq |z|^\nu \cdot ||q||_{\delta,\nu,W} \leq \delta^\nu ||q||_{\delta,\nu,W}. \tag{4.5.a}$$

In particular, q vanishes to order ν at the origin.

If $\Gamma \in \mathcal{H}_{\delta,\nu}(S^2 \otimes V)$, define:

$$\mathcal{L}(\Gamma)_{ijk}{}^l := \partial_{z_i}\Gamma_{jk}{}^l - \partial_{z_j}\Gamma_{ik}{}^l \in \mathcal{H}_\delta(\mathfrak{A}),$$
$$(\Gamma \star \Gamma)_{ijk}{}^l := \Gamma_{in}{}^l\Gamma_{jk}{}^n - \Gamma_{jn}{}^l\Gamma_{ik}{}^n \in \mathcal{H}_\delta(\mathfrak{A}).$$

We note that since $\mathcal{L}$ involves derivatives, $\mathcal{L}(\Gamma)$ need not belong to $\mathcal{H}_{\delta,\mu}(\mathfrak{A})$ for any μ. However, we have $\Gamma \star \Gamma \in \mathcal{H}_{\delta,2\nu}(\mathfrak{A})$ since we can estimate

$$||\Gamma \star \Gamma||_{\delta,2\nu,\mathfrak{A}} \leq C_1 ||\Gamma||^2_{\delta,\nu,S^2} \quad \text{for} \quad C_1 = C_1(m, |_{\mathfrak{A}}, |_{S^2 \otimes V}).$$

The constant C_1 depends, as will many other constants, upon the dimension m, and upon the norm $|_{\mathfrak{A}}$ chosen on $\mathfrak{A}$. It also will also depend on the norm

$|_{S^2\otimes V}$ chosen on $S^2 \otimes V$. If ∇^Γ has Christoffel symbol Γ, then

$$\mathcal{R}^\Gamma = \mathcal{L}(\Gamma) + \Gamma \star \Gamma.$$

Let ρ_s be the symmetric part of the Ricci tensor.

Lemma 4.5.1 *Let $m \geq 3$.*

(1) $\mathcal{H}_{\delta,\nu}(W)$ is a Banach space with respect to the norm $||_{\delta,\nu,W}$.
(2) If $\Theta \in \mathcal{H}_{\delta,\nu}(S^2)$, then there exists $\Gamma \in \mathcal{H}_{\delta,\nu+1}(S^2 \otimes V)$ so

(a) $||\Gamma||_{\delta,\nu+1,S^2\otimes V} \leq C_2||\Theta||_{\delta,\nu,S^2}$ for $C_2 = C_2(m, |_{S^2\otimes V}, |_{S^2})$.
(b) $\Gamma_{ij}{}^j = 0$ for any i.
(c) $\rho(\mathcal{L}(\Gamma)) = \Theta$.

Proof. Let q_n be a sequence of elements of $\mathcal{H}_{\delta,\nu}(W)$ forming a Cauchy sequence with respect to the norm $||_{\delta,\nu,W}$. By Equation (4.5.a), q_n converges uniformly to a continuous $W \otimes_{\mathbb{R}} \mathbb{C}$ valued function q on B_δ. The Cauchy integral formula then shows q is holomorphic on B_δ. It is now immediate that $q \in \mathcal{H}_{\delta,\nu}(W)$ and that $q_n \to q$ in $||_{\delta,\nu,W}$; Assertion (1) now follows.

If $q \in \mathcal{H}_{\delta,\nu}(W)$, define:

$$\left(\int_k q\right)(z) := z_k \int_0^1 q(z_1, \ldots, z_{k-1}, tz_k, z_{k+1}, \ldots, z_m)dt.$$

We may then estimate:

$$\begin{aligned}
&\left|\left(\int_k q\right)(z)\right|_W \\
&\leq C_3|z_k| \int_0^1 ||q||_{\delta,\nu,W} \cdot |(z_1, \ldots, z_{k-1}, tz_k, z_{k+1}, \ldots, z_m)|^\nu dt \\
&\leq C_3|z|^{\nu+1}||q||_{\delta,\nu,W}
\end{aligned}$$

where $C_3 = C_3(m, |_W)$. It then follows that

$$\left|\left|\int_k q\right|\right|_{\delta,\nu+1,W} \leq C_3||q||_{\delta,\nu,W}. \qquad (4.5.b)$$

By letting $w = tz_k$, we may express

$$\int_k q = \int_0^{z_k} q(z_1, \ldots, z_{k-1}, w, z_{k+1}, \ldots, z_m)dw.$$

This shows that

$$\partial_{z_k} \int_k q = q. \qquad (4.5.c)$$

We have assumed throughout that $m \geq 3$. For each pair of indices $\{i, j\}$, not necessarily distinct, choose $\iota_{ij} = \iota_{ji}$ distinct from i and from j. Let

$$\Gamma_{ij}{}^{l} := \left\{ \begin{array}{lll} \int_l \Theta_{ij} & \text{if} & l = \iota_{ij} \\ 0 & \text{if} & l \neq \iota_{ij} \end{array} \right\}.$$

Assertion (2a) then follows from Equation (4.5.b) for a suitably chosen constant C_2. Since ι_{ij} is distinct from i and j, Assertion (2b) follows. Furthermore, $\Gamma(x)$ is real if x is real. Finally, we use Equation (4.5.c) to establish Assertion (2c) by computing:

$$\rho(\mathcal{L}(\Gamma))_{ij} = \sum_l \partial_{x_l} \Gamma_{ij}{}^{l} = \Theta_{ij}.$$

□

Given $\Gamma_\nu \in \mathcal{H}_{\delta,\nu}(S^2 \otimes V)$ and $\Gamma_\mu \in \mathcal{H}_{\delta,\mu}(S^2 \otimes V)$, we polarize $\star$ to define:

$$\begin{aligned} &\{\Gamma_\nu \star \Gamma_\mu\}_{ijk}{}^{l} \\ &= \tfrac{1}{2}\left\{\Gamma_{\nu,in}{}^{l}\Gamma_{\mu,jk}{}^{n} + \Gamma_{\mu,in}{}^{l}\Gamma_{\nu,jk}{}^{n} - \Gamma_{\nu,jn}{}^{l}\Gamma_{\mu,ik}{}^{n} - \Gamma_{\mu,jn}{}^{l}\Gamma_{\nu,ik}{}^{n}\right\}. \end{aligned}$$

There exists $C_4 = C_4(m, |_{S^2\otimes V}, |_{S^2})$ satisfying:

$$||\rho_s(\Gamma_\nu \star \Gamma_\mu)||_{\delta,\nu+\mu,S^2} \leq C_4 ||\Gamma_\nu||_{\delta,\nu,S^2\otimes V} \cdot ||\Gamma_\mu||_{\delta,\mu,S^2\otimes V}. \qquad (4.5.d)$$

Let $\Gamma_\ell \in \mathcal{H}_{\delta,\nu_\ell}(S^2 \otimes V)$ be a sequence for $1 \leq \ell \leq \ell_0$. Define a connection ∇ with Christoffel symbol $\Gamma_1 + \cdots + \Gamma_{\ell_0}$; the curvature is then:

$$\mathcal{R} = \sum_{\ell=1}^{\ell_0} \mathcal{L}(\Gamma_\ell) + \sum_{\ell=1}^{\ell_0}\sum_{n=1}^{\ell_0} \Gamma_\ell \star \Gamma_n.$$

The following is a crucial technical result:

Lemma 4.5.2 *Let $\mathcal{A} \in \mathfrak{A}$. Let $\Xi \in \mathcal{H}_1(S^2)$ satisfy*

$$\Xi - \rho_s(\mathcal{A}) \in \mathcal{H}_{1,2}(S^2).$$

Let $\Gamma_{1,uv}{}^{l} := \frac{1}{3}(A_{wuv}{}^{l} + A_{wvu}{}^{l})x^w$. There exists $\delta = \delta(\mathcal{A}) > 0$, a constant $C = C(m, \mathcal{A}, |_{S^2}, |_{S^2\otimes V}, |_{\mathfrak{A}}) > 0$, and a sequence $\Gamma_\ell \in \mathcal{H}_{\delta,2\ell-1}(S^2 \otimes V)$ for $\ell = 2, 3, \ldots$ satisfying:

(1) $||\Gamma_\ell||_{\delta,2\ell-1,S^2\otimes V} \leq C^{2\ell-1}$ for $1 \leq \ell$.
(2) $\Gamma_{\ell,ij}{}^{j} = 0$ for $2 \leq \ell$ and any index i.

(3) If ∇^ℓ has Christoffel symbol $\Gamma_1 + \cdots + \Gamma_\ell$, then

$$||\rho_s(\mathcal{R}^\ell) - \Xi||_{\delta,2\ell,S^2} \leq C^{2\ell}.$$

Proof. Let $\mathcal{A} \in \mathfrak{A}$. Since Γ_1 is a homogeneous linear polynomial, there is a constant $C_5 = C_5(\mathcal{A}, |_{S^2\otimes V}, |_{\mathfrak{A}}) > 0$ so that for any $\delta > 0$, we have the estimate

$$||\Gamma_1||_{\delta,1,S^2\otimes V} \leq C_5.$$

By Theorem 1.3.4, $\mathcal{R}^1(0) = \mathcal{A}$. Since Γ^1 is linear, $\mathcal{L}(\Gamma^1)$ is constant and $\mathcal{L}(\Gamma^1) = \mathcal{A}$. Consequently, $\mathcal{R}^1 - \mathcal{A} = \Gamma_1 \star \Gamma_1$ is a quadratic polynomial in z. Thus $\rho_s(\mathcal{R}^1) - \rho_s(\mathcal{A}) = \rho_s(\Gamma_1 \star \Gamma_1) \in \mathcal{H}_{1,2}(S^2)$. For any $\delta \leq 1$, we have:

$$\begin{aligned}
&||\rho_s(\mathcal{R}^1) - \Xi||_{\delta,2,S^2} \leq ||\rho_s(\mathcal{R}^1) - \Xi||_{1,2,S^2} \\
&\leq ||\rho_s(\mathcal{R}^1) - \rho_s(\mathcal{A})||_{1,2,S^2} + ||\rho_s(\mathcal{A}) - \Xi||_{1,2,S^2} \\
&= ||\rho_s(\Gamma_1 \star \Gamma_1)||_{1,2,S^2} + ||\rho_s(\mathcal{A}) - \Xi||_{1,2,S^2} \\
&= C_6^2 < \infty
\end{aligned}$$

for $C_6 = C_6(\mathcal{A}, \Xi, |_{S^2})$. Let $C := \max(C_2, C_5, C_6, 3C_4C_5) + 1$. Choose $0 < \delta < 1$ satisfing:

$$C\delta^2 < 1 \quad \text{and} \quad \frac{C^3\delta^2}{1 - C^2\delta^2} \leq C_5. \tag{4.5.e}$$

If $\ell = 1$, Assertion (2) of Lemma 4.5.2 holds vacuously and Assertions (1) and (3) hold since $C > C_5$ and $C > C_6$. We assume inductively that $\Gamma_1, \ldots, \Gamma_\ell$ have been chosen with the desired properties. Let

$$\Theta_\ell := \Xi - \rho_s(\mathcal{R}^\ell) \in \mathcal{H}_{\delta,2\ell}(S^2).$$

Apply Lemma 4.5.1 to choose $\Gamma_{\ell+1} \in \mathcal{H}_{\delta,2\ell+1}(S^2 \otimes V)$ satisfying:

$$\begin{aligned}
&||\Gamma_{\ell+1}||_{\delta,2\ell+1,S^2\otimes V} \leq C_2||\Theta_\ell||_{\delta,2\nu,S^2} \leq C^{2\ell+1}, \\
&\Gamma_{\ell+1,ij}{}^j = 0 \quad \text{for any } i, \\
&\rho(\mathcal{L}(\Gamma_{\ell+1})) = \Theta_\ell.
\end{aligned}$$

We estimate:

$$\begin{aligned}
&\left|\rho_s\left[\mathcal{R}^{\ell+1}\right](z) - \Xi\right|_{S^2} \\
&= \left|\rho_s\left[\mathcal{R}^\ell + \mathcal{L}(\Gamma_{\ell+1}) + 2\textstyle\sum_{\mu\leq\ell}\Gamma_\ell \star \Gamma_{\ell+1} + \Gamma_{\ell+1}\Gamma_{\ell+1}\right](z) - \Xi\right|_{S^2} \\
&= \left|\rho_s\left[2\textstyle\sum_{\mu\leq\ell}\Gamma_\ell \star \Gamma_{\ell+1} + \Gamma_{\ell+1}\Gamma_{\ell+1}\right](z)\right|_{S^2}
\end{aligned}$$

$$\begin{aligned}
&\leq 2C_4\{C_5|z| + C^3|z|^3 + \cdots + C^{2\ell+1}|z|^{2\ell+1}\}C^{2\ell+1}|z|^{2\ell+1}\\
&\leq 2C_4\{C_5 + C^3\delta^2 + \cdots + C^{2\ell+1}\delta^{2\ell}\}C^{2\ell+1}|z|^{2\ell+2}\\
&\leq 2C_4\{C_5 + C^3\delta^2/(1 - C^2\delta^2)\}C^{2\ell+1}|z|^{2\ell+2}\\
&\leq 3C_4C_5C^{2\ell+1}|z|^{2\ell+2} \leq C^{2\ell+2}|z|^{2\ell+2}. \qquad \square
\end{aligned}$$

Lemma 4.5.2 deals with the symmetric Ricci tensor. The following is a useful observation concerning the anti-symmetric Ricci tensor ρ_a. This observation is closely related to Theorem 3.2.1.

Lemma 4.5.3 *Let Γ be the Christoffel symbols of an affine connection. Let $\mathcal{R}$ be the associated curvature operator.*

(1) $\rho_a(\mathcal{R}) = \rho_a(\mathcal{L}(\Gamma))$.
(2) If $\Gamma_{ij}{}^j = 0$ for each index i, then $\rho_a(\mathcal{L}(\Gamma)) = 0$.

Proof. We may decompose $\mathcal{R} = \mathcal{L}(\Gamma) + \Gamma \star \Gamma$. We compute:

$$\rho(\Gamma \star \Gamma)_{jk} = \Gamma_{in}{}^i\Gamma_{jk}{}^n - \Gamma_{jn}{}^i\Gamma_{ik}{}^n.$$

Since this is symmetric in $\{j, k\}$, Assertion (1) follows. We also compute:

$$\rho(\mathcal{L}(\Gamma)) = \partial_{x_i}\Gamma_{jk}{}^i - \partial_{x_j}\Gamma_{ik}{}^i.$$

This is symmetric in $\{j, k\}$ if the assumptions of Assertion (2) hold. $\square$

We continue our sequence of technical results:

Lemma 4.5.4 *Let $M_\delta := B_\delta \cap \mathbb{R}^m$, let $\mathcal{A} \in \mathfrak{A}$, and let $\Xi \in \mathcal{H}_1(S^2)$ satisfy $\Xi - \rho_s(\mathcal{A}) \in \mathcal{H}_{1,2}(S^2)$. Then there exists $\epsilon > 0$ and there exists a real analytic affine connection ∇ on M_ϵ so $\mathcal{R}_0 = \mathcal{A}$, and so for any $P \in M_\epsilon$,*

$$\rho_a(\mathcal{R})(P) = \rho(\mathcal{A}) \quad \textit{and} \quad \rho_s(\mathcal{R})(P) = \Xi(P).$$

Proof. We apply Lemma 4.5.2. Choose $\epsilon < \delta$ so $C\epsilon < 1$. Assertion (1) of Lemma 4.5.2 shows

$$|\Gamma_\ell(z)|_{S^2} \leq C^{2\ell-1}|z|^{2\ell-1} < (C\epsilon)^{2\ell-1}$$

for any $z \in B_\epsilon$. Consequently the series

$$\Gamma(z) := \Gamma_1(z) + \cdots + \Gamma_\ell(z) + \cdots$$

converges uniformly on B_ϵ. Thus the associated connection ∇ defined by Γ is a real analytic affine connection on M_ϵ. Since uniform convergence in the holomorphic context implies the uniform convergence on compact subsets of all derivatives,

$$\mathcal{R}(z) = \lim_{\ell\to\infty} \mathcal{R}^\ell(z) \quad \text{for any} \quad z \in B_\epsilon.$$

Note that

$$\Gamma_1 + \cdots + \Gamma_\ell = \Gamma_1 + O(|x|^3),$$

Theorem 1.3.4 now yields $\mathcal{R}_0^\ell = \mathcal{R}_0^1 = \mathcal{A}$. Consequently

$$\mathcal{R}_0 = \lim_{\ell\to\infty} \mathcal{R}_0^\ell = \mathcal{A}.$$

Furthermore, Assertion (3) of Lemma 4.5.2 together with the estimate of Equation (4.5.a) shows that one has the estimate:

$$\begin{aligned} |\rho_s(\mathcal{R})(z) - \Xi(z)|_{S^2} &= \lim_{\ell\to\infty} |\rho_s(\mathcal{R}^\ell)(z) - \Xi(z)|_{S^2} \\ &\leq \lim_{\ell\to\infty} (C\epsilon)^\ell = 0 \quad \text{for any} \quad z \in B_\epsilon. \end{aligned}$$

We apply Lemma 4.5.3 to see

$$\rho_a(\mathcal{R})(z) = \lim_{\ell\to\infty} \rho_a(\mathcal{L}(\Gamma_1) + \cdots + \mathcal{L}(\Gamma_\ell))(z).$$

By Lemma 4.5.2, ${\Gamma_{\ell,ij}}^j = 0$ for $\ell \geq 2$. Thus by Lemma 4.5.3, $\rho_a(\mathcal{L}(\Gamma_\ell)) = 0$ for $\ell \geq 2$. It now follows that $\rho_a(\mathcal{R}) = \rho_a(\mathcal{L}(\Gamma_1)) = \rho_a(\mathcal{A})$ is constant on the coordinate frame as well. □

Remark 4.5.1 In Lemma 4.5.4, we can take $\Xi(P) = \rho_s(\mathcal{A})$ to be constant. We then have relative to the coordinate frame that

$$\rho(\mathcal{R})(P)_{ij} = \rho(\mathcal{A})_{ij}$$

for any $P \in M_\epsilon$. This establishes Remark 1.3.1.

To apply Lemma 4.5.4 to establish the results of Section 1.4, we must first find a suitable Ξ. To do this, we need the following technical result:

Lemma 4.5.5 *Let $M_\delta := B_\delta \cap \mathbb{R}^m$, let $\mathcal{A} \in \mathfrak{A}$, and let g be the germ of a real analytic metric on M_δ for some $\delta > 0$ so that $g = g(0) + O(|z|^2)$. Then there exists $\varepsilon > 0$, a real analytic orthonormal frame $\vec{e}$ for TM_ε, and $\Xi \in \mathcal{H}_\varepsilon(S^2)$ so that*

(1) $e_i = \partial_{z_i} + O(|z|^2)$.
(2) $\Xi(e_i, e_j) = \rho_s(\mathcal{A})_{ij}$ and $\Xi(\partial_{z_i}, \partial_{z_j}) = \rho_s(\mathcal{A})_{ij} + O(|z|^2)$.

Proof. By making a linear change of coordinates, we may assume that $\{\partial_{x_i}\}$ is an orthonormal frame at 0. Let $\varepsilon_{ij} = g(\partial_{x_i}, \partial_{x_j})(0)$;

$$\varepsilon_{ij} = \left\{ \begin{array}{c} 0 \text{ if } i \neq j \\ \pm 1 \text{ if } i = j \end{array} \right\}.$$

The requisite holomorphic frame is then defined using the generalized Gram–Schmidt process:

$$\begin{aligned}
f_1(z) &= \partial_{z_1}, & e_1(z) &= \frac{f_1(z)}{\sqrt{\varepsilon_{11} g(f_1(z), f_1(z))}}, \\
f_2(z) &= \partial_{z_2} - \varepsilon_{11} g(\partial_{z_2}, e_1(z)) e_1(z), & e_2(z) &= \frac{f_2(z)}{\sqrt{\varepsilon_{22} g(f_2(z), f_2(z))}}, \\
&\cdots\cdots & &\cdots\cdots \\
f_m(z) &= \partial_{z_m} - \sum_{j<m} \varepsilon_{jj} g(\partial_{z_m}, e_j(z)) e_j(z), & e_m(z) &= \frac{f_m(z)}{\sqrt{\varepsilon_{mm} g(f_m(z), f_m(z))}}.
\end{aligned}$$

Assertion (1) now follows. We define $\Xi(e_i, e_j) = \rho_s(\mathcal{A})_{ij}$ to establish Assertion (2). □

We use Lemma 4.5.4 to establish the following result from which Theorem 1.4.3 follows:

Theorem 4.5.1 *Let $M_\delta := B_\delta \cap \mathbb{R}^m$, let $\mathcal{A} \in \mathfrak{A}$, and let g be a real analytic pseudo-Riemannian metric on M_δ for some $\delta > 0$. Then:*

(1) If $\mathcal{A}$ is Ricci symmetric, there exists $\epsilon > 0$ and there exists a Ricci symmetric connection ∇ on M_ϵ so that $\mathcal{R}_0 = \mathcal{A}$ and so that $\mathcal{R}$ has constant scalar curvature.

(2) If $\mathcal{A}$ is Ricci anti-symmetric, there exists $\epsilon > 0$ and there exists a Ricci anti-symmetric connection ∇ on M_ϵ so that $\mathcal{R}_0 = \mathcal{A}$ and so that $\mathcal{R}$ has constant scalar curvature 0.

(3) If $\mathcal{A}$ is Ricci flat, there exists $\epsilon > 0$ and there exists a Ricci flat connection ∇ on M_ϵ so that $\mathcal{R}_0 = \mathcal{A}$ and so that $\mathcal{R}$ has constant scalar curvature 0.

Proof. We use the construction of Lemma 3.3.1 to choose real analytic coordinates on V so that $g = g(0) + O(|x|^2)$. We then use Lemma 4.5.5 to construct Ξ; by rescaling the coordinate system, we may assume the domain of definition of g, $\vec{e}$, and of Ξ includes the ball of radius 2 in $\mathbb{C}^m$. We then apply Lemma 4.5.4 to construct ∇ and $\mathcal{R}$. Since the matrix of $\rho_s(\mathcal{R})$ is constant with respect to the moving orthonormal frame $\vec{e}$, $\mathcal{R}$ has constant scalar curvature; the remaining properties are now immediate since the matrix of $\rho_a(\mathcal{R})$ relative to the coordinate frame is constant as well and is given by $\rho_a(\mathcal{A})$. $\square$

The space of *Weyl projective curvature operators* is $\ker(\rho) \cap \mathfrak{A}$. Let σ_a and σ_s be as given in Definition 4.4.1:

$$(\sigma_a(\psi))(x,y)z = -\tfrac{1}{1+m}\{2\psi(x,y)z + \psi(x,z)y - \psi(y,z)x\} \text{ for } \psi \in \Lambda^2,$$
$$(\sigma_s(\phi))(x,y)z = \tfrac{1}{1-m}\{\phi(x,z)y - \phi(y,z)x\} \text{ for } \phi \in S^2.$$

Then the projection $\mathcal{P}$ on $\ker(\rho) \cap \mathfrak{A}$ is given by:

$$\mathcal{P}(\mathcal{A}) = \mathcal{A} - \sigma_s(\rho_s(\mathcal{A})) - \sigma_a(\rho_a(\mathcal{A})). \tag{4.5.f}$$

Following [Simon, Schwenk-Schellschmidt, and Viesel (1991)] one says that $\mathcal{A} \in \mathfrak{A}$ is *projectively flat* if $\mathcal{P}(\mathcal{A}) = 0$. One says that ∇ is projectively flat if the associated curvature operator is projectively flat for all $P \in M$.

Theorem 4.5.2 *Let $M_\delta := B_\delta \cap \mathbb{R}^m$ and let $\mathcal{A} \in \mathfrak{A}$.*

(1) If $\mathcal{A}$ is projectively flat, then there exists a projectively flat connection on M_δ for some $\delta > 0$ so that $\mathcal{R}_0 = \mathcal{A}$.

(2) If $\mathcal{A}$ is projectively flat and Ricci symmetric, then there exists a projectively flat and Ricci symmetric connection on M_δ for some $\delta > 0$ so that $\mathcal{R}_0 = \mathcal{A}$.

Proof. If $\Theta \in V^* \otimes V^*$, set

$$H(\Theta)(x,y)z := \Theta(x,y)z - \Theta(y,x)z + \Theta(x,z)y - \Theta(y,z)x.$$

Clearly $H(\Theta)(x,y) = -H(\Theta)(y,x)$. One verifies that the Bianchi identity is satisfied and thus $H(\Theta) \in \mathfrak{A}$ by computing:

$$\begin{aligned}
&H(\Theta)(x,y)z + H(\Theta)(y,z)x + H(\Theta)(z,x)y \\
&= \quad \Theta(x,y)z - \Theta(y,x)z + \Theta(x,z)y - \Theta(y,z)x \\
&\quad + \Theta(y,z)x - \Theta(z,y)x + \Theta(y,x)z - \Theta(z,x)y \\
&\quad + \Theta(z,x)y - \Theta(x,z)y + \Theta(z,y)x - \Theta(x,y)z \\
&= 0.
\end{aligned}$$

We examine the associated Ricci tensor:

$$\begin{aligned}
&\rho(H(\Theta))(y,z) \\
&= e^i\{\Theta(e_i,y)z - \Theta(y,e_i)z + \Theta(e_i,z)y - \Theta(y,z)e_i\} \\
&= \Theta(z,y) - \Theta(y,z) + \Theta(y,z) - m\Theta(y,z)
\end{aligned}$$

$$= \tfrac{1-m}{2}\{\Theta(z,y)+\Theta(y,z)\} + \tfrac{1+m}{2}\{\Theta(z,y)-\Theta(y,z)\}$$
$$= (1-m)\Theta_s(y,z) - (1+m)\Theta_a(y,z).$$

So modulo a suitable renormalization, H can be used to split the short exact sequence of Lemma 4.4.1. In particular, $\mathcal{A} \in \mathfrak{A}$ is projectively flat if and only if $\mathcal{A} \in \text{Range}(H)$.

Let $\theta = \theta_i(x)dx^i$ be a 1-form. We define a connection ∇^θ by requiring that ∇^θ on coordinate vector fields satisfies:

$$\nabla^\theta_x y = \theta(x)y + \theta(y)x.$$

In a system of local coordinates, this is equivalent to the condition

$$\nabla_{\partial_{x_i}} \partial_{x_j} = \theta_i \partial_{x_j} + \theta_j \partial_{x_i}.$$

Set $\Psi(x,y) = x\theta(y)$. If $\mathcal{R}^\theta$ is the curvature operator of ∇^θ, then:

$$\begin{aligned}
\mathcal{R}^\theta(x,y)z &= \nabla_x(\theta(y)z + \theta(z)y) - \nabla_y(\theta(x)z + \theta(z)x)\\
&= \theta(x)\theta(y)z + \theta(x)\theta(z)y + \theta(z)\theta(y)x + \theta(y)\theta(z)x\\
&\quad -\theta(y)\theta(x)z - \theta(y)\theta(z)x - \theta(z)\theta(x)y - \theta(x)\theta(z)y\\
&\quad +x(\theta(y))z + x(\theta(z))y - y(\theta(x))z - y(\theta(z))x\\
&= \theta(y)\theta(z)x - \theta(x)\theta(z)y\\
&\quad +x(\theta(y))z + x(\theta(z))y - y(\theta(x))z - y(\theta(z))x\\
&= H(-\theta\otimes\theta + \Psi).
\end{aligned}$$

Consequently ∇^θ is projectively flat. Let $\Theta \in V^* \otimes V^*$. Set $\theta = x_i\Theta_{ij}dx^j$; the associated Christoffel symbols are then linear functions on V. Then $\theta(0) = 0$ and $\Psi = \Theta$ so

$$\rho(\mathcal{R}^\theta)(0) = (1-m)\Theta_s - (m+1)\Theta_a.$$

Assertion (1) now follows. Furthermore, if Θ is symmetric, then Ψ is symmetric for any point $P \in V$ and $\mathcal{R}^\theta$ is Ricci symmetric. This establishes Assertion (2). □

There is only one natural GL equivariant geometric realization question uncovered by the forgoing results. It is answered, in the negative, by the following result:

Theorem 4.5.3 *Let $m \geq 3$. If ∇ is a projectively flat, Ricci anti-symmetric, affine connection, then ∇ is flat. Thus if $0 \neq \mathcal{A} \in \mathfrak{A}$ is projectively flat and Ricci anti-symmetric, then $\mathcal{A}$ is not geometrically realizable by a projectively flat, Ricci anti-symmetric, affine connection.*

Proof. Suppose $\rho(\mathcal{R}) \in \Lambda^2$. Let $\omega_{ij} := -\frac{1}{m+1}\rho(\partial_{x_i}, \partial_{x_j})$. Since $\mathcal{R}$ is projectively flat and since ρ is alternating, $\mathcal{R} = \sigma_a(\rho_a(\mathcal{A}))$. Thus by Equation (4.5.f),

$$\mathcal{R}(\partial_{x_i}, \partial_{x_j})\partial_{x_k} = 2\omega_{ij}\partial_{x_k} + \omega_{ik}\partial_{x_j} - \omega_{jk}\partial_{x_i}.$$

Let $\mathcal{R}(x, y; z)w$ be the covariant derivative of the curvature operator. We have:

$$\begin{aligned}
\mathcal{R}(\partial_{x_i}, \partial_{x_j}; \partial_{x_\ell})\partial_{x_k} &= 2\omega_{ij;\ell}\partial_{x_k} + \omega_{ik;\ell}\partial_{x_j} - \omega_{jk;\ell}\partial_{x_i},\\
\mathcal{R}(\partial_{x_j}, \partial_{x_\ell}; \partial_{x_i})\partial_{x_k} &= 2\omega_{j\ell;i}\partial_{x_k} + \omega_{jk;i}\partial_{x_\ell} - \omega_{\ell k;i}\partial_{x_j},\\
\mathcal{R}(\partial_{x_\ell}, \partial_{x_i}; \partial_{x_j})\partial_{x_k} &= 2\omega_{\ell i;j}\partial_{x_k} + \omega_{\ell k;j}\partial_{x_i} - \omega_{ik;j}\partial_{x_\ell}.
\end{aligned}$$

Summing and applying the second Bianchi identity of Lemma 3.1.2 yields:

$$\begin{aligned}
0 = {} & (2\omega_{ij;\ell} + 2\omega_{j\ell;i} + 2\omega_{\ell i;j})\partial_{x_k} \\
& +(\omega_{ik;\ell} - \omega_{\ell k;i})\partial_{x_j} + (\omega_{\ell k;j} - \omega_{jk;\ell})\partial_{x_i} + (\omega_{jk;i} - \omega_{ik;j})\partial_{x_\ell}.
\end{aligned} \tag{4.5.g}$$

Let $\{i, j, \ell\}$ be distinct indices. Set $k = i$. Examining the coefficient of ∂_{x_j} in Equation (4.5.g) yields

$$0 = \omega_{ii;\ell} - \omega_{\ell i;i} = \omega_{i\ell;i} \quad \text{(do not sum over i).}$$

Polarizing this identity then yields

$$\omega_{i\ell;j} + \omega_{j\ell;i} = 0.$$

Next we set $k = \ell$ and examine the coefficient of ∂_{x_k} in Equation (4.5.g):

$$\begin{aligned}
0 &= 2\omega_{ij;k} + 2\omega_{jk;i} + 2\omega_{ki;j} + \omega_{jk;i} - \omega_{ik;j}\\
&= 2\omega_{ij;k} + 3\omega_{jk;i} + 3\omega_{ki;j} = -2\omega_{kj;i} + 3\omega_{jk;i} - 3\omega_{kj;i} = 8\omega_{jk;i}.
\end{aligned}$$

Thus if $\{x, y, z\}$ are linearly independent vectors, then $\nabla_x\omega(y, z) = 0$; since the set of all triples of linearly independent vectors is dense in the set of all triples, this relation holds by continuity for all $\{x, y, z\}$ and thus $\nabla\omega = 0$. We compute:

$$\begin{aligned}
0 &= \{(\nabla_x\nabla_y - \nabla_y\nabla_x - \nabla_{[x,y]})\omega\}(z, w)\\
&= \omega(\mathcal{R}(x, y)z, w) + \omega(z, \mathcal{R}(x, y)w)\\
&= 4\omega(x, y)\omega(z, w) + 2\omega(x, z)\omega(y, w) - 2\omega(x, w)\omega(y, z).
\end{aligned}$$

Set $x = z$ and $y = w$ to see that $6\omega(x, y)^2 = 0$. Consequently $\omega = 0$ so $\mathcal{R} = 0$. □

4.6 Decomposition of $\mathfrak{A}$ as an Orthogonal Module

All of the factors in the orthogonal module decomposition of $\mathfrak{A}$ appear with multiplicity one except for S_0^2 and Λ^2. Thus it is only the identification of these factors which is at issue in any given decomposition. There are two different decompositions of $\mathfrak{A}$ as an orthogonal module given in [Bokan (1990)]. In the first decomposition, we have:

$$\begin{array}{lll} W_1^B = \sigma_1(\mathbb{R}), & W_2^B \approx W_5^B \approx S_0^2, & W_3^B \approx W_4^B \approx \Lambda^2, \\ W_6^B = W_6^{\mathcal{O}}, & W_7^B = W_7^{\mathcal{O}}, & W_8^B = W_8^{\mathcal{O}}. \end{array}$$

The decomposition is characterized by the fact that

$$\begin{aligned} \ker(\rho) &= W_4^B \oplus W_5^B \oplus W_6^B \oplus W_7^B \oplus W_8^B, \\ \ker(\rho)^\perp &= W_1^B \oplus W_2^B \oplus W_3^B. \end{aligned}$$

We adopt the notation of Equation (4.2.a) and apply Lemma 4.2.1 to see:

$$W_5^B = \xi_S^{\mathfrak{A}}(m-1,1) \quad \text{and} \quad W_4^B = \xi_\Lambda^{\mathfrak{A}}(-1,m+1).$$

By Lemma 4.2.2, the orthogonal complements are given by:

$$W_2^B = \xi_S^{\mathfrak{A}}(0,1) \quad \text{and} \quad W_3^B = \xi_\Lambda^{\mathfrak{A}}(1,0).$$

This yields, naturally enough, the splitting discussed in Definition 4.4.1.

There is another decomposition given in [Bokan (1990)] of importance. It takes the form:

$$\begin{array}{lll} Z_1^B = \sigma_1(\mathbb{R}), & Z_2^B \approx Z_4^B \approx S_0^2, & Z_5^B \approx Z_7 \approx \Lambda^2, \\ Z_3^B = W_6^{\mathcal{O}}, & Z_6^B = W_7^{\mathcal{O}}, & Z_8^B = W_8^{\mathcal{O}}. \end{array}$$

This decomposition is characterized by the fact that

$$\begin{aligned} &\mathfrak{R} = Z_1^B \oplus Z_2^B \oplus Z_3^B, \\ &Z_4^B \oplus Z_5^B \subset \{A \in \mathfrak{A} : A(x,y,z,w) = A(x,y,w,z)\}. \end{aligned}$$

It now follows that

$$\begin{array}{ll} Z_2^B = \xi_S^{\mathfrak{A}}(1,1), & Z_3^B = \xi_S^{\mathfrak{A}}(1,-1), \\ Z_5^B = \xi_\Lambda^{\mathfrak{A}}(1,1), & Z_7^B = \xi_\Lambda^{\mathfrak{A}}(1,-3). \end{array}$$

Chapter 5

Affine Kähler Geometry

In Chapter 5, we study (para)-complex affine geometry. We adopt the notation established in Section 1.5 and apply the results of Section 3.4. Let $(V, \langle\cdot,\cdot\rangle, J_\pm)$ be a para-Hermitian vector space $(+)$ or a pseudo-Hermitian vector space $(-)$. In Section 5.1, we study the space of quadratic invariants for

$$\begin{aligned}\mathfrak{K}_\pm^{\mathfrak{A}} &= \{A \in \mathfrak{A} : A(x,y,z,w) = \mp A(x,y,J_\pm z, J_\pm w)\}\\ &= \{\mathcal{A} \in \mathfrak{A} : J_\pm \mathcal{A}(x,y) = \mathcal{A}(x,y)J_\pm\}.\end{aligned}$$

In Section 5.2, we study the Ricci tensors ρ and ρ_{13}. In Section 5.3, we give an algorithm for constructing (para)-Kähler affine manifolds and the associated (para)-Kähler affine curvature tensors. In Section 5.4, we establish Theorem 1.5.2 giving the decomposition of the space $\mathfrak{K}_-^{\mathfrak{A}}$ of Kähler affine curvature operators as a module over $\mathcal{U}_-$ and $\mathcal{U}_-^\star$ [Matzeu and Nikčević (1991)]. We also establish the geometric realization result of Theorem 1.5.3 in the complex setting. In Section 5.5 we describe the corresponding decomposition of the para-Kähler affine curvature operators $\mathfrak{K}_+^{\mathfrak{A}}$ as a module for the group $\mathcal{U}_+^\star$ also given in Theorem 1.5.2. Theorem 1.5.3 in the para-complex setting then follows as a consequence of this discussion. In Section 5.6, we give the structure of $\mathfrak{K}_\pm^{\mathfrak{A}}$ as a module for the group $\mathrm{GL}_\pm^\star$.

5.1 Affine Kähler Curvature Tensor Quadratic Invariants

The invariants $\{\psi_1^{\mathcal{O}}, ..., \psi_{10}^{\mathcal{O}}\}$ were defined in Equation (4.1.h) and are orthogonal quadratic invariants of $\mathfrak{A}$; they restrict, therefore, to orthogonal quadratic invariants of $\mathfrak{K}_\pm^{\mathfrak{A}}$. We recall their definition and define some additional invariants:

Definition 5.1.1 Set:

$$\psi_1^{\mathcal{O}} = \mathcal{I}\{(1,2,2,1)(3,4,4,3)\}, \qquad \psi_2^{\mathcal{O}} = \mathcal{I}\{(1,2,2,3)(1,4,4,3)\},$$
$$\psi_3^{\mathcal{O}} = \mathcal{I}\{(1,2,2,3)(3,4,4,1)\}, \qquad \psi_4^{\mathcal{O}} = \mathcal{I}\{(1,2,2,3)(1,4,3,4)\},$$
$$\psi_5^{\mathcal{O}} = \mathcal{I}\{(1,2,2,3)(3,4,1,4)\}, \qquad \psi_6^{\mathcal{O}} = \mathcal{I}\{(1,2,3,2)(1,4,3,4)\},$$
$$\psi_7^{\mathcal{O}} = \mathcal{I}\{(1,2,3,2)(3,4,1,4)\}, \qquad \psi_8^{\mathcal{O}} = \mathcal{I}\{(1,2,3,4)(1,3,2,4)\},$$
$$\psi_9^{\mathcal{O}} = \mathcal{I}\{(1,2,3,4)(1,3,4,2)\}, \qquad \psi_{10}^{\mathcal{O}} = \mathcal{I}\{(1,2,3,4)(1,4,3,2)\},$$
$$\psi_{11}^{\mathfrak{K}\pm} := \mathcal{I}\{(1,2,3,4)(J_\pm 1, J_\pm 2, 3, 4)\},$$
$$\psi_{12}^{\mathfrak{K}\pm} := \mathcal{I}\{(1,2,3,1)(4, J_\pm 2, J_\pm 3, 4)\},$$
$$\psi_{13}^{\mathfrak{K}\pm} := \mathcal{I}\{(1,2,3,1)(4, J_\pm 3, J_\pm 2, 4)\},$$
$$\psi_{14}^{\mathfrak{K}\pm} := \mathcal{I}\{(1,2,J_\pm 2,1)(3,4,J_\pm 4,3)\},$$
$$\sigma_1^{\mathfrak{K}\pm} := \mathcal{I}\{(1,2,3,4)(1,3,4,J_\pm 2)\}, \quad \sigma_2^{\mathfrak{K}\pm} := \mathcal{I}\{(1,2,3,1)(4,2,J_\pm 3,4)\},$$
$$\sigma_3^{\mathfrak{K}\pm} := \mathcal{I}\{(1,3,1,2)(4,2,4,J_\pm 3)\}, \quad \sigma_4^{\mathfrak{K}\pm} := \mathcal{I}\{(1,2,3,1)(4,J_\pm 2,4,3)\},$$
$$\sigma_5^{\mathfrak{K}\pm} := \mathcal{I}\{(1,2,3,1)(4,J_\pm 3,4,2)\}, \quad \sigma_6^{\mathfrak{K}\pm} := \mathcal{I}\{(1,2,2,1)(3,4,J_\pm 4,3)\}.$$

Lemma 5.1.1

(1) The invariants $\{\psi_1^{\mathcal{O}},...,\psi_{10}^{\mathcal{O}},\psi_{11}^{\mathfrak{K}\pm},...,\psi_{14}^{\mathfrak{K}\pm},\sigma_1^{\mathfrak{K}\pm},...,\sigma_6^{\mathfrak{K}\pm}\}$ *span* $\mathcal{I}_2^{\mathcal{U}-}(\mathfrak{K}_-^{\mathfrak{A}})$.

(2) The invariants $\{\psi_1^{\mathcal{O}},...,\psi_{10}^{\mathcal{O}},\psi_{11}^{\mathfrak{K}\pm},...,\psi_{14}^{\mathfrak{K}\pm}\}$ *span* $\mathcal{I}_2^{\mathcal{U}_\pm^\star}(\mathfrak{K}_-^{\mathfrak{A}})$.

(3) We have $\dim\{\mathcal{I}_2^{\mathcal{U}-}(\mathfrak{K}_-^{\mathfrak{A}})\} \le 20$ *and* $\dim\{\mathcal{I}_2^{\mathcal{U}_\pm^\star}(\mathfrak{K}_\pm^{\mathfrak{A}})\} \le 14$.

Remark 5.1.1 It will follow from Theorem 5.4.1 that in fact equality holds in Assertion (3) if $m \ge 6$ and thus the invariants in Assertion (1) and in Assertion (2) form a basis for the associated spaces of invariants if $m \ge 6$.

Proof. We argue as follows and at some length to establish Assertion (1) and Assertion (2). Assertion (3) is immediate from Assertion (1) and from Assertion (2). Adopt the notation of Definition 2.3.2. Choose invariants involving a minimal number of decorations with $J_\pm$ to represent invariants and to create a spanning set. Let $\mathcal{S}_i \subset \mathcal{I}_2^{\mathcal{U}_\pm}(\mathfrak{K}_\pm^{\mathfrak{A}})$ be the linear span of the space of all invariants corresponding to invariants with at most i decorations by $J_\pm$. If S_1 and S_2 belong to $\mathcal{S}_{i+1}$, we write $S_1 \sim S_2$ if $S_1 - S_2 \in \mathcal{S}_i$. Since $S_1 - S_2$ can be written in terms with fewer decorations, either will do. If $S_1 \sim 0$, then S_1 is unnecessary.

(1) We begin with some general observations. Let $(S) = (\star,\star,\star,\star)$ be a generic quadruple; we do not change (S) during the argument.

(a) Let $1 \leq \mu \leq 4$. By replacing the basis $\{e^\mu_{i_\mu}\}$ by the basis $\{J_\pm e^\mu_{i_\mu}\}$, we see that we do not need to consider invariants involving two copies of $J_\pm\mu$. Thus every index is either undecorated with $J_\pm$ or is decorated with exactly one $J_\pm$. It was to make such arguments precise that we chose to have four distinct bases which were not linked.

(b) Using the Kähler identity, we have $\mathcal{I}\{(*, *, J_\pm\mu, J_\pm\nu)(S)\} = \mp\mathcal{I}\{(*, *, \mu, \nu)(S)\}$. We also have $\mathcal{I}\{(*, *, J_\pm\mu, \nu)(S)\} = -\mathcal{I}\{(*, *, \mu, J_\pm\nu)(S)\}$.

(c) If ν is a decorated index, by making the change of basis described in (1a), we can interchange the roles of ν and $J_\pm\nu$ in the invariant; we change the sign since J_+^* is an anti-isometry and since $J_-^2 = -\operatorname{Id}$.

(d) The Bianchi identity implies $\mathcal{I}\{(J_\pm\sigma, J_\pm\mu, \delta, J_\pm\nu)(S)\} \sim 0$; we can express this invariant as a sum of invariants involving fewer decorations by $J_\pm$.

(e) We can permute the roles of $\{1, 2, 3, 4\}$ indexing the bases.

(f) If $\mu \neq \nu$ are decorated indices, then $\mathcal{I}\{(*, *, \mu, \nu)(S)\} \sim 0$, $\mathcal{I}\{(*, *, J_\pm\mu, \nu)(S)\} \sim 0$, $\mathcal{I}\{(*, *, \mu, J_\pm\nu)(S)\} \sim 0$, and $\mathcal{I}\{(*, *, J_\pm\mu, J_\pm\nu)(S)\} \sim 0$.

(g) We may express $\mathcal{I}\{(4, J_\pm 2, 4, 3)(S)\} = -\mathcal{I}\{(4, 2, 4, J_\pm 3)(S)\}$. To see this, we argue:

$$\begin{aligned}
&\mathcal{I}\{(4, J_\pm 2, 4, 3)(S)\} = \mp\mathcal{I}\{(4, J_\pm 2, J_\pm 4, J_\pm 3)(S)\} \\
&= \pm\mathcal{I}\{(J_\pm 2, J_\pm 4, 4, J_\pm 3)(S)\} \pm \mathcal{I}\{(J_\pm 4, 4, J_\pm 2, J_\pm 3)\} \\
&= \pm\mathcal{I}\{(J_\pm 2, J_\pm 4, 4, J_\pm 3)(S)\} - \mathcal{I}\{(J_\pm 4, 4, 2, 3)(S)\} \\
&= \mathcal{I}\{(J_\pm 2, J_\pm 4, J_\pm J_\pm 4, J_\pm 3)(S)\} + \mathcal{I}\{(4, 2, J_\pm 4, 3)(S)\} \\
&\qquad + \mathcal{I}\{(2, J_\pm 4, 4, 3)(S)\} \\
&= \mathcal{I}\{(J_\pm 2, 4, 4, 3)(S)\} \pm 2\mathcal{I}\{(4, 2, J_\pm 4, J_\pm J_\pm 3)(S)\} \\
&= -\mathcal{I}\{(4, J_\pm 2, 4, 3)(S)\} - 2\mathcal{I}\{(4, 2, 4, J_\pm 3)(S)\}.
\end{aligned}$$

(2) Suppose an invariant involves the quadruple $(*, *, *, *)$ where every index is decorated. We apply (1e) and (1f) to replace the invariant by an equivalent invariant involving $(*, *, 1, J_\pm 1)$. Applying the Bianchi identity yields two invariants involving quadruples of the form $(*, 1, \mu, J_\pm 1)$ where μ is a decorated index; this contradicts (1f).

(3) If an invariant involves four decorations by $J_\pm$, then this contradicts (2). Thus no minimal invariant involves four decorations by $J_\pm$.

(4) Suppose the invariant involves three decorations by $J_\pm$. Let "1" be the index which does not involve $J_\pm$. Then by (2), the index "1" appears once in each quadruple of the invariant.

(a) We apply (1) to see that the invariant involves one of the three possible quadruples: $(3, 1, J_\pm 2, 2)$, $(*, *, 1, 2)$, or $(*, *, 2, 1)$. If the invariant involves the quadruple $(3, 1, J_\pm 2, 2)$, then applying the Bianchi identity yields two invariants. The first involves the quadruple $(1, J_\pm 2, 3, 2)$. It also involves the quadruple $(J_\pm 2, 3, 1, 2)$. The first possibility contradicts (1f) and may therefore be ignored. Thus we conclude the invariant involves the quadruples $(*, *, 1, *)$ or $(*, *, *, 1)$. Applying a similar argument to the other quadruple means the other variable can also be taken to be $(*, *, 1, *)$ or $(*, *, *, 1)$. Thus we have $\mathcal{I}\{(*, *, 1, *)(*, *, 1, *)\}$ or $\mathcal{I}\{(*, *, 1, *)(*, *, *, 1)\}$ or $\mathcal{I}\{(*, *, *, 1)(*, *, *, 1)\}$.

(b) Consider $\mathcal{I}\{(*, *, 2, 1)(*, *, 3, 1)\}$ or $\mathcal{I}\{(*, *, 2, 1)(*, *, 1, 3)\}$ or $\mathcal{I}\{(*, *, 1, 2)(*, *, 1, 3)\}$. We apply (1b) and (1c) to choose equivalent invariants of the form, respectively:
$\mathcal{I}\{(*, *, J_\pm 2, 1)(*, *, J_\pm 3, 1)\} = \mp\mathcal{I}\{(*, *, 2, 1)(*, *, 3, 1)\} \sim 0$,
$\mathcal{I}\{(*, *, J_\pm 2, 1)(*, *, 1, J_\pm 3)\} = \mp\mathcal{I}\{(*, *, 2, 1)(*, *, 1, 3)\} \sim 0$,
$\mathcal{I}\{(*, *, 1, J_\pm 2)(*, *, 1, J_\pm 3)\} = \mp\mathcal{I}\{(*, *, 1, 2)(*, *, 1, 3)\} \sim 0$.

(c) The possible invariants remaining to be considered are
$\mathcal{I}\{(*, *, 1, 2)(*, *, 1, J_\pm 2)\}$, $\mathcal{I}\{(*, *, 1, 2)(*, *, J_\pm 2, 1)\}$,
$\mathcal{I}\{(*, *, 2, 1)(*, *, J_\pm 2, 1)\}$.
We use the Kähler identity and permute the indices to see
$\mathcal{I}\{(*, *, 1, 2)(*, *, 1, J_\pm 2)\} = \mathcal{I}\{(*, *, J_\pm 2, 1)(*, *, 2, 1)\}$.
Thus this case may be ignored.

(d) If the invariant has the form $\mathcal{I}\{(*, *, 2, 1)(*, *, J_\pm 2, 1)\}$ or $\mathcal{I}\{(*, *, 2, 1)(*, *, 1, J_\pm 2)\}$, then the remaining indices are different from "1" and from "2". We can apply the Bianchi identity to the quadruple $(*, *, 2, 1)$ to create invariants $\mathcal{I}\{(*, *, 3, 1)(*, *, 2, 1)\}$ or $\mathcal{I}\{(*, *, 3, 1)(*, *, 1, 2)\}$. This contradicts (4b).

We conclude therefore we do not need to consider invariants with three decorations by $J_\pm$.

(5) Suppose that there are two decorations by $J_\pm$. We suppose the indices "1" and "2" are decorated and the indices "3" and "4" are not decorated.

(a) We have some general remarks about this case. Suppose the invariant is $\mathcal{I}\{(*, *, \tilde{1}, 3)(*, *, \tilde{2}, 3)\}$, or $\mathcal{I}\{(*, *, \tilde{1}, 3)(*, *, 3, \tilde{2})\}$, or $\mathcal{I}\{(*, *, 3, \tilde{1})(*, *, 3, \tilde{2})\}$ where either $\tilde{1} = 1$ or $\tilde{1} = J_\pm 1$ and where $\tilde{2} = 2$ or $\tilde{2} = J_\pm 2$. By applying (1c), we can assume $\tilde{1} = J_\pm 1$ and $\tilde{2} = J_\pm 2$. We can then apply (1a) to replace "3" by "$J_\pm 3$" and use

(1d) to reduce the number of decorations by $J_\pm$. Thus it is not necessary to consider these possibilities.

(b) Suppose all four indices occur in each quadruple.

i. Express $\mathcal{I}\{(*,*,\nu,J_\pm 1)(S)\} = -\mathcal{I}\{(*,*,J_\pm\nu,1)(S)\}$. Since the index "1" is no longer decorated, we can permute the coordinate indices to write this in the form $(*,*,*,4)$. Thus we may assume that $(*,*,*,4)$ is a quadruple of the invariant.

ii. We wish to ensure the final index in the other quadruple is not decorated. We suppose the contrary.

A. If $\mu \neq 4$, consider $\mathcal{I}\{(*,*,*,4)(*,*,\mu,J_\pm 1)\}$
$= -\mathcal{I}\{(*,*,*,4)(*,*,J_\pm\mu,1)\}$ where "4" is not decorated and "1" is no longer decorated.

B. $\mathcal{I}\{(*,*,*,4)(\tilde\nu,\tilde\mu,4,J_\pm 1)\}$ (where $\{\tilde\nu,\tilde\mu\} \subset \{2,3,J_\pm 2,J_\pm 3\}$)
$= -\mathcal{I}\{(*,*,*,4)(\tilde\mu,4,\tilde\nu,J_\pm 1)\} - \mathcal{I}\{(*,*,*,4)(4,\tilde\nu,\tilde\mu,J_\pm 1)\}$
$= \mathcal{I}\{(*,*,*,4)(\tilde\mu,4,J_\pm\tilde\nu,1)\} + \mathcal{I}\{(*,*,*,4)(4,\tilde\nu,J_\pm\tilde\mu,1)\}$
where "1" and "4" are not decorated.

Thus we can assume the final indices are not decorated. This gives rise to the two cases which are discussed below:

iii. We have $\mathcal{I}\{(*,*,*,4)(*,*,*,4)\}$.

A. $\psi_{11}^{\mathfrak{K}\pm} = \mathcal{I}\{(1,2,3,4)(J_\pm 1,J_\pm 2,3,4)\}$.

B. $\Psi_1 := \mathcal{I}\{(1,2,3,4)(J_\pm 1,3,J_\pm 2,4)\}$
$= -\mathcal{I}\{(1,2,3,4)(3,J_\pm 2,J_\pm 1,4)\}$
$\quad -\mathcal{I}\{(1,2,3,4)(J_\pm 2,J_\pm 1,3,4)\}$
$= -\mathcal{I}\{(1,2,J_\pm 3,J_\pm 4)(J_\pm 3,J_\pm 2,J_\pm 1,J_\pm 4)\} + \psi_{11}^{\mathfrak{K}\pm}$
$= -\mathcal{I}\{(1,2,3,4)(J_\pm 3,J_\pm 2,1,4)\} + \psi_{11}^{\mathfrak{K}\pm}$
$= -\mathcal{I}\{(3,2,1,4)(J_\pm 1,J_\pm 2,3,4)\} + \psi_{11}^{\mathfrak{K}\pm}$
$= -\mathcal{I}\{(3,J_\pm 1,J_\pm 2,4)(2,1,3,4)\} + \psi_{11}^{\mathfrak{K}\pm}$
$= -\mathcal{I}\{(J_\pm 1,3,J_\pm 2,4)(1,2,3,4)\} + \psi_{11}^{\mathfrak{K}\pm} = -\Psi_1 + \psi_{11}^{\mathfrak{K}\pm}$.
Thus $\Psi_1 = \frac{1}{2}\psi_{11}^{\mathfrak{K}\pm}$.

C. $\Psi_2 := \mathcal{I}\{(1,3,2,4)(J_\pm 1,3,J_\pm 2,4)\}$
$= -\mathcal{I}\{(2,1,3,4)(J_\pm 1,3,J_\pm 2,4)\}$
$\quad -\mathcal{I}\{(3,2,1,4)(J_\pm 1,3,J_\pm 2,4)\}$
$= \Psi_1 + \mathcal{I}\{(3,2,J_\pm 1,4)(1,3,J_\pm 2,4)\} \sim \Psi_1$.

D. $\Psi_3 = \mathcal{I}\{(1,3,2,4)(J_\pm 2,3,J_\pm 1,4)\}$. This contradicts (5a).

iv. We have $\mathcal{I}\{(*,*,*,4)(*,*,*,3)\}$. Clear the previous notation.

A. $\Psi_1 := \mathcal{I}\{(1,2,3,4)(J_\pm 1,J_\pm 2,4,3)\}$
$= \mathcal{I}\{(2,3,1,4)(J_\pm 2,4,J_\pm 1,3)\}$

$$+\mathcal{I}\{(2,3,1,4)(4,J_\pm 1,J_\pm 2,3)\}$$
$$+\mathcal{I}\{(3,1,2,4)(J_\pm 2,4,J_\pm 1,3)\}$$
$$+\mathcal{I}\{(3,1,2,4)(4,J_\pm 1,J_\pm 2,3)\}$$
$$= 2\mathcal{I}\{(2,3,1,4)(J_\pm 2,4,J_\pm 1,3)\}$$
$$+2\mathcal{I}\{(2,3,1,4)(4,J_\pm 1,J_\pm 2,3)\}$$
$= 2\Psi_4 - 2\Psi_3$. See C and D below.

B. $\Psi_2 := \mathcal{I}\{(1,2,3,4)(J_\pm 1,4,J_\pm 2,3)\}$
$$= -\mathcal{I}\{(1,2,3,4)(4,J_\pm 2,J_\pm 1,3)\}$$
$$-\mathcal{I}\{(1,2,3,4)(J_\pm 2,J_\pm 1,4,3)\}$$
$= -\Psi_2 + \Psi_1$. Thus $\Psi_2 = \frac{1}{2}\Psi_1$.

C. $\Psi_3 := \mathcal{I}\{(1,3,2,4)(J_\pm 2,4,J_\pm 1,3)\}$
$$= \mathcal{I}\{(1,J_\pm 3,2,4)(J_\pm 2,4,1,3)\}$$
$$= -\mathcal{I}\{(1,J_\pm 3,2,4)(4,1,J_\pm 2,3)\}$$
$$-\mathcal{I}\{(1,J_\pm 3,2,4)(1,J_\pm 2,4,3)\}$$
$$= \mathcal{I}\{(1,3,2,4)(4,1,2,3)\} - \mathcal{I}\{(1,J_\pm 3,2,4)(1,J_\pm 2,4,3)\}$$
$$\sim \pm\mathcal{I}\{(1,3,J_\pm 2,J_\pm 4)(1,2,J_\pm 4,J_\pm 3)\} \sim 0.$$

D. $\Psi_4 := \mathcal{I}\{(1,3,2,4)(J_\pm 1,4,J_\pm 2,3)\}$
$$= \mathcal{I}\{(1,J_\pm 3,2,4)(J_\pm 1,4,2,3)\}$$
$$= -\mathcal{I}\{(1,J_\pm 3,2,4)(4,2,J_\pm 1,3)\}$$
$$-\mathcal{I}\{(1,J_\pm 3,2,4)(2,J_\pm 1,4,3)\}$$
$$= \mp\mathcal{I}\{(1,3,2,4)(4,2,1,3)\}$$
$$+\mathcal{I}\{(1,3,J_\pm 2,J_\pm 4)(J_\pm 2,J_\pm 1,J_\pm 4,J_\pm 3)\}$$
$$\sim \mathcal{I}\{(1,3,2,4)(J_\pm 2,J_\pm 1,4,3)\}$$
$$= -\mathcal{I}\{(J_\pm 1,J_\pm 2,3,4)(1,4,2,3)\}$$
$$= -\mathcal{I}\{(1,2,3,4)(J_\pm 1,4,J_\pm 2,3)\}$$
$$= -\Psi_2 = -\tfrac{1}{2}\Psi_1.$$

Thus $\Psi_1 = 2\Psi_4 - 2\Psi_3 \sim -\Psi_1$ so $\Psi_1 \sim 0$, $\Psi_2 \sim 0$, $\Psi_3 \sim 0$, and $\Psi_4 \sim 0$. Consequently, this case does not appear.

(c) Suppose only three indices appear in each quadruple. We assume the index "1" appears twice in the first quadruple and that the index "4" appears twice in the second quadruple. By applying the Bianchi identity, we can ensure that $(*,*,1,1)$, $(*,*,1,J_\pm 1)$, and $(1,J_\pm 1,*,*)$ do not appear. Furthermore, we can use the Kähler identity to replace $(*,*,2,J_\pm 1)$ by $-(*,*,J_\pm 2,1)$. We have:
$\psi_{12}^{\mathfrak{K}_\pm} = \mathcal{I}\{(1,2,3,1)(4,J_\pm 2,J_\pm 3,4)\}$, and
$\psi_{13}^{\mathfrak{K}_\pm} = \mathcal{I}\{(1,2,3,1)(4,J_\pm 3,J_\pm 2,4)\}$.
The indices "1" and "4" must appear either in the third or the fourth position since the appearance of both in either the fi-

nal two or in the first two positions is impossible. Thus any "$J_\pm 1$" can be moved to an index "2" or "3". We clear the previous notation and consider the remaining possible invariants: $\mathcal{I}\{(1,2,3,1)(4,J_\pm 2,4,J_\pm 3)\}$,
$\mathcal{I}\{(1,3,2,1)(4,J_\pm 2,4,J_\pm 3)\}$, $\mathcal{I}\{(1,2,1,3)(4,J_\pm 2,4,J_\pm 3)\}$,
$\mathcal{I}\{(1,3,1,2)(4,J_\pm 2,4,J_\pm 3)\}$. By (1g),
$(4,J_\pm 2,4,J_\pm 3) = -(4,2,4,J_\pm J_\pm 3) = (4,2,4,3) \sim 0$ so these invariants do not appear.

(d) Suppose only two indices appear in each quadruple. We apply the Bianchi identity to eliminate $(1,J_\pm 1,*,*)$ and $(*,*,1,J_\pm 1)$. It must then have the form $\mathcal{I}\{(1,2,\tilde{1},\tilde{2})(3,4,\tilde{4},\tilde{3})\}$ where $\tilde{i}$ is either i or $J_\pm i$. Since we do not have $(1,2,J_\pm 2,J_\pm 1)$, using the $\mathbb{Z}_2$ symmetry in "1" and "2", we see that the invariant in question is:
$\psi_{14}^{\mathfrak{K}\pm} = \mathcal{I}\{(1,2,J_\pm 2,1)(3,4,J_\pm 4,3)\}$.

(6) Exactly one index $J_\pm$ appears in the invariant. We choose the notation so that the decorated index is "1" and distinguish cases:

(a) The invariant involves two quadruples where every index appears once. We assume the first quadruple is (1234). We apply the Bianchi identity to assume that in the second quadruple the indices (12) do not touch each other in either the first or the last pair of indices and similarly that the indices (34) do not touch each other in either the first or last pair of indices. This leads to four basic types. We clear the previous notation.

i. $\mathcal{I}\{(1,2,3,4)(1,3,2,4)\}$. Insert a $J_\pm$.

A. $\Psi_1 := \mathcal{I}\{(1,2,3,4)(1,3,2,J_\pm 4)\}$
$= -\mathcal{I}\{(1,2,3,J_\pm 4)(1,3,2,4)\}$
$= -\mathcal{I}\{(1,3,2,4)(1,2,3,J_\pm 4)\}$
$= -\mathcal{I}\{(1,2,3,4)(1,3,2,J_\pm 4)\} = -\Psi_1$. Thus $2\Psi_1 = 0$.

B. $\Psi_2 := \mathcal{I}\{(1,2,3,4)(1,J_\pm 3,2,4)\}$
$= -\mathcal{I}\{(1,2,J_\pm 3,4)(1,3,2,4)\}$
$= \pm\mathcal{I}\{(1,2,J_\pm 3,J_\pm 4)(1,3,2,J_\pm 4)\}$
$= -\mathcal{I}\{(1,2,3,4)(1,3,2,J_\pm 4)\}$
$= -\Psi_1 = 0$.

C. $\Psi_3 := \mathcal{I}\{(1,2,3,4)(J_\pm 1,3,2,4)\}$
$= -\mathcal{I}\{(J_\pm 1,2,3,4)(1,3,2,4)\}$
$= -\mathcal{I}\{(1,3,2,4)(J_\pm 1,2,3,4)\}$
$= -\mathcal{I}\{(1,2,3,4)(J_\pm 1,3,2,4)\} = -\Psi_3$. Thus $2\Psi_3 = 0$.

D. $\Psi_4 := \mathcal{I}\{(1,2,3,4)(1,3,J_\pm 2,4)\}$

$= -\mathcal{I}\{(1,2,3,4)(1,3,2,J_\pm 4)\} = -\Psi_1.$

ii. $\mathcal{I}\{(1,2,3,4)(1,3,4,2)\}$. Insert a $J_\pm$.

A. $\sigma_1^{\mathfrak{K}\pm} = \mathcal{I}\{(1,2,3,4)(1,3,4,J_\pm 2)\}$.

B. $\Psi_5 := \mathcal{I}\{(1,2,3,4)(1,J_\pm 3,4,2)\}$
$= -\mathcal{I}\{(1,2,J_\pm 3,4)(1,3,4,2)\}$
$= \pm\mathcal{I}\{(1,2,J_\pm 3,J_\pm 4)(1,3,J_\pm 4,2)\}$
$= -\mathcal{I}\{(1,2,3,4)(1,3,J_\pm 4,2)\}$
$= \mathcal{I}\{(1,2,3,4)(1,3,4,J_\pm 2)\} = \sigma_1^{\mathfrak{K}\pm}$.

C. $\Psi_6 := \mathcal{I}\{(1,2,3,4)(J_\pm 1,3,4,2)\}$
$= -\mathcal{I}\{(1,2,3,4)(3,4,J_\pm 1,2)\}$
$\quad -\mathcal{I}\{(1,2,3,4)(4,J_\pm 1,3,2)\}$
$= \mathcal{I}\{(1,2,3,4)(3,4,1,J_\pm 2)\} + \Psi_{10}$
$= -\mathcal{I}\{(1,2,3,4)(1,3,4,J_\pm 2)\}$
$\quad -\mathcal{I}\{(1,2,3,4)(4,1,3,J_\pm 2)\} + \Psi_{10}$
$= -\sigma_1^{\mathfrak{K}\pm} + \Psi_8 + \Psi_{10}$. See below for Ψ_8 and Ψ_{10}.

D. $\Psi_7 := \mathcal{I}\{(1,2,3,4)(1,3,J_\pm 4,2)\}$
$= -\mathcal{I}\{(1,2,3,4)(1,3,4,J_\pm 2)\} = -\sigma_1^{\mathfrak{K}\pm}$.

iii. $\mathcal{I}\{(1,2,3,4)(1,4,2,3)\}$. Insert a $J_\pm$. We permute the indices to express this in the form already considered in (6)(a)(ii) above: $= (1,\bar{2},\bar{3},\bar{4})(1,\bar{3},\bar{4},\bar{2})$.

iv. $\mathcal{I}\{(1,2,3,4)(1,4,3,2)\}$. Insert a $J_\pm$.

A. $\Psi_8 := \mathcal{I}\{(1,2,3,4)(1,4,3,J_\pm 2)\}$
$= \mathcal{I}\{(1,2,J_\pm 3,J_\pm 4)(1,J_\pm 4,J_\pm 3,J_\pm 2)\}$
$= \mathcal{I}\{(1,2,3,4)(1,J_\pm 4,3,2)\} = \Psi_9$.

B. $\Psi_9 := \mathcal{I}\{(1,2,3,4)(1,J_\pm 4,3,2)\}$
$= \mathcal{I}\{(1,J_\pm 4,3,2)(1,2,3,4)\}$
$= \mathcal{I}\{(1,J_\pm 2,3,4)(1,4,3,2)\}$
$= -\mathcal{I}\{(1,2,3,4)(1,4,3,J_\pm 2)\} = -\Psi_8$.
Thus $\Psi_8 = \Psi_9 = 0$.

C. $\Psi_{10} := \mathcal{I}\{(1,2,3,4)(J_\pm 1,4,3,2)\}$
$= -\mathcal{I}\{(1,4,3,2)(J_\pm 1,2,3,4)\}$
$= -\mathcal{I}\{(1,2,3,4)(J_\pm 1,4,3,2)\} = -\Psi_{10}$. Thus $\Psi_{10} = 0$.

D. $\Psi_{11} := \mathcal{I}\{(1,2,3,4)(1,4,J_\pm 3,2)\}$
$= -\mathcal{I}\{(1,2,3,4)(1,4,3,J_\pm 2)\} = -\Psi_9 = 0$.

(b) By applying the Bianchi identity, we may assume that $(1,J_\pm 1,*,*)$ and $(*,*,1,J_\pm 1)$ do not appear. We continue our study.

(c) The invariant involves two quadruples where one index is repeated in each quadruple. We assume the notation is chosen so the indices

"2" and "3" appear in each quadruple. We clear the previous notation. We use the Bianchi identity to avoid quadruples $(*,*,1,1)$, $(*,*,J_\pm 1,1)$, $(*,*,1,J_\pm 1)$, $(1,J_\pm 1,*,*)$. Thus the invariants have the following form; insert a $J_\pm$.

i. $\mathcal{I}\{(1,2,3,1)(4,2,3,4)\}$. Insert a $J_\pm$.

A. $\Psi_1 := \mathcal{I}\{(1,2,3,J_\pm 1)(4,2,3,4)\}$
$= -\mathcal{I}\{(1,2,J_\pm 3,J_\pm 1)(4,2,J_\pm 3,J_\pm J_\pm 4)\}$
$= -\mathcal{I}\{(1,2,3,1)(4,2,3,J_\pm 4)\} = -\Psi_1$. Thus $\Psi_1 = 0$.

B. $\Psi_2 := \mathcal{I}\{(1,J_\pm 2,3,1)(4,2,3,4)\}$
$= -\mathcal{I}\{(1,2,3,1)(4,J_\pm 2,3,4)\}$
$= -\Psi_2$. Thus $\Psi_2 = 0$.

ii. $\mathcal{I}\{(1,2,3,1)(4,3,2,4)\}$. Insert a $J_\pm$.

A. $\sigma_2^{\mathfrak{K}\pm} = -\mathcal{I}\{(1,2,3,J_\pm 1)(4,3,2,4)\}$.

B. $\Psi_3 := \mathcal{I}\{(1,J_\pm 2,3,1)(4,3,2,4)\}$
$= \mathcal{I}\{(1,J_\pm 2,3,1)(4,3,J_\pm J_\pm 2,J_\pm J_\pm 4)\}$
$= \mp\mathcal{I}\{(1,2,3,1)(4,3,J_\pm 2,J_\pm J_\pm 4)\}$
$= \mathcal{I}\{(1,2,3,1)(4,3,2,J_\pm 4)\} = -\sigma_2^{\mathfrak{K}\pm}$.

iii. $\mathcal{I}\{(1,2,3,1)(4,2,4,3)\}$. Insert a $J_\pm$. We apply (1g).

A. $\Psi_4 := \mathcal{I}\{(1,2,3,J_\pm 1)(4,2,4,3)\}$
$= -\mathcal{I}\{(1,2,3,1)(4,2,4,J_\pm 3)\}$
$= \mathcal{I}\{(1,2,3,1)(4,J_\pm 2,4,3)\} = \sigma_4^{\mathfrak{K}\pm}$.

B. $\sigma_4^{\mathfrak{K}\pm} = -\mathcal{I}\{(1,J_\pm 2,3,1)(4,2,4,3)\}$.

iv. $\mathcal{I}\{(1,2,3,1)(4,3,4,2)\}$. Insert a $J_\pm$. We apply (1g).

A. $\sigma_5^{\mathfrak{K}\pm} = \mathcal{I}\{(1,2,3,J_\pm 1)(4,3,4,2)\}$.

B. $\Psi_5 := \mathcal{I}\{(1,J_\pm 2,3,1)(4,3,4,2)\}$
$= -\mathcal{I}\{(1,2,3,1)(4,3,4,J_\pm 2)\}$
$= \mathcal{I}\{(1,2,3,1)(4,J_\pm 3,4,2)\}$
$= -\mathcal{I}\{(1,2,J_\pm 3,1)(4,3,4,2)\}$
$= \mathcal{I}\{(1,2,3,J_\pm 1)(4,3,4,2)\} = \sigma_5^{\mathfrak{K}\pm}$.

v. $\mathcal{I}\{(1,2,1,3)(4,2,4,3)\}$. Insert a $J_\pm$.

A. $\Psi_6 := \mathcal{I}\{(1,2,1,J_\pm 3)(4,2,4,3)\}$
$= -\mathcal{I}\{(1,2,1,3)(4,2,4,J_\pm 3)\} = -\Psi_6$. Thus $\Psi_6 = 0$.

B. $\Psi_7 := \mathcal{I}\{(1,J_\pm 2,1,3)(4,2,4,3)\}$
$= -\mathcal{I}\{(1,2,1,3)(4,J_\pm 2,4,3)\} = -\Psi_7$. Thus $\Psi_7 = 0$.

C. $\Psi_8 := \mathcal{I}\{(J_\pm 1,2,1,3)(4,2,4,3)\}$
$= -\mathcal{I}\{(1,2,J_\pm 1,3)(1,2,4,3)\}$

$$= \mathcal{I}\{(1,2,1,J_\pm 3)(I,2,4,3)\} = \Psi_6 = 0.$$

vi. $\mathcal{I}\{(1,2,1,3)(4,3,4,2)\}$. Insert a $J_\pm$.

A. $\sigma_3^{\mathfrak{K}_\pm} = \mathcal{I}\{(1,2,1,J_\pm 3)(4,3,4,2)\}$.

B. $\Psi_9 := \mathcal{I}\{(1,J_\pm 2,1,3)(4,3,4,2)\}$
$= -\mathcal{I}\{(1,2,1,3)(4,3,4,J_\pm 2)\} = -\sigma_3^{\mathfrak{K}_\pm}$.

vii. The invariant involves two quadruples each of which involves two indices. Such invariants yield, up to sign, $\sigma_6^{\mathfrak{K}_\pm} = \mathcal{I}\{(1,2,2,1)(3,4,J_\pm 4,3)\}$.

(7) No index is decorated with $J_\pm$. These invariants are $\mathcal{O}$ invariants. By Lemma 4.1.4, $\dim\{\mathcal{I}_2^{\mathcal{O}}(\mathfrak{A})\} \le 10$ and hence there are at most ten such invariants; they are enumerated in Equation (4.1.h).

Lemma 5.1.1 follows since we have constructed a spanning set with twenty elements if the structure group is $\mathcal{U}_-$ and a spanning set with fourteen elements if the structure group is $\mathcal{U}_\pm^\star$. □

5.2 The Ricci Tensor for a Kähler Affine Connection

Lemma 5.2.1 gives the appropriate equivariance properties for ρ and for ρ_{13}. Recall that

$$\begin{aligned}
\mathfrak{K}^{\mathfrak{A}}_{\pm,+} &= \{A \in \mathfrak{K}^{\mathfrak{A}}_\pm : A(J_\pm x, J_\pm y, z, w) = A(x,y,z,w)\},\\
\mathfrak{K}^{\mathfrak{A}}_{\pm,-} &= \{A \in \mathfrak{K}^{\mathfrak{A}}_\pm : A(J_\pm x, J_\pm y, z, w) = -A(x,y,z,w)\},\\
\Omega_\pm &= \langle e_a, J_\pm e_b\rangle e^a \otimes e^b, \quad \text{and} \quad \tau^{\mathfrak{A}}_\pm = \varepsilon^{il}\varepsilon^{jk} A(e_i, J_\pm e_j, e_k, e_l).
\end{aligned}$$

Lemma 5.2.1 *Let $A \in \mathfrak{K}^{\mathfrak{A}}_\pm$. Let $\rho_{13} = \rho_{13}(A)$ and $\rho = \rho(A)$.*

(1) $\langle \rho, \Omega_\pm\rangle = \tau^{\mathfrak{A}}_\pm$ and $\langle \rho_{13}, \Omega_\pm\rangle = \tau^{\mathfrak{A}}_\pm$.

(2) $J_\pm^*\rho_{13} = \mp\rho_{13}$.

(3) If $A \in \mathfrak{K}^{\mathfrak{A}}_{\pm;\mp}$, then $J_\pm^*\rho_{13} = \mp\rho_{13}$ and $J_\pm^*\rho = \mp\rho$.

(4) If $A \in \mathfrak{K}^{\mathfrak{A}}_{\pm;\pm}$, then $\rho_{13} = 0$ and $J_\pm^*\rho = \pm\rho$.

Proof. We establish Assertion (1) by computing:

$$\begin{aligned}
\langle \rho, \Omega_\pm\rangle &= \varepsilon^{il} A(e_i, e_j, e_k, e_l)\langle e_a, J_\pm e_b\rangle\langle e^j, e^a\rangle\langle e^k, e^b\rangle\\
&= \varepsilon^{il} A(e_i, e_j, J_\pm e_k, e_l)\langle e_a, J_\pm J_\pm e_b\rangle\langle e^j, e^a\rangle\langle J_\pm e^k, J_\pm e^b\rangle\\
&= -\varepsilon^{il} A(e_i, e_j, J_\pm e_k, e_l)\langle e_a, e_b\rangle\langle e^j, e^a\rangle\langle e^k, e^b\rangle\\
&= -\varepsilon^{il}\varepsilon^{jk} A(e_i, e_j, J_\pm e_k, e_l) = \varepsilon^{il}\varepsilon^{jk} A(e_i, J_\pm e_j, e_k, e_l) = \tau^{\mathfrak{A}}_\pm,
\end{aligned}$$

$$\begin{aligned}\langle\rho_{13},\Omega_\pm\rangle &= \varepsilon^{ik}A(e_i,e_j,e_k,e_l)\langle e_a,J_\pm e_b\rangle\langle e^j,e^a\rangle\langle e^l,e^b\rangle\\ &= \varepsilon^{ik}A(e_i,J_\pm e_j,e_k,e_l)\langle J_\pm e_a,J_\pm e_b\rangle\langle J_\pm e^j,J_\pm e^a\rangle\langle e^l,e^b\rangle\\ &= \varepsilon^{ik}A(e_i,J_\pm e_j,e_k,e_l)\langle e_a,e_b\rangle\langle e^j,e^a\rangle\langle e^l,e^b\rangle\\ &= \varepsilon^{ik}\varepsilon^{jl}A(e_i,J_\pm e_j,e_k,e_l)\\ &= \varepsilon^{ik}\varepsilon^{jl}A(J_\pm e_i,J_\pm J_\pm e_j,J_\pm e_k,J_\pm e_l)\\ &= -\varepsilon^{ik}\varepsilon^{jl}A(J_\pm e_i,e_j,e_k,e_l)=\varepsilon^{ik}\varepsilon^{jl}A(e_j,J_\pm e_i,e_k,e_l)=\tau_\pm^{\mathfrak{A}}.\end{aligned}$$

To prove Assertion (2), we use the Bianchi identity and replace the frame $\{e_i\}$ by the equivalent frame $\{J_\pm e_i\}$. Since $J_\pm^*\varepsilon=\mp\varepsilon$, we have:

$$\begin{aligned}0 &= \varepsilon^{ij}\{A(J_\pm e_i,J_\pm x,e_j,J_\pm y)+A(J_\pm x,e_j,J_\pm e_i,J_\pm y)\\ &\qquad +A(e_j,J_\pm e_i,J_\pm x,J_\pm y)\}\\ &= \varepsilon^{ij}\{-A(e_i,J_\pm x,J_\pm e_j,J_\pm y)+A(J_\pm x,e_j,J_\pm e_i,J_\pm y)\\ &\qquad +A(e_j,J_\pm e_i,J_\pm x,J_\pm y)\}\\ &= \varepsilon^{ij}\{2A(J_\pm x,e_j,J_\pm e_i,J_\pm y)+A(e_j,J_\pm e_i,J_\pm x,J_\pm y)\},\end{aligned}$$

$$\begin{aligned}0 &= \varepsilon^{ij}\{A(e_j,J_\pm e_i,x,y)+A(J_\pm e_i,x,e_j,y)+A(x,e_j,J_\pm e_i,y)\}\\ &= \varepsilon^{ij}\{A(e_j,J_\pm e_i,x,y)-A(e_i,x,J_\pm e_j,y)+A(x,e_j,J_\pm e_i,y)\}\\ &= \varepsilon^{ij}\{A(e_j,J_\pm e_i,x,y)+2A(x,e_i,J_\pm e_j,y)\}.\end{aligned}$$

We use these identities to see that:

$$\begin{aligned}\rho_{13}(J_\pm x,y) &= \varepsilon^{ij}A(e_j,J_\pm x,e_i,y)=\mp\varepsilon^{ij}A(e_j,J_\pm x,J_\pm e_i,J_\pm y)\\ &= \mp\tfrac{1}{2}\varepsilon^{ij}A(e_j,J_\pm e_i,J_\pm x,J_\pm y)=\tfrac{1}{2}\varepsilon^{ij}A(e_j,J_\pm e_i,x,y)\\ &= \varepsilon^{ij}A(e_i,x,J_\pm e_j,y)=-\varepsilon^{ij}A(e_i,x,e_j,J_\pm y)=-\rho_{13}(x,J_\pm y).\end{aligned}$$

Replacing x by $J_\pm x$ then shows $\rho_{13}(x,y)=\mp\rho_{13}(J_\pm x,J_\pm y)$. This proves Assertion (2). Let $A\in\mathfrak{K}_{\pm;\delta}^{\mathfrak{A}}$. We compute:

$$\begin{aligned}\rho(J_\pm x,J_\pm y) &= \varepsilon^{ij}A(e_i,J_\pm x,J_\pm y,e_j)\\ &= \mp\delta\varepsilon^{ij}A(J_\pm e_i,J_\pm J_\pm x,J_\pm J_\pm y,J_\pm e_j)\\ &= \mp\delta\varepsilon^{ij}A(J_\pm e_i,x,y,J_\pm e_j)=\delta\varepsilon^{ij}A(e_i,x,y,e_j)=\delta\rho(x,y),\end{aligned}$$

$$\begin{aligned}\rho_{13}(J_\pm x,J_\pm y) &= \varepsilon^{ij}A(e_i,J_\pm x,e_j,J_\pm y)\\ &= \mp\delta\varepsilon^{ij}A(J_\pm e_i,J_\pm J_\pm x,J_\pm e_j,J_\pm J_\pm y)\\ &= \mp\delta A\epsilon^{ij}(J_\pm e_i,x,J_\pm e_j,y)=\delta\rho_{13}(x,y).\end{aligned}$$

Since $J_\pm^*\rho_{13}=\mp\rho_{13}=\delta\rho_{13}$, $\rho_{13}=0$ on $\mathfrak{K}_{\pm,\pm}^{\mathfrak{A}}$ since the signs differ. □

5.3 Constructing Affine (Para)-Kähler Manifolds

We use the formalism described in Lemma 3.1.3 to define affine (para)-Kähler connections. Let $(V, \langle\cdot,\cdot\rangle, J_\pm)$ be a para-Hermitian vector space $(+)$ or a pseudo-Hermitian vector space $(-)$. We consider the following subspaces of $\otimes^3 V^* \otimes V$ consisting of all tensors satisfying the following identities:

$$\begin{aligned}\mathfrak{Z}_\pm &:= \{\Theta_\pm \in \otimes^3 V^* \otimes V : \Theta_\pm(x,y,z) = \Theta_\pm(x,z,y),\\ &\qquad \Theta_\pm(x, J_\pm y, z) = \Theta_\pm(x, y, J_\pm z) = J_\pm \Theta_\pm(x,y,z)\}, &(5.3.a)\\ \mathfrak{Z}_{\pm,h} &:= \{\Theta_\pm \in \mathfrak{Z}_\pm : \Theta_\pm(J_\pm x, y, z) = J_\pm \Theta_\pm(x,y,z)\}, &(5.3.b)\\ \mathfrak{Z}_{\pm,a} &:= \{\Theta_\pm \in \mathfrak{Z}_\pm : \Theta_\pm(J_\pm x, y, z) = -J_\pm \Theta_\pm(x,y,z)\}. &(5.3.c)\end{aligned}$$

The notation "h" for (para)-holomorphic and "a" for anti-(para)-holomorphic arises as follows. If we fix (y, z), then the map $x \to \Theta(x, y, z)$ is a linear map from V to V. Such a map is (para)-holomorphic if and only if it is (para)-complex linear and such a map is anti-(para)-holomorphic if and only if it is anti-(para)-complex linear. Thus $\mathfrak{Z}_{\pm,h}$ consists of functions which are (para)-holomorphic in the first argument while $\mathfrak{Z}_{\pm,a}$ consists of functions which are anti-(para)-holomorphic in the first argument. Define a linear map ϱ from $\mathfrak{Z}_\pm$ to itself by setting:

$$\varrho(\Theta_\pm)(x,y,z) := J_\pm^{-1}\Theta_\pm(J_\pm x, y, z).$$

It is immediate that $\varrho^2 = \mathrm{Id}$, that ϱ preserves the inner product on $\mathfrak{Z}_\pm$, that $\mathfrak{Z}_{\pm,h}$ is the $+1$ eigenspace of ϱ, and that $\mathfrak{Z}_{\pm,a}$ is the -1 eigenspace of ϱ. This shows that we have an orthogonal direct sum decomposition

$$\mathfrak{Z}_\pm = \mathfrak{Z}_{\pm,h} \oplus \mathfrak{Z}_{\pm,a}.$$

Let $\{u^1, \ldots, u^m\}$ be the dual coordinates on V induced by some basis for V. Recall that $x = c^i \partial_{u_i}$ is said to be a *coordinate vector field* if the coefficients c^i are constant; this notion is independent of the particular basis chosen. If $\Theta_\pm \in \mathfrak{Z}_\pm$, then $\Theta_\pm$ defines a connection $\nabla = \nabla^{\Theta_\pm}$; it may be characterized as follows. Let x and y be coordinate vector fields and let $P \in V$. Then

$$(\nabla_x^{\Theta_\pm} y)(P) := \Theta_\pm(P, x, y).$$

Lemma 5.3.1 *Let $\Theta_\pm \in \mathfrak{Z}_\pm$.*

(1) $(V, J_\pm, \nabla^{\Theta_\pm})$ is a (para)-Kähler manifold with curvature operator

$$\begin{aligned}\mathcal{R}^{\Theta_\pm}(P)(x,y)z &= \Theta_\pm(x,y,z) - \Theta_\pm(y,x,z)\\ &+\Theta_\pm(P,x,\Theta_\pm(P,y,z)) - \Theta_\pm(P,y,\Theta_\pm(P,x,z)).\end{aligned}$$

(2) If $\Theta_\pm \in \mathfrak{Z}_{\pm,h}$, then $\mathcal{R}^{\Theta_\pm}(P) \in \mathfrak{K}^{\mathfrak{A}}_{\pm,\pm}$ for all $P \in V$.

(3) If $\Theta_\pm \in \mathfrak{Z}_{\pm,a}$, then $\mathcal{R}^{\Theta_\pm}(0) \in \mathfrak{K}^{\mathfrak{A}}_{\pm,\mp}$.

Proof. Let x, y, and z be coordinate vector fields. We use the first relation of Equation (5.3.a) to show that the torsion tensor of $\nabla = \nabla^{\Theta_\pm}$ vanishes by computing:

$$\{\nabla_y z - \nabla_z y - [y,z]\}(P) = \Theta_\pm(P,y,z) - \Theta_\pm(P,z,y) - 0 = 0.$$

We use the second relation of Equation (5.3.a) to show $\nabla J_\pm = J_\pm \nabla$ by computing:

$$\{\nabla_y J_\pm z\}(P) = \Theta_\pm(P,y,J_\pm z) = J_\pm\Theta_\pm(P,y,z) = \{J_\pm\nabla_y z\}(P).$$

Because $\Theta_\pm$ is a multi-linear map, $x\Theta_\pm(\cdot,y,z) = \Theta_\pm(x,y,z)$. We complete the proof of Assertion (1) by observing:

$$\begin{aligned}\mathcal{R}^{\Theta_\pm}(P)(x,y)z &= \{x\Theta_\pm(\cdot,y,z) - y\Theta_\pm(\cdot,x,z)\}(P)\\ &+\Theta_\pm(P,x,\Theta_\pm(P,y,z)) - \Theta_\pm(P,y,\Theta_\pm(P,x,z)).\end{aligned}$$

Suppose that $\Theta_\pm \in \mathfrak{Z}_{\pm,h}$. We use Assertion (1), the first relation of Equation (5.3.a), and Equation (5.3.b) to prove Assertion (2) by computing:

$$\begin{aligned}\mathcal{R}^{\Theta_\pm}(P)(J_\pm x, J_\pm y)z &= \Theta_\pm(J_\pm x, J_\pm y, z) - \Theta_\pm(J_\pm y, J_\pm x, z)\\ &+\Theta_\pm(P,J_\pm x,\Theta_\pm(P,J_\pm y,z)) - \Theta_\pm(P,J_\pm y,\Theta_\pm(P,J_\pm x,z))\\ &= J_\pm^2\Theta_\pm(x,y,z) - J_\pm^2\Theta_\pm(y,x,z)\\ &+J_\pm^2\Theta_\pm(P,x,\Theta_\pm(P,y,z)) - J_\pm^2\Theta_\pm(P,y,(P,x,z))\\ &= \pm\mathcal{R}^{\Theta_\pm}(P)(x,y)z.\end{aligned}$$

If $\Theta_\pm \in \mathfrak{Z}_{\pm,a}$, a similar calculation using Equation (5.3.c) yields

$$\begin{aligned}\mathcal{R}^{\Theta_\pm}(P)(J_\pm x, J_\pm y)z &= -J_\pm^2\Theta_\pm(x,y,z) + J_\pm^2\Theta_\pm(y,x,z)\\ &+J_\pm^2\Theta_\pm(P,x,\Theta_\pm(P,y,z)) - J_\pm^2\Theta_\pm(P,y,(P,x,z)).\end{aligned}$$

The linear terms and the quadratic terms now have opposite signs and no longer yield $\mathcal{R}^{\Theta_\pm}$ at an arbitrary point P of V. If we take $P=0$, then the quadratic terms vanish and we obtain $\mp\mathcal{R}^{\Theta_\pm}(0)$ as desired. □

Let $\Xi\Theta_\pm := \mathcal{R}^{\Theta_\pm}(0)$. By Lemma 5.3.1,

$$(\Xi\Theta_\pm)(x,y)z = \Theta_\pm(x,y,z) - \Theta_\pm(y,x,z). \tag{5.3.d}$$

If $T \in \mathrm{GL}$, then the natural action T^* on $\otimes^3 V^* \otimes V$ of Equation (1.1.b) is:

$$(T^*\Theta)(x,y,z) := T^{-1}\Theta(Tx,Ty,Tz). \tag{5.3.e}$$

We verify this action defines the structure of a module for the group $\mathrm{GL}_\pm^\star$ on the spaces $\mathfrak{Z}_\pm$, $\mathfrak{Z}_{\pm,h}$, $\mathfrak{Z}_{\pm,a}$, $\mathfrak{K}_\pm^{\mathfrak{A}}$, $\mathfrak{K}_{\pm,+}^{\mathfrak{A}}$, and $\mathfrak{K}_{\pm,-}^{\mathfrak{A}}$ and that the map Ξ of Equation (5.3.d) is a morphism of modules for the group $\mathrm{GL}_\pm$:

Lemma 5.3.2 *Let $T \in \mathrm{GL}_\pm$.*

(1) $\Xi T^* = T^*\Xi$.

(2) $T^*\mathfrak{Z}_\pm = \mathfrak{Z}_\pm$, $\quad T^*\mathfrak{Z}_{\pm,h} = \mathfrak{Z}_{\pm,h}$, $\quad$ *and* $T^*\mathfrak{Z}_{\pm,a} = \mathfrak{Z}_{\pm,a}$.

(3) $T^*\mathfrak{K}_\pm^{\mathfrak{A}} = \mathfrak{K}_\pm^{\mathfrak{A}}$, $\quad T^*\mathfrak{K}_{\pm,+}^{\mathfrak{A}} = \mathfrak{K}_{\pm,+}^{\mathfrak{A}}$, *and* $T^*\mathfrak{K}_{\pm,-}^{\mathfrak{A}} = \mathfrak{K}_{\pm,-}^{\mathfrak{A}}$.

(4) $\Xi\mathfrak{Z}_\pm \subset \mathfrak{K}_\pm^{\mathfrak{A}}$, $\quad \Xi\mathfrak{Z}_{\pm,h} \subset \mathfrak{K}_{\pm,\pm}^{\mathfrak{A}}$, *and* $\Xi\mathfrak{Z}_{\pm,a} \subset \mathfrak{K}_{\pm,\mp}^{\mathfrak{A}}$.

Proof. Let $\Theta_\pm \in \mathfrak{Z}_\pm$. We use Equation (5.3.d) and Equation (5.3.e) to prove Assertion (1) by computing:

$$\begin{aligned}
&\{T^*(\Xi\Theta_\pm)\}(x,y,z) = T^{-1}(\Xi\Theta_\pm)(Tx,Ty,Tz)\\
&\quad = T^{-1}\Theta_\pm(Tx,Ty,Tz) - T^{-1}\Theta_\pm(Ty,Tx,Tz)\\
&\quad = (T^*\Theta_\pm)(x,y,z) - (T^*\Theta_\pm)(y,x,z)\\
&\quad = \{\Xi(T^*\Theta_\pm)\}(x,y,z).
\end{aligned}$$

Since $T \in \mathrm{GL}_\pm$, we have $TJ_\pm = \delta J_\pm$ for $\delta = \pm 1$. The only matter at issue in the proof of Assertions (2) and (3) is the presence of the sign $\pm$ represented by δ and this always cancels out. We show that $T^*\Theta_\pm \in \mathfrak{Z}_\pm$ by verifying:

$$\begin{aligned}
&(T^*\Theta_\pm)(x,y,z) = T^{-1}\Theta_\pm(Tx,Ty,Tz) = T^{-1}\Theta_\pm(Tx,Tz,Ty)\\
&\quad = (T^*\Theta_\pm)(x,z,y),\\
&(T^*\Theta_\pm)(x,J_\pm y,z) = T^{-1}\Theta_\pm(Tx,TJ_\pm y,Tz)\\
&\quad = \delta T^{-1}\Theta_\pm(Tx,J_\pm Ty,Tz) = \delta T^{-1}J_\pm\Theta_\pm(Tx,Ty,Tz)\\
&\quad = \delta^2 J_\pm T^{-1}\Theta_\pm(Tx,Ty,Tz) = J_\pm(T^*\Theta_\pm)(x,y,z),\\
&(T^*\Theta_\pm)(x,y,J_\pm z) = (T^*\Theta_\pm)(x,J_\pm z,y)\\
&\quad = J_\pm(T^*\Theta_\pm)(x,z,y) = J_\pm(T^*\Theta_\pm)(x,y,z).
\end{aligned}$$

Let $\Theta_{\pm,h} \in \mathfrak{Z}_{\pm,h}$ and let $\Theta_{\pm,a} \in \mathfrak{Z}_{\pm,a}$. We show that $T^*\Theta_{\pm,h} \in \mathfrak{Z}_{\pm,h}$ and that $T^*\Theta_{\pm,a} \in \mathfrak{Z}_{\pm,a}$ by verifying the defining relations are satisfied:

$$\begin{aligned}
&(T^*\Theta_{\pm,h})(J_\pm x,y,z) = T^{-1}\Theta_{\pm,h}(TJ_\pm x,Ty,Tz)\\
&\quad= \delta T^{-1}\Theta_{\pm,h}(J_\pm Tx,Ty,Tz) = \delta T^{-1}J_\pm\Theta_{\pm,h}(Tx,Ty,Tz)\\
&\quad= \delta^2 J_\pm T^{-1}\Theta_{\pm,h}(Tx,Ty,Tz) = J_\pm(T^*\Theta_{\pm,h})(x,y,z),
\end{aligned}$$

$$\begin{aligned}
&(T^*\Theta_{\pm,a})(J_\pm x,y,z) = T^{-1}\Theta_{\pm,a}(TJ_\pm x,Ty,Tz)\\
&\quad= \delta T^{-1}\Theta_{\pm,a}(J_\pm Tx,Ty,Tz) = -\delta T^{-1}J_\pm\Theta_{\pm,a}(Tx,Ty,Tz)\\
&\quad= -\delta^2 J_\pm T^{-1}\Theta_{\pm,a}(Tx,Ty,Tz) = -J_\pm(T^*\Theta_{\pm,a})(x,y,z).
\end{aligned}$$

Let $\mathcal{A}\in\mathfrak{K}_\pm^{\mathfrak{A}}$. By Lemma 3.1.3, $T^*\mathcal{A}\in\mathfrak{A}$ so only the equivariance with respect to $J_\pm$ is at issue. Since $(T^*\mathcal{A})(x,y) = T^{-1}\mathcal{A}(Tx,Ty)T$, we show $T^*\mathcal{A}\in\mathfrak{K}_\pm^{\mathfrak{A}}$ by computing:

$$\begin{aligned}
&(T^*\mathcal{A})(x,y)J_\pm = T^{-1}\mathcal{A}(Tx,Ty)TJ_\pm = \delta T^{-1}\mathcal{A}(Tx,Ty)J_\pm T\\
&\quad= \delta T^{-1}J_\pm\mathcal{A}(Tx,Ty)T = \delta^2 J_\pm T^{-1}\mathcal{A}(Tx,Ty)T\\
&\quad= J_\pm(T^*\mathcal{A})(x,y).
\end{aligned}$$

Let $\mathcal{A}_+\in\mathfrak{K}_{\pm,+}^{\mathfrak{A}}$ and let $\mathcal{A}_-\in\mathfrak{K}_{\pm,-}^{\mathfrak{A}}$. We show that $T^*\mathcal{A}_+\in\mathfrak{K}_{\pm,+}^{\mathfrak{A}}$ and that $T^*\mathcal{A}_-\in\mathfrak{K}_{\pm,-}^{\mathfrak{A}}$ by checking the defining relations are satisfied:

$$\begin{aligned}
&(T^*\mathcal{A}_+)(J_\pm x,J_\pm y) = T^{-1}\mathcal{A}_+(TJ_\pm x,TJ_\pm y)T\\
&\quad= \delta^2 T^{-1}\mathcal{A}_+(J_\pm Tx,J_\pm Ty)T = T^{-1}\mathcal{A}_+(Tx,Ty)T\\
&\quad= (T^*\mathcal{A}_+)(x,y),
\end{aligned}$$

$$\begin{aligned}
&(T^*\mathcal{A}_-)(J_\pm x,J_\pm y) = T^{-1}\mathcal{A}_-(TJ_\pm x,TJ_\pm y)T\\
&\quad= \delta^2 T^{-1}\mathcal{A}_-(J_\pm Tx,J_\pm Ty)T = -T^{-1}\mathcal{A}_-(Tx,Ty)T\\
&\quad= -(T^*\mathcal{A}_-)(x,y).
\end{aligned}$$

Although Assertion (4) follows from Lemma 5.3.1, a direct computation is instructive:

$$\begin{aligned}
&(\Xi\Theta_\pm)(x,y)J_\pm z = \Theta_\pm(x,y,J_\pm z) - \Theta_\pm(y,x,J_\pm z)\\
&\quad= J_\pm\Theta_\pm(x,y,z) - J_\pm\Theta_\pm(y,x,z) = J_\pm(\Xi\Theta_\pm)(x,y)z,
\end{aligned}$$

$$\begin{aligned}
&(\Xi\Theta_{\pm,h})(J_\pm x,J_\pm y)z = \Theta_{\pm,h}(J_\pm x,J_\pm y,z) - \Theta_{\pm,h}(J_\pm y,J_\pm x,z)\\
&\quad= J_\pm^2\Theta_{\pm,h}(x,y,z) - J_\pm^2\Theta_{\pm,h}(y,x,z) = \pm(\Xi\Theta_{\pm,h})(x,y,z),
\end{aligned}$$

$$\begin{aligned}
&(\Xi\Theta_{\pm,a})(J_\pm x,J_\pm y)z = \Theta_{\pm,a}(J_\pm x,J_\pm y,z) - \Theta_{\pm,a}(J_\pm y,J_\pm x,z)\\
&\quad= -J_\pm^2\Theta_{\pm,a}(x,y,z) + J_\pm^2\Theta_{\pm,a}(y,x,z) = \mp(\Xi\Theta_{\pm,a})(x,y,z). \quad\square
\end{aligned}$$

5.4 Affine Kähler Curvature Operators

In Section 5.4, we will discuss the decomposition [Matzeu and Nikčević (1991)] of $\mathfrak{K}_-^{\mathfrak{A}}$ as a module with structure group $\mathcal{U}_-$ or $\mathcal{U}_-^\star$. There are twelve different components in this decomposition if $n \geq 6$ and ten components if $n = 4$. We also refer to related work in [Nikčević (1992)]; see also [Nikčević (1994)]. We adopt the notation of Section 5.3; any element in Range(Ξ) is geometrically realizable in the appropriate context. The fact that $\Xi : \mathfrak{Z}_- \to \mathfrak{K}_-^{\mathfrak{A}}$ is a module morphism plays an important role.

We assume that $\{e_i, f_i\}_{i=1}^{\bar{m}}$ is an orthonormal basis for V with $J_- e_i = f_i$ and $J_- f_i = -e_i$. Let $\Theta_- \in \mathfrak{Z}_-$. We expand

$$\Theta_-(\cdot, e_i, e_j) = u_{ij}{}^k(\cdot)e_k + v_{ij}{}^k(\cdot)f_k$$

and as a convenient notation set:

$$\Theta_{-,ij}{}^k(\cdot) := u_{ij}{}^k(\cdot) + \sqrt{-1}v_{ij}{}^k(\cdot).$$

Conversely given $\{\Theta_{-,ij}{}^k\} \in V^* \otimes V \otimes \mathbb{C}$ with $\Theta_{-,ij}{}^k = \Theta_{-,ji}{}^k$, we can recover Θ_- by using the second identity of Equation (5.3.a) to define

$$\Theta_-(\cdot, f_i, e_j) = \Theta_-(\cdot, e_j, f_i) := J_\pm\Theta_-(\cdot, e_i, e_j),$$
$$\Theta_-(\cdot, f_i, f_j) = -\Theta_-(\cdot, e_i, e_j).$$

Then Θ_- belongs to $\mathfrak{Z}_{-,h}$ if and only if the components $\Theta_{-,ij}{}^k$ are holomorphic; Θ_- belongs to $\mathfrak{Z}_{-,a}$ if and only if the components $\Theta_{-,ij}{}^k$ are anti-holomorphic. Since

$$\tau_-^{\mathfrak{A}} = \varepsilon^{il}\varepsilon^{jk}A(e_i, J_-e_j, e_k, e_l)$$

involves only one decoration by J_-, it takes values in the representation space χ and not $\mathbb{R}$.

Lemma 5.4.1 *Let $m \geq 4$. Then:*

(1) $(\tau \oplus \tau_-^{\mathfrak{A}}) \circ \Xi : \mathfrak{Z}_{-,a} \to \mathbb{R} \oplus \chi \to 0.$

(2) $\rho \circ \Xi : \mathfrak{Z}_{-,h} \to S_-^2 \oplus \Lambda_-^2 \to 0.$

(3) $(\rho \oplus \rho_{13}) \circ \Xi : \mathfrak{Z}_{-,a} \to S_{0,+}^{2,\mathcal{U}_-} \oplus \Lambda_{0,+}^{2,\mathcal{U}_-} \oplus S_{0,+}^{2,\mathcal{U}_-} \oplus \Lambda_{0,+}^{2,\mathcal{U}_-} \to 0.$

Proof. Given $\Theta_- \in \mathfrak{Z}_-$, we shall apply Lemma 5.3.1 to construct a Kähler affine connection $\nabla = \nabla^{\Theta_-}$ with associated curvature $\mathcal{A} := \Xi(\Theta_-)$ to be the curvature operator of ∇ at the origin. We first consider

$$\Theta_{-,11}{}^1 = \varrho_1(x^1 - \sqrt{-1}y^1) \quad \text{and} \quad \Theta_{-,12}{}^2 = \Theta_{-,21}{}^2 = \varrho_2(y^1 + \sqrt{-1}x^1).$$

Since Θ_- is anti-holomorphic, $\mathcal{A} \in \mathfrak{K}^{\mathfrak{A}}_{-,+}$. We have:

$$\nabla_{e_1}e_1 = -\nabla_{f_1}f_1 = \varrho_1(x^1e_1 - y^1f_1),\ \nabla_{e_1}f_1 = \nabla_{f_1}e_1 = \varrho_1(y^1e_1 + x^1f_1),$$
$$\nabla_{e_1}e_2 = -\nabla_{f_1}f_2 = \varrho_2(y^1e_2 + x^1f_2),\ \nabla_{e_2}e_1 = -\nabla_{f_2}f_1 = \varrho_2(y^1e_2 + x^1f_2),$$
$$\nabla_{e_1}f_2 = \nabla_{f_1}e_2 = \varrho_2(-x^1e_2 + y^1f_2),\ \nabla_{e_2}f_1 = \nabla_{f_2}e_1 = \varrho_2(-x^1e_2 + y^1f_2),$$

$$\begin{array}{ll}
\mathcal{A}(e_1,f_1)e_1 = 2\varrho_1f_1, & \mathcal{A}(e_1,f_1)f_1 = -2\varrho_1e_1,\\
\mathcal{A}(e_1,e_2)e_1 = -\mathcal{A}(e_1,f_2)f_1 = \varrho_2f_2, & \mathcal{A}(e_1,e_2)f_1 = \mathcal{A}(e_1,f_2)e_1 = -\varrho_2e_2,\\
\mathcal{A}(f_1,e_2)e_1 = -\mathcal{A}(f_1,f_2)f_1 = \varrho_2e_2, & \mathcal{A}(f_1,e_2)f_1 = \mathcal{A}(f_1,f_2)e_1 = \varrho_2f_2,\\
\mathcal{A}(e_1,f_1)f_2 = -2\varrho_2f_2, & \mathcal{A}(e_1,f_1)e_2 = -2\varrho_2e_2,\\
\rho(e_1,e_1) = \rho(f_1,f_1) = -2\varrho_1, & \rho(e_1,f_1) = -\rho(f_1,e_1) = 2\varrho_2,\\
\tau = -4\varrho_1\varepsilon_{11}, & \tau^{\mathfrak{A}}_- = -4\varrho_2\varepsilon_{11}.
\end{array}$$

Assertion (1) follows since ϱ_1 and ϱ_2 are arbitrary.

Next take $\Theta_{-,11}{}^1 = \varrho_1(x^2 + \sqrt{-1}y^2)$ and $\Theta_{-,22}{}^2 = \varrho_2(x^1 + \sqrt{-1}y^1)$. Since Θ_- is holomorphic, $\mathcal{A} \in \mathfrak{K}^{\mathfrak{A}}_{-,-}$. Then:

$$\nabla_{e_1}e_1 = -\nabla_{f_1}f_1 = \varrho_1(x^2e_1 + y^2f_1),\ \nabla_{e_1}f_1 = \nabla_{f_1}e_1 = \varrho_1(-y^2e_1 + x^2f_1),$$
$$\nabla_{e_2}e_2 = -\nabla_{f_2}f_2 = \varrho_2(x^1e_2 + y^1f_2),\ \nabla_{e_2}f_2 = \nabla_{f_2}e_2 = \varrho_2(-y^1e_2 + x^1f_2),$$

$$\begin{array}{ll}
\mathcal{A}(e_2,e_1)e_1 = -\mathcal{A}(e_2,f_1)f_1 = \varrho_1e_1, & \mathcal{A}(f_2,e_1)e_1 = -\mathcal{A}(f_2,f_1)f_1 = \varrho_1f_1,\\
\mathcal{A}(e_2,e_1)f_1 = \mathcal{A}(e_2,f_1)e_1 = \varrho_1f_1, & \mathcal{A}(f_2,e_1)f_1 = \mathcal{A}(f_2,f_1)e_1 = -\varrho_1e_1,\\
\mathcal{A}(e_1,e_2)e_2 = -\mathcal{A}(e_1,f_2)f_2 = \varrho_2e_2, & \mathcal{A}(f_1,e_2)e_2 = -\mathcal{A}(f_1,f_2)f_2 = \varrho_2f_2,\\
\mathcal{A}(e_1,e_2)f_2 = \mathcal{A}(e_1,f_2)e_2 = \varrho_2f_2, & \mathcal{A}(f_1,e_2)f_2 = \mathcal{A}(f_1,f_2)e_2 = -\varrho_2e_2,\\
\rho(e_2,e_1) = -2\varrho_1, & \rho(f_2,f_1) = 2\varrho_1,\\
\rho(e_1,e_2) = -2\varrho_2, & \rho(f_1,f_2) = 2\varrho_2.
\end{array}$$

If we take $\varrho_1 = \varrho_2$, then $\rho \in S^2_-$; if we take $\varrho_1 = -\varrho_2$, then $\rho \in \Lambda^2_-$. Since S^2_- and Λ^2_- are inequivalent irreducible modules for the group $\mathcal{U}^\star_-$ and since Ξ is a morphism of modules for the group $\mathcal{U}^\star_-$ from $\mathfrak{Z}_{-,a}$ to $\mathfrak{K}^{\mathfrak{A}}_{-,-}$, Assertion (2) follows.

We begin the proof of Assertion (3) by taking:

$$\Theta_{-,11}{}^1 = \varrho_1(x^1 - \sqrt{-1}y^1) + \varrho_2(x^2 - \sqrt{-1}y^2),$$
$$\Theta_{-,22}{}^2 = \varrho_3(x^2 - \sqrt{-1}y^2) + \varrho_4(x^1 - \sqrt{-1}y^1).$$

Since Θ_- is anti-holomorphic, $\mathcal{A} \in \mathfrak{K}^{\mathfrak{A}}_{-,-}$. We then have

$$\nabla_{e_1}e_1 = -\nabla_{f_1}f_1 = (\varrho_1x^1 + \varrho_2x^2)e_1 - (\varrho_1y^1 + \varrho_2y^2)f_1,$$
$$\nabla_{e_1}f_1 = \nabla_{f_1}e_1 = (\varrho_1y^1 + \varrho_2y^2)e_1 + (\varrho_1x^1 + \varrho_2x^2)f_1,$$

$$\nabla_{e_2}e_2 = -\nabla_{f_2}f_2 = (\varrho_3 x^2 + \varrho_4 x^1)e_2 - (\varrho_3 y^2 + \varrho_4 y^1)f_2,$$
$$\nabla_{e_2}f_2 = \nabla_{f_2}e_2 = (\varrho_3 y^2 + \varrho_4 y^1)e_2 + (\varrho_3 x^2 + \varrho_4 x^1)f_2.$$

The components of the curvature operator are described by:

$$\begin{array}{ll}
\mathcal{A}(e_1,f_1)f_1 = -2\varrho_1 e_1, & \mathcal{A}(e_1,f_1)e_1 = 2\varrho_1 f_1,\\
\mathcal{A}(e_2,e_1)e_1 = -\mathcal{A}(e_2,f_1)f_1 = \varrho_2 e_1, & \mathcal{A}(e_2,e_1)f_1 = \mathcal{A}(e_2,f_1)e_1 = \varrho_2 f_1,\\
\mathcal{A}(f_2,e_1)e_1 = -\mathcal{A}(f_2,f_1)f_1 = -\varrho_2 f_1, & \mathcal{A}(f_2,e_1)f_1 = \mathcal{A}(f_2,f_1)e_1 = \varrho_2 e_1,\\
\mathcal{A}(e_2,f_2)f_2 = -2\varrho_3 e_2, & \mathcal{A}(e_2,f_2)e_2 = 2\varrho_3 f_2,\\
\mathcal{A}(e_1,e_2)e_2 = -\mathcal{A}(e_1,f_2)f_2 = \varrho_4 e_2, & \mathcal{A}(e_1,e_2)f_2 = \mathcal{A}(e_1,f_2)e_2 = \varrho_4 f_2,\\
\mathcal{A}(f_1,e_2)e_2 = -\mathcal{A}(f_1,f_2)f_2 = -\varrho_4 f_2, & \mathcal{A}(f_1,e_2)f_2 = \mathcal{A}(f_1,f_2)e_2 = \varrho_4 e_2,
\end{array}$$

$$\begin{array}{ll}
\rho(e_1,e_1) = \rho(f_1,f_1) = -2\varrho_1, & \rho(e_1,e_2) = \rho(f_1,f_2) = -2\varrho_4,\\
\rho(e_2,e_2) = \rho(f_2,f_2) = -2\varrho_3, & \rho(e_2,e_1) = \rho(f_2,f_1) = -2\varrho_2,\\
\rho_{13}(e_1,e_1) = \rho_{13}(f_1,f_1) = 2\varrho_1, & \rho_{13}(e_1,e_2) = \rho_{13}(f_1,f_2) = 0,\\
\rho_{13}(e_2,e_2) = \rho_{13}(f_2,f_2) = 2\varrho_3, & \rho_{13}(e_2,e_1) = \rho_{13}(f_2,f_1) = 0,\\
\tau = -4\varrho_1\varepsilon_{11} - 4\varrho_3\varepsilon_{22}, & \tau_-^{\mathfrak{A}} = 0.
\end{array}$$

Note that $\tau = 0$ implies $\langle \rho, g\rangle = \langle \rho_{13}, g\rangle = 0$. Let $\vec{\varrho} = (\varrho_1, \varrho_2, \varrho_3, \varrho_4)$. By Lemma 5.2.1, if $\tau_-^{\mathfrak{A}} = 0$, then $\langle \rho, \Omega_-\rangle = \langle \rho_{13}, \Omega_-\rangle = 0$.

(1) If $\vec{\varrho} = (0,1,0,1)$, then $\tau = 0$, $0 \neq \rho(A) \in S_{0,+}^{2,\mathcal{U}_-}$, and $\rho_{13} = 0$.
(2) If $\vec{\varrho} = (0,1,0,-1)$, then $\tau_-^{\mathfrak{A}} = 0$, $0 \neq \rho(A) \in \Lambda_{0,+}^{2,\mathcal{U}_-}$, and $\rho_{13} = 0$.
(3) If $\vec{\varrho} = (\varepsilon_{22}, 0, -\varepsilon_{11}, 0)$, then $\tau = 0$ and $0 \neq \rho_{13}(A) \in S_{0,+}^{2,\mathcal{U}_-}$.

Next take

$$\Theta_{-,12}{}^2 = \Theta_{-,21}{}^2 = \varrho_5(x^2 - \sqrt{-1}y^2).$$

Since Θ_- is anti-holomorphic, $\mathcal{A} \in \mathfrak{K}_{-,-}^{\mathfrak{A}}$. We have:

$$\begin{array}{ll}
\nabla_{e_1}e_2 = -\nabla_{f_1}f_2 = \varrho_5(x^2 e_2 - y^2 f_2), & \nabla_{e_1}f_2 = \nabla_{f_1}e_2 = \varrho_5(y^2 e_2 + x^2 f_2),\\
\nabla_{e_2}e_1 = -\nabla_{f_2}f_1 = \varrho_5(x^2 e_2 - y^2 f_2), & \nabla_{f_2}e_1 = \nabla_{e_2}f_1 = \varrho_5(y^2 e_2 + x^2 f_2),\\
\mathcal{A}(e_2,e_1)e_2 = -\mathcal{A}(e_2,f_1)f_2 = \varrho_5 e_2, & \mathcal{A}(f_2,e_1)e_2 = -\mathcal{A}(f_2,f_1)f_2 = -\varrho_5 f_2,\\
\mathcal{A}(e_2,e_1)f_2 = \mathcal{A}(e_2,f_1)e_2 = \varrho_5 f_2, & \mathcal{A}(f_2,e_1)f_2 = \mathcal{A}(f_2,f_1)e_2 = \varrho_5 e_2,\\
\mathcal{A}(e_2,f_2)e_1 = 2\varrho_5 f_2, & \mathcal{A}(e_2,f_2)f_1 = -2\varrho_5 e_2,\\
\rho_{13}(e_1,e_2) = 2\varrho_5, & \rho_{13}(f_1,f_2) = 2\varrho_5,\\
\tau = 0, & \tau_-^{\mathfrak{A}} = 0.
\end{array}$$

We have $\rho_{13} \in S_{0,+}^{2,\mathcal{U}_-} \oplus \Lambda_{0,+}^{2,\mathcal{U}_-}$. Since ρ_{13} is not symmetric, ρ_{13} has a non-zero component in $\Lambda_{0,+}^{2,\mathcal{U}_-}$. Assertion (3) follows because $S_{0,+}^{2,\mathcal{U}_-}$ and $\Lambda_{0,+}^{2,\mathcal{U}_-}$

are inequivalent and irreducible modules for the group $\mathcal{U}_-^\star$ and because Ξ is a morphism of modules for the group $\mathcal{U}_-^\star$. $\square$

It follows from Lemma 5.4.1 that we have surjective maps

$$0 \to \ker(\rho) \cap \mathfrak{K}_{-,-}^{\mathfrak{A}} \to \mathfrak{K}_{-,-}^{\mathfrak{A}} \to S_-^2 \oplus \Lambda_-^2 \to 0,$$
$$0 \to \ker(\rho) \cap \ker(\rho_{13}) \cap \mathfrak{K}_{-,+}^{\mathfrak{A}} \to \mathfrak{K}_{-,+}^{\mathfrak{A}} \to 2 \cdot (S_{0,+}^{2,\mathcal{U}_-} \oplus \Lambda_{0,+}^{2,\mathcal{U}_-} \oplus \mathbb{R}) \to 0.$$

Recall that we have defined:

$$\begin{aligned}
W_{-,9}^{\mathfrak{A}} &:= \{A \in \mathfrak{K}_{-,+}^{\mathfrak{A}} : A(x,y,z,w) = -A(x,y,w,z)\} \cap \ker(\rho), \\
W_{-,10}^{\mathfrak{A}} &:= \{A \in \mathfrak{K}_{-,+}^{\mathfrak{A}} : A(x,y,z,w) = A(x,y,w,z)\} \cap \ker(\rho), \\
W_{-,11}^{\mathfrak{A}} &:= \mathfrak{K}_{-,+}^{\mathfrak{A}} \cap (W_{-,9}^{\mathfrak{A}})^\perp \cap (W_{-,10}^{\mathfrak{A}})^\perp \cap \ker(\rho_{13}) \cap \ker(\rho), \\
W_{-,12}^{\mathfrak{A}} &:= \mathfrak{K}_{-,-}^{\mathfrak{A}} \cap \ker(\rho).
\end{aligned}$$

It is then clear that

$$W_{-,9}^{\mathfrak{A}} = \mathfrak{K}_{-,+}^{\mathfrak{A}} \cap W_6^{\mathcal{O}}, \quad W_{-,10}^{\mathfrak{A}} = \mathfrak{K}_{-,+}^{\mathfrak{A}} \cap W_7^{\mathcal{O}}, \quad W_{-,11}^{\mathfrak{A}} = \mathfrak{K}_{-,+}^{\mathfrak{A}} \cap W_8^{\mathcal{O}}.$$

Lemma 5.4.2 *Let $m \geq 4$.*

(1) If $m \geq 6$, then $W_{-,12}^{\mathfrak{A}} \cap \Xi\mathfrak{Z}_{-,h} \neq \{0\}$.
(2) $W_{-,9}^{\mathfrak{A}} \cap \Xi\mathfrak{Z}_{-,a} \neq \{0\}$.
(3) $W_{-,10}^{\mathfrak{A}} \cap \Xi\mathfrak{Z}_{-,a} \neq \{0\}$.
(4) If $m \geq 6$, then $W_{-,11}^{\mathfrak{A}} \cap \Xi\mathfrak{Z}_{-,a} \neq \{0\}$.
(5) The spaces $W_{-,9}^{\mathfrak{A}}$ and $W_{-,10}^{\mathfrak{A}}$ are isomorphic modules for the group $\mathcal{U}_-$.

Proof. To prove Assertion (1), we take ${\Theta_{-,11}}^2 = x^3 + \sqrt{-1}y^3$; we shall let $\mathcal{A} := \Xi\Theta_-$ be the resulting curvature operator at 0. Since Θ_- is holomorphic, $\mathcal{A} \in \mathfrak{K}_{-,-}^{\mathfrak{A}}$. We have:

$$\begin{aligned}
&\nabla_{e_1}e_1 = -\nabla_{f_1}f_1 = x^3e_2 + y^3f_2, \ \nabla_{e_1}f_1 = \nabla_{f_1}e_1 = -y^3e_2 + x^3f_2, \\
&\mathcal{A}(e_3,e_1)e_1 = -\mathcal{A}(e_3,f_1)f_1 = e_2, \ \mathcal{A}(f_3,e_1)e_1 = -\mathcal{A}(f_3,f_1)f_1 = f_2, \\
&\mathcal{A}(e_3,e_1)f_1 = \mathcal{A}(e_3,f_1)e_1 = f_2, \quad \mathcal{A}(f_3,e_1)f_1 = \mathcal{A}(f_3,f_1)e_1 = -e_2.
\end{aligned}$$

Since $\rho = 0$, $0 \neq \mathcal{A} \in W_{-,12}^{\mathfrak{A}}$; this establishes Assertion (1).

Next, we consider:

$$\begin{aligned}
&{\Theta_{-,11}}^2 = \varrho_1(x^1 - \sqrt{-1}y^1), \qquad {\Theta_{-,11}}^1 = \varrho_3(x^2 - \sqrt{-1}y^2), \\
&{\Theta_{-,12}}^1 = {\Theta_{-,21}}^1 = \varrho_2(x^1 - \sqrt{-1}y^1).
\end{aligned}$$

This implies:

$$\begin{aligned}
&\nabla_{e_1}e_1 = -\nabla_{f_1}f_1 = \varrho_1(x^1e_2 - y^1f_2) + \varrho_3(x^2e_1 - y^2f_1),\\
&\nabla_{f_1}e_1 = \nabla_{e_1}f_1 = \varrho_1(y^1e_2 + x^1f_2) + \varrho_3(y^2e_1 + x^2f_1),\\
&\nabla_{e_1}e_2 = -\nabla_{f_1}f_2 = \nabla_{e_2}e_1 = -\nabla_{f_2}f_1 = \varrho_2(x^1e_1 - y^1f_1),\\
&\nabla_{f_1}e_2 = \nabla_{e_1}f_2 = \nabla_{e_2}f_1 = \nabla_{f_2}e_1 = \varrho_2(y^1e_1 + x^1f_1).
\end{aligned}$$

Since Θ_- is anti-holomorphic, $\mathcal{A} \in \mathfrak{K}^{\mathfrak{A}}_{-,+}$. We have:

$$\begin{aligned}
&\mathcal{A}(e_1,f_1)e_1 = 2\varrho_1f_2, && \mathcal{A}(e_1,f_1)f_1 = -2\varrho_1e_2,\\
&\mathcal{A}(e_1,f_1)e_2 = 2\varrho_2f_1, && \mathcal{A}(e_1,f_1)f_2 = -2\varrho_2e_1,\\
&\mathcal{A}(e_1,e_2)e_1 = \varrho_2e_1 - \varrho_3e_1, && \mathcal{A}(e_1,e_2)f_1 = \varrho_2f_1 - \varrho_3f_1,\\
&\mathcal{A}(e_1,f_2)e_1 = \varrho_2f_1 + \varrho_3f_1, && \mathcal{A}(e_1,f_2)f_1 = -\varrho_2e_1 - \varrho_3e_1,\\
&\mathcal{A}(f_1,f_2)e_1 = \varrho_2e_1 - \varrho_3e_1, && \mathcal{A}(f_1,f_2)f_1 = \varrho_2f_1 - \varrho_3f_1,\\
&\mathcal{A}(f_1,e_2)f_1 = \varrho_2e_1 + \varrho_3e_1, && \mathcal{A}(f_1,e_2)e_1 = -\varrho_2f_1 - \varrho_3f_1.
\end{aligned}$$

Consequently:

$$\begin{aligned}
&A(e_1,f_1,e_1,f_2) = -A(e_1,f_1,f_1,e_2) = 2\varrho_1\varepsilon_{22},\\
&A(e_1,f_1,e_2,f_1) = -A(e_1,f_1,f_2,e_1) = 2\varrho_2\varepsilon_{11},\\
&A(e_1,e_2,e_1,e_1) = A(e_1,e_2,f_1,f_1) = (\varrho_2 - \varrho_3)\varepsilon_{11},\\
&A(e_1,f_2,e_1,f_1) = -A(e_1,f_2,f_1,e_1) = (\varrho_2 + \varrho_3)\varepsilon_{11},\\
&A(f_1,f_2,e_1,e_1) = A(f_1,f_2,f_1,f_1) = (\varrho_2 - \varrho_3)\varepsilon_{11},\\
&A(f_1,e_2,f_1,e_1) = -\mathcal{A}(f_1,e_2,e_1,f_1) = (\varrho_2 + \varrho_3)\varepsilon_{11}.
\end{aligned}$$

We take $\vec{\varrho} = (-\frac{1}{2}\varepsilon_{22}, -\frac{1}{2}\varepsilon_{11}, -\frac{1}{2}\varepsilon_{11})$ to create A_1 with:

$$\begin{aligned}
&A_1(e_1,f_1,e_1,f_2) = -A_1(e_1,f_1,f_1,e_2) = -1,\\
&A_1(e_1,f_1,e_2,f_1) = -A_1(e_1,f_1,f_2,e_1) = -1,\\
&A_1(e_1,e_2,e_1,e_1) = A_1(e_1,e_2,f_1,f_1) = 0,\\
&A_1(e_1,f_2,e_1,f_1) = -A_1(e_1,f_2,f_1,e_1) = -1,\\
&A_1(f_1,f_2,e_1,e_1) = A_1(f_1,f_2,f_1,f_1) = 0,\\
&A_1(f_1,e_2,f_1,e_1) = -A_1(f_1,e_2,e_1,f_1) = -1.
\end{aligned}$$

This yields:

$$\begin{aligned}
&\rho(A_1)(e_1,e_2) = \rho(A_1)(e_2,e_1) = \varepsilon_{11},\\
&\rho(A_1)(f_1,f_2) = \rho(A_1)(f_2,f_1) = \varepsilon_{11}.
\end{aligned}$$

Interchanging the roles of the indices "1" and "2" then creates A_2 with:

$$\begin{aligned}
&A_2(e_2,f_2,e_2,f_1) = -A_2(e_2,f_2,f_2,e_1) = -1,\\
&A_2(e_2,f_2,e_1,f_2) = -A_2(e_2,f_2,f_1,e_2) = -1,\\
&A_2(e_2,e_1,e_2,e_2) = A_2(e_2,e_1,f_2,f_2) = 0,\\
&A_2(e_2,f_1,e_2,f_2) = -A_2(e_2,f_1,f_2,e_2) = -1,\\
&A_2(f_2,f_1,e_2,e_2) = A_2(f_2,f_1,f_2,f_2) = 0,\\
&A_2(f_2,e_1,f_2,e_2) = -A_2(f_2,e_1,e_2,f_2) = -1.
\end{aligned}$$

This yields:

$$\begin{aligned}
&\rho(A_2)(e_2,e_1) = \rho(A_2)(e_1,e_2) = \varepsilon_{22},\\
&\rho(A_2)(f_2,f_1) = \rho(A_2)(f_1,f_2) = \varepsilon_{22}.
\end{aligned}$$

These tensors are anti-symmetric in the last two indices so $\rho_{13} = -\rho$. We prove Assertion (2) by verifying that:

$$0 \neq A_1 - \varepsilon_{11}\varepsilon_{22}A_2 \in W^{\mathfrak{A}}_{-,9}.$$

Next, we take $\varrho = (\frac{1}{2}\varepsilon_{22}, -\frac{1}{2}\varepsilon_{11}, \frac{1}{2}\varepsilon_{11})$ to create a tensor with

$$\begin{aligned}
&A_3(e_1,f_1,e_1,f_2) = -A_3(e_1,f_1,f_1,e_2) = 1,\\
&A_3(e_1,f_1,e_2,f_1) = -A_3(e_1,f_1,f_2,e_1) = -1,\\
&A_3(e_1,e_2,e_1,e_1) = A_3(e_1,e_2,f_1,f_1) = -1,\\
&A_3(e_1,f_2,e_1,f_1) = -A_3(e_1,f_2,f_1,e_1) = 0,\\
&A_3(f_1,f_2,e_1,e_1) = A_3(f_1,f_2,f_1,f_1) = -1,\\
&A_3(f_1,e_2,f_1,e_1) = -A_3(f_1,e_2,e_1,f_1) = 0.
\end{aligned}$$

This yields:

$$\begin{aligned}
&\rho(A_3)(e_1,e_2) = \rho(A_3)(f_1,f_2) = \varepsilon_{11},\\
&\rho(A_3)(e_2,e_1) = \rho(A_3)(f_2,f_1) = -\varepsilon_{11}.
\end{aligned}$$

We interchange the roles of the indices "1" and "2" to create A_4 with:

$$\begin{aligned}
&A_4(e_2,f_2,e_2,f_1) = -A_4(e_2,f_2,f_2,e_1) = 1,\\
&A_4(e_2,f_2,e_1,f_2) = -A_4(e_2,f_2,f_1,e_2) = -1,\\
&A_4(e_2,e_1,e_2,e_2) = A_4(e_2,e_1,f_2,f_2) = -1,\\
&A_4(e_2,f_1,e_2,f_2) = -A_4(e_2,f_1,f_2,e_2) = 0,\\
&A_4(f_2,f_1,e_2,e_2) = A_4(f_2,f_1,f_2,f_2) = -1,\\
&A_4(f_2,e_1,f_2,e_2) = -A_4(f_2,e_1,e_2,f_2) = 0.
\end{aligned}$$

This yields:

$$\rho(A_4)(e_2, e_1) = \rho(A_4)(f_2, f_1) = \varepsilon_{22},$$
$$\rho(A_4)(e_1, e_2) = \rho(A_4)(f_1, f_2) = -\varepsilon_{22}.$$

These two tensors are symmetric in the last two indices so $\rho_{13} = \rho$. We complete the proof of Assertion (3) by checking:

$$0 \neq A_3 + \varepsilon_{11}\varepsilon_{22}A_4 \in W^{\mathfrak{A}}_{-,10}.$$

Let $m \geq 6$. Let $\Theta_{-,11}{}^2 = x^3 - \sqrt{-1}y^3$ define $\mathcal{A}_5 := \Xi\Theta_-$. Since Θ_- is anti-holomorphic, $\mathcal{A} \in \mathfrak{K}^{\mathfrak{A}}_{-,+}$. We have:

$$\begin{aligned}
&\nabla_{e_1}e_1 = -\nabla_{f_1}f_1 = x^3e_2 - y^3f_2, \quad \nabla_{e_1}f_1 = \nabla_{f_1}e_1 = y^3e_2 + x^3f_2,\\
&\mathcal{A}_5(e_3, e_1)e_1 = -\mathcal{A}_5(e_3, f_1)f_1 = e_2, \mathcal{A}_5(f_3, e_1)e_1 = -\mathcal{A}_5(f_3, f_1)f_1 = -f_2,\\
&\mathcal{A}_5(e_3, e_1)f_1 = \mathcal{A}_5(e_3, f_1)e_1 = f_2, \quad \mathcal{A}_5(f_3, e_1)f_1 = \mathcal{A}_5(f_3, f_1)e_1 = e_2,\\
&\rho_{13}(\mathcal{A}_5) = \rho(\mathcal{A}_5) = 0.
\end{aligned}$$

Consequently, we may conclude that:

$$\mathcal{A}_5 \in W^{\mathfrak{A}}_{-,9} \oplus W^{\mathfrak{A}}_{-,10} \oplus W^{\mathfrak{A}}_{-,11}.$$

To prove Assertion (4), it suffices to show $\mathcal{A}_5$ has a non-zero component in $W^{\mathfrak{A}}_{-,11}$. Suppose to the contrary that $\mathcal{A}_5 \in W^{\mathfrak{A}}_{-,9} \oplus W^{\mathfrak{A}}_{-,10}$. We may then decompose $\mathcal{A}_5 = \mathcal{A}_9 + \mathcal{A}_{10}$ where $\mathcal{A}_9 \in W^{\mathfrak{A}}_{-,9}$ and $\mathcal{A}_{10} \in W^{\mathfrak{A}}_{-,10}$. We lower indices to define A_5, A_9, and A_{10}. We then have

$$\begin{aligned}
A_5(x, y, z, w) + A_5(x, y, w, z) &= A_9(x, y, z, w) + A_9(x, y, w, z)\\
&\quad + A_{10}(x, y, z, w) + A_{10}(x, y, w, z)\\
&= 2A_{10}(x, y, z, w).
\end{aligned}$$

This yields a contradiction; the Bianchi identity is not satisfied by this tensor:

$$\begin{aligned}
&A_{10}(f_3, f_1, e_2, e_1) + A_{10}(f_1, e_2, f_3, e_1) + A_{10}(e_2, f_3, f_1, e_1)\\
&= \tfrac{1}{2}\{A_5(f_3, f_1, e_2, e_1) + A_5(f_3, f_1, e_1, e_2) + A_5(f_1, e_2, f_3, e_1)\\
&\quad + A_5(f_1, e_2, e_1, f_3) + A_5(e_2, f_3, f_1, e_1) + A_5(e_2, f_3, e_1, f_1)\}\\
&= 0 + \tfrac{1}{2} + 0 + 0 + 0 + 0 \neq 0.
\end{aligned}$$

Assertion (4) now follows.

Let $A \in \mathfrak{K}^{\mathfrak{A}}_{-,+}$. We define $A^*(x, y, z, w) := A(x, y, z, J_-w)$. Clearly A^* satisfies Equation (4.1.a) and Equation (4.1.b) so $A^* \in \mathfrak{A}$. Since $A \in \mathfrak{K}^{\mathfrak{A}}_{-,+}$,

$$\begin{aligned}
&A^*(x, y, J_-z, J_-w) = A(x, y, J_-z, J_-J_-w) = A(x, y, z, J_-w)\\
&\quad = A^*(x, y, z, w),
\end{aligned}$$

$$J_-^* A^*(x,y,z,w) = A(J_-x, J_-y, J_-z, J_-J_-w) = A(x,y,z,J_-w)$$
$$= A^*(x,y,z,w).$$

This shows that $A^* \in \mathfrak{K}^{\mathfrak{A}}_{-,+}$ as well. Clearly $\rho_{13}(A) = 0$ if and only if $\rho_{13}(A^*) = 0$. Since $A^*(x,y,z,w) = -A(x,y,J_-z,w)$,

$$\rho(A^*)(y,z) = -\rho(A)(y,J_-z)$$

and thus $\rho(A) = 0$ if and only if $\rho(A^*) = 0$.

Suppose $A_9 \in W^{\mathfrak{A}}_{-,9}$ and $A_{10} \in W^{\mathfrak{A}}_{-,10}$. We compute:

$$A_9^*(x,y,w,z) = A_9(x,y,w,J_-z) = -A_9(x,y,J_-w,z)$$
$$= A_9(x,y,z,J_-w) = A_9^*(x,y,z,w),$$
$$A_{10}^*(x,y,w,z) = A_{10}(x,y,w,J_-z) = -A_{10}(x,y,J_-w,z)$$
$$= -A_{10}(x,y,z,J_-w) = -A_{10}^*(x,y,z,w).$$

Thus $A_9^* \in W^{\mathfrak{A}}_{-,10}$ and $A_{10}^* \in W^{\mathfrak{A}}_{-,9}$. This shows that the map $A \to A^*$ is an endomorphism of modules for the group $\mathcal{U}_-$ intertwining the modules $W^{\mathfrak{A}}_{-,9}$ and $W^{\mathfrak{A}}_{-,10}$; Assertion (5) now follows. □

The Assertions of Theorem 1.5.2 for the groups $\mathcal{U}_-$ and $\mathcal{U}_-^\star$ for $m \geq 6$ will follow from the following result. We restrict to the generic case $m \geq 6$ to avoid the minor technical fuss in dealing with dimension $m = 4$; the geometrical realization results in the complex Kähler setting of Theorem 1.5.3 also follow:

Theorem 5.4.1 *Let $m \geq 6$.*

(1) We have isomorphisms of modules for the group $\mathcal{U}_-$:

$$\mathfrak{K}^{\mathfrak{A}}_{-,-} \approx S_-^{2,\mathcal{U}_-} \oplus \Lambda_-^{2,\mathcal{U}_-} \oplus W^{\mathfrak{A}}_{-,12},$$
$$\mathfrak{K}^{\mathfrak{A}}_{-,+} \approx 2\cdot\mathbb{R} \oplus 4\cdot S_{0,+}^{2,\mathcal{U}_-} \oplus 2\cdot W^{\mathfrak{A}}_{-,9} \oplus W^{\mathfrak{A}}_{-,11}.$$

(2) The modules $\{S_-^{2,\mathcal{U}_-}, \Lambda_-^{2,\mathcal{U}_-}, W^{\mathfrak{A}}_{-,12}, \mathbb{R}, S_{0,+}^{2,\mathcal{U}_-}, W^{\mathfrak{A}}_{-,9}, W^{\mathfrak{A}}_{-,11}\}$ are inequivalent and irreducible modules for the group $\mathcal{U}_-$.

(3) We have isomorphisms of the modules for the group $\mathcal{U}_-^\star$:

$$\mathfrak{K}^{\mathfrak{A}}_{-,-} = S_-^{2,\mathcal{U}_-} \oplus \Lambda_-^{2,\mathcal{U}_-} \oplus W^{\mathfrak{A}}_{-,12},$$
$$\mathfrak{K}^{\mathfrak{A}}_{-,+} = \mathbb{R} \oplus \chi \oplus 2\cdot S_{0,+}^{2,\mathcal{U}_-} \oplus 2\cdot \Lambda_{0,+}^{2,\mathcal{U}_-} \oplus W^{\mathfrak{A}}_{-,9} \oplus W^{\mathfrak{A}}_{-,10} \oplus W^{\mathfrak{A}}_{-,11}.$$

(4) The modules $\{S_-^{2,\mathcal{U}_-}, \Lambda_-^{2,\mathcal{U}_-}, W^{\mathfrak{A}}_{-,12}, \mathbb{R}, \chi, S_{0,+}^{2,\mathcal{U}_-}, \Lambda_{0,+}^{2,\mathcal{U}_-}, W^{\mathfrak{A}}_{-,9}, W^{\mathfrak{A}}_{-,10}, W^{\mathfrak{A}}_{-,11}\}$ are irreducible and inequivalent modules for the group $\mathcal{U}_-^\star$.

(5) *Let $\mathcal{A} \in \mathfrak{K}_-^{\mathfrak{A}}$. We can choose linear functions $\{u_{ij}{}^k, v_{ij}{}^k\}$ on $\mathbb{R}^m$ with $u_{ij}{}^k = u_{ji}{}^k$ and $v_{ij}{}^k = v_{ji}{}^k$ defining an affine connection ∇ on $\mathbb{R}^m$ so*

$$\begin{aligned}\nabla_{e_i} e_j &= -\nabla_{f_i} f_j = u_{ij}{}^k e_k + v_{ij}{}^k f_k,\\ \nabla_{f_i} e_j &= \nabla_{e_i} f_j = -v_{ij}{}^k e_k + u_{ij}{}^k f_k.\end{aligned}$$

Then ∇ is Kähler affine with $\mathcal{R}^\nabla(0) = \mathcal{A}$. Furthermore, if $\mathcal{A}$ belongs to $\mathfrak{K}_{-,-}^{\mathfrak{A}}$, then $\{u, v\}$ can be chosen to be holomorphic so that $\mathcal{R}(P) \in \mathfrak{K}_{-,-}^{\mathfrak{A}}$ for all $P \in \mathbb{R}^m$.

Proof. We have the following isomorphisms of modules for the group $\mathcal{U}_-$:

$$\mathbb{R} \cdot \tau \approx \mathbb{R} \cdot \tau_-^{\mathfrak{A}}, \quad S_{0,+}^{2,\mathcal{U}_-} \approx \Lambda_{0,+}^{2,\mathcal{U}_-}, \quad W_{-,9}^{\mathfrak{A}} \approx W_{-,10}^{\mathfrak{A}}.$$

The decomposition of Assertion (1) follows from Lemma 5.4.1 and Lemma 5.4.2. Lemma 2.2.2 then permits us to estimate:

$$\dim\{\mathcal{I}_2^{\mathcal{U}_-}(\mathfrak{K}_-^{\mathfrak{A}})\} \geq 3 + 3 + 10 + 3 + 1 = 20.$$

On the other hand, by Lemma 5.1.1, we have the reverse inequality; this establishes Remark 5.1.1 for the group $\mathcal{U}_-$. The fact that the modules

$$\{S_-^2, \Lambda_-^2, W_{-,12}^{\mathfrak{A}}, \mathbb{R}, S_{0,+}^{2,\mathcal{U}_-}, W_{-,9}^{\mathfrak{A}}, W_{-,11}^{\mathfrak{A}}\}$$

are inequivalent irreducible modules for the group $\mathcal{U}_-$ then follows from Lemma 2.2.2. This proves Assertion (2).

Since $\tau_-^{\mathfrak{A}}$ changes sign if $J_-T = -TJ_-$, $\tau_-^{\mathfrak{A}}$ defines a representation space of dimension $r = 1$ given by the non-trivial $\mathbb{Z}_2$ character χ. Thus τ and $\tau_-^{\mathfrak{A}}$ define inequivalent representations. Similarly, we have

$$\Lambda_{0,+}^{2,\mathcal{U}_-} \approx S_{0,+}^{2,\mathcal{U}_-} \otimes \chi \quad \text{and} \quad W_{-,10}^{\mathfrak{A}} \approx W_{-,9}^{\mathfrak{A}} \otimes \chi$$

as modules for the group $\mathcal{U}_-^\star$. The decomposition of Assertion (3) now follows and we may estimate thereby

$$\dim\{\mathcal{I}_2^{\mathcal{U}_-^\star}(\mathfrak{K}_-^{\mathfrak{A}})\} \geq 3 + 2 + 6 + 3 = 14.$$

Again, the reverse inequality is provided by Lemma 5.1.1; Assertion (4) now follows from Lemma 2.2.2. This also completes the proof of Remark 5.1.1 for the group $\mathcal{U}_-^\star$.

We have in fact shown that $\Xi(\mathfrak{Z}_-) = \mathfrak{K}_-^{\mathfrak{A}}$. If $A \in \mathfrak{K}_{-,-}^{\mathfrak{A}}$, we took $u + \sqrt{-1}v$ to be holomorphic. We used Lemma 5.3.1 to ensure $\mathcal{R}$ belongs to $\mathfrak{K}_{-,-}^{\mathfrak{A}}$ for every point of $\mathbb{R}^m$. Assertion (5) follows. □

5.5 Affine Para-Kähler Curvature Operators

We now pass to the para-complex setting. Let $(V, \langle\cdot,\cdot\rangle, J_+)$ be a para-Hermitian vector space. We take a basis so $J_+e_i = f_i$ and $J_+f_i = e_i$. As in the pseudo-Hermitian setting, let Θ_+ be defined by the data $(u_{ij}{}^k, v_{ij}{}^k)$ where u and v are linear functions with $u_{ij}{}^k = u_{ji}{}^k$ and $v_{ij}k = v_{ji}{}^k$. We define a connection $\nabla = \nabla^{\Theta_+}$ where

$$\nabla_{e_i}e_j = \nabla_{f_i}f_j = u_{ij}{}^k e_k + v_{ij}{}^k f_k,$$
$$\nabla_{e_i}f_j = \nabla_{f_i}e_j = v_{ij}{}^k e_k + u_{ij}{}^k f_k.$$

This connection is para-Kähler and the map Ξ sending Θ_+ to the curvature of ∇ at 0 is a map of $\mathcal{U}_+^\star$ representation spaces by Lemma 5.3.1. If Θ_+ is para-holomorphic, the curvature lies in $\mathfrak{K}_{+,+}^{\mathfrak{A}}$ for all points of $\mathbb{R}^m$, while if Θ_+ is anti-para-holomorphic, then the curvature lies in $\mathfrak{K}_{+,-}^{\mathfrak{A}}$ at the origin. We extend Lemma 5.4.1 from the pseudo-Hermitian to the para-Hermitian setting.

Lemma 5.5.1

(1) $(\tau \oplus \tau_+^{\mathfrak{A}}) \circ \Xi : \mathfrak{Z}_{+,a} \to \mathbb{R} \oplus \chi \to 0.$

(2) $\rho \circ \Xi : \mathfrak{Z}_{+,h} \to S_+^{2,\mathcal{U}_+} \oplus \Lambda_+^{2,\mathcal{U}_+} \to 0.$

(3) $(\rho \oplus \rho_{13}) \circ \Xi : \mathfrak{Z}_{+,a} \to S_{0,-}^{2,\mathcal{U}_+} \oplus \Lambda_{0,-}^{2,\mathcal{U}_+} \oplus S_{0,-}^{2,\mathcal{U}_+} \oplus \Lambda_{0,-}^{2,\mathcal{U}_+} \to 0.$

Proof. We shall always let the background metric be given by:

$$g = dx_1^2 - dy_1^2 + \ldots + dx_{\bar{m}}^2 - dy_{\bar{m}}^2.$$

We first take

$$\Theta_{+,11}{}^1 = \varrho_1(x^1 - \iota y^1) \quad \text{and} \quad \Theta_{+,12}{}^2 = \Theta_{21}{}^2 = \varrho_2(y^1 - \iota x^1).$$

Let $\mathcal{A} := \Xi\Theta_+ \in \mathfrak{K}_{+,-}^{\mathfrak{A}}$. We may then establish Assertion (1) by computing:

$$\begin{array}{ll}
\nabla_{e_1}e_1 = \nabla_{f_1}f_1 = \varrho_1(x^1e_1 - y^1f_1), & \nabla_{e_1}f_1 = \nabla_{f_1}e_1 = \varrho_1(-y^1e_1 + x^1f_1), \\
\nabla_{e_1}e_2 = \nabla_{f_1}f_2 = \varrho_2(y^1e_2 - x^1f_2), & \nabla_{e_2}e_1 = \nabla_{f_2}f_1 = \varrho_2(y^1e_2 - x^1f_2), \\
\nabla_{e_1}f_2 = \nabla_{f_1}e_2 = \varrho_2(-x^1e_2 + y^1f_2), & \nabla_{e_2}f_1 = \nabla_{f_2}e_1 = \varrho_2(-x^1e_2 + y^1f_2), \\
\mathcal{A}(e_1,f_1)e_1 = 2\varrho_1f_1, & \mathcal{A}(e_1,f_1)f_1 = 2\varrho_1e_1, \\
\mathcal{A}(e_1,e_2)e_1 = \mathcal{A}(e_1,f_2)f_1 = -\varrho_2f_2, & \mathcal{A}(e_1,e_2)f_1 = \mathcal{A}(e_1,f_2)e_1 = -\varrho_2e_2, \\
\mathcal{A}(f_1,e_2)e_1 = \mathcal{A}(f_1,f_2)f_1 = \varrho_2e_2, & \mathcal{A}(f_1,e_2)f_1 = \mathcal{A}(f_1,f_2)e_1 = \varrho_2f_2, \\
\mathcal{A}(e_1,f_1)f_2 = -2\varrho_2f_2, & \mathcal{A}(e_1,f_1)e_2 = -2\varrho_2e_2, \\
\tau = -4\varrho_1, & \tau_+^{\mathfrak{A}} = -4\varrho_2.
\end{array}$$

Next take $\Theta_{+,11}{}^1 = \varrho_1(x^2 + \iota y^2)$ and $\Theta_{+,22}{}^2 = \varrho_2(x^1 + \iota y^1)$. Again, let $\mathcal{A} := \Xi\Theta_+ \in \mathfrak{K}^{\mathfrak{A}}_{+,+}$. Then:

$$\nabla_{e_1}e_1 = \nabla_{f_1}f_1 = \varrho_1(x^2e_1 + y^2f_1),\ \nabla_{e_1}f_1 = \nabla_{f_1}e_1 = \varrho_1(y^2e_1 + x^2f_1),$$
$$\nabla_{e_2}e_2 = \nabla_{f_2}f_2 = \varrho_2(x^1e_2 + y^1f_2),\ \nabla_{e_2}f_2 = \nabla_{f_2}e_2 = \varrho_2(y^1e_2 + x^1f_2),$$

$$\begin{aligned}
&\mathcal{A}(e_2,e_1)e_1 = \mathcal{A}(e_2,f_1)f_1 = \varrho_1e_1, && \mathcal{A}(f_2,e_1)e_1 = \mathcal{A}(f_2,f_1)f_1 = \varrho_1f_1,\\
&\mathcal{A}(e_2,e_1)f_1 = \mathcal{A}(e_2,f_1)e_1 = \varrho_1f_1, && \mathcal{A}(f_2,e_1)f_1 = \mathcal{A}(f_2,f_1)e_1 = \varrho_1e_1,\\
&\mathcal{A}(e_1,e_2)e_2 = \mathcal{A}(e_1,f_2)f_2 = \varrho_2e_2, && \mathcal{A}(f_1,e_2)e_2 = \mathcal{A}(f_1,f_2)f_2 = \varrho_2f_2,\\
&\mathcal{A}(e_1,e_2)f_2 = \mathcal{A}(e_1,f_2)e_2 = \varrho_2f_2, && \mathcal{A}(f_1,e_2)f_2 = \mathcal{A}(f_1,f_2)e_2 = \varrho_2e_2,
\end{aligned}$$

$$\begin{aligned}
&\rho(e_2,e_1) = -2\varrho_1, && \rho(f_2,f_1) = -2\varrho_1,\\
&\rho(e_1,e_2) = -2\varrho_2, && \rho(f_1,f_2) = -2\varrho_2.
\end{aligned}$$

If we take $\varrho_1 = \varrho_2$, then $\rho \in S^{2,\mathcal{U}_+}_+$; if we take $\varrho_1 = -\varrho_2$, then $\rho \in \Lambda^{2,\mathcal{U}_+}_+$. This proves Assertion (2).

We begin the proof of Assertion (3) by taking:

$$\begin{aligned}
\Theta_{+,11}{}^1 &= \varrho_1(x^1 - \iota y^1) + \varrho_2(x^2 - \iota y^2),\\
\Theta_{+,22}{}^2 &= \varrho_3(x^2 - \iota y^2) + \varrho_4(x^1 - \iota y^1).
\end{aligned}$$

We then have

$$\begin{aligned}
\nabla_{e_1}e_1 &= \nabla_{f_1}f_1 = (\varrho_1x^1 + \varrho_2x^2)e_1 - (\varrho_1y^1 + \varrho_2y^2)f_1,\\
\nabla_{e_1}f_1 &= \nabla_{f_1}e_1 = -(\varrho_1y^1 + \varrho_2y^2)e_1 + (\varrho_1x^1 + \varrho_2x^2)f_1,\\
\nabla_{e_2}e_2 &= \nabla_{f_2}f_2 = (\varrho_3x^2 + \varrho_4x^1)e_2 - (\varrho_3y^2 + \varrho_4y^1)f_2,\\
\nabla_{e_2}f_2 &= \nabla_{f_2}e_2 = -(\varrho_3y^2 + \varrho_4y^1)e_2 + (\varrho_3x^2 + \varrho_4x^1)f_2.
\end{aligned}$$

Let $\mathcal{A} := \Xi\Theta_+ \in \mathfrak{K}^{\mathfrak{A}}_{+,-}$. Then taking into account the fact that $\{e_1, e_2\}$ are spacelike while $\{f_1, f_2\}$ are timelike, we compute:

$$\begin{aligned}
&\mathcal{A}(e_1,f_1)f_1 = 2\varrho_1e_1, && \mathcal{A}(e_1,f_1)e_1 = 2\varrho_1f_1,\\
&\mathcal{A}(e_2,e_1)e_1 = \mathcal{A}(e_2,f_1)f_1 = \varrho_2e_1, && \mathcal{A}(e_2,e_1)f_1 = \mathcal{A}(e_2,f_1)e_1 = \varrho_2f_1,\\
&\mathcal{A}(f_2,e_1)e_1 = \mathcal{A}(f_2,f_1)f_1 = -\varrho_2f_1, && \mathcal{A}(f_2,e_1)f_1 = \mathcal{A}(f_2,f_1)e_1 = -\varrho_2e_1,\\
&\mathcal{A}(e_2,f_2)f_2 = 2\varrho_3e_2, && \mathcal{A}(e_2,f_2)e_2 = 2\varrho_3f_2,\\
&\mathcal{A}(e_1,e_2)e_2 = \mathcal{A}(e_1,f_2)f_2 = \varrho_4e_2, && \mathcal{A}(e_1,e_2)f_2 = \mathcal{A}(e_1,f_2)e_2 = \varrho_4f_2,\\
&\mathcal{A}(f_1,e_2)e_2 = \mathcal{A}(f_1,f_2)f_2 = -\varrho_4f_2, && \mathcal{A}(f_1,e_2)f_2 = \mathcal{A}(f_1,f_2)e_2 = -\varrho_4e_2,
\end{aligned}$$

$$\begin{aligned}
&\rho(f_1,f_1)=-\rho(e_1,e_1)=2\varrho_1, &&\rho(e_2,e_1)=-\rho(f_2,f_1)=-2\varrho_2,\\
&\rho(f_2,f_2)=-\rho(e_2,e_2)=2\varrho_3, &&\rho(e_1,e_2)=-\rho(f_1,f_2)=-2\varrho_4,\\
&\rho_{13}(e_1,e_1)=-\rho_{13}(f_1,f_1)=2\varrho_1, &&\rho_{13}(e_2,e_1)=\rho_{13}(f_2,f_1)=0,\\
&\rho_{13}(e_2,e_2)=-\rho_{13}(f_2,f_2)=2\varrho_3, &&\rho_{13}(e_1,e_2)=\rho_{13}(f_1,f_2)=0,\\
&\tau=-4\varrho_1-4\varrho_3, &&\tau_+^{\mathfrak{A}}=0.
\end{aligned}$$

Note that $\tau=0$ implies $\langle\rho,g\rangle=\langle\rho_{13},g\rangle=0$. Let $\vec{\varrho}:=(\varrho_1,\varrho_2,\varrho_3,\varrho_4)$. Similarly, by Lemma 5.2.1, if $\tau_+^{\mathfrak{A}}=0$, then $\langle\rho,\Omega_+\rangle=\langle\rho_{13},\Omega_+\rangle=0$.

(1) If $\vec{\varrho}=(0,1,0,1)$, then $\tau=0$, $0\neq\rho(A)\in S^{2,\mathcal{U}_+}_{0,-}$, and $\rho_{13}=0$.
(2) If $\vec{\varrho}=(0,1,0,-1)$, then $\tau_+^{\mathfrak{A}}=0$, $0\neq\rho(A)\in\Lambda^{2,\mathcal{U}_+}_{0,-}$, and $\rho_{13}=0$.
(3) If $\vec{\varrho}=(1,0,-1,0)$, then $\tau=0$ and $0\neq\rho_{13}(A)\in S^{2,\mathcal{U}_+}_{0,-}$.

Thus we complete the proof of Assertion (3) by constructing an example where ρ_{13} has a non-zero component in $\Lambda^{2,\mathcal{U}_+}_{0,-}$. We take

$$\Theta_{+,12}{}^2=\Theta_{+,21}{}^2=\varrho_5(x^2-\iota y^2).$$

Set $\mathcal{A}:=\Xi\Theta_+\in\mathfrak{K}^{\mathfrak{A}}_{+,-}$. Then:

$$\begin{aligned}
&\nabla_{e_1}e_2=\nabla_{f_1}f_2=\varrho_5(x^2e_2-y^2f_2), &&\nabla_{e_1}f_2=\nabla_{f_1}e_2=\varrho_5(x^2f_2-y^2e_2),\\
&\nabla_{e_2}e_1=\nabla_{f_2}f_1=\varrho_5(x^2e_2-y^2f_2), &&\nabla_{f_2}e_1=\nabla_{e_2}f_1=\varrho_5(x^2f_2-y^2e_2),
\end{aligned}$$

$$\begin{aligned}
&\mathcal{A}(e_2,e_1)e_2=\mathcal{A}(e_2,f_1)f_2=\varrho_5e_2, &&\mathcal{A}(f_2,e_1)e_2=\mathcal{A}(f_2,f_1)f_2=-\varrho_5f_2,\\
&\mathcal{A}(e_2,e_1)f_2=\mathcal{A}(e_2,f_1)e_2=\varrho_5f_2, &&\mathcal{A}(f_2,e_1)f_2=\mathcal{A}(f_2,f_1)e_2=-\varrho_5e_2,\\
&\mathcal{A}(e_2,f_2)e_1=2\varrho_5f_2, &&\mathcal{A}(e_2,f_2)f_1=2\varrho_5e_2,
\end{aligned}$$

$$\rho_{13}(e_1,e_2)=2\varrho_5,\qquad\qquad \rho_{13}(f_1,f_2)=-2\varrho_5.$$

Since $\tau=0$ and $\tau_+^{\mathfrak{A}}=0$, $\rho(A)\in S^2_{0,-}\oplus\Lambda^2_{0,-}$. It is not symmetric and thus has a non-zero component in $\Lambda^{2,\mathcal{U}_+}_{0,-}$. Assertion (3) follows. □

We continue our discussion by extending Lemma 5.4.2 to this context. We use the notation of Definition 1.5.1:

Lemma 5.5.2

(1) If $m\geq 6$, then $W^{\mathfrak{A}}_{+,12}\cap\Xi\mathfrak{Z}_{+,h}\neq\{0\}$.
(2) $W^{\mathfrak{A}}_{+,9}\cap\Xi\mathfrak{Z}_{+,a}\neq\{0\}$.
(3) $W^{\mathfrak{A}}_{+,10}\cap\Xi\mathfrak{Z}_{+,a}\neq\{0\}$.
(4) If $m\geq 6$, then $W^{\mathfrak{A}}_{+,11}\cap\Xi\mathfrak{Z}_{+,a}\neq\{0\}$.

Proof. To prove Assertion (1), we set $\Theta_{+,11}{}^2=x^3+\iota y^3$. We then have

$$\begin{aligned}
&\nabla_{e_1}e_1=\nabla_{f_1}f_1=x^3e_2+y^3f_2, \quad \nabla_{e_1}f_1=\nabla_{f_1}e_1=y^3e_2+x^3f_2,\\
&\mathcal{A}(e_3,e_1)e_1=\mathcal{A}(e_3,f_1)f_1=e_2, \quad \mathcal{A}(f_3,e_1)e_1=\mathcal{A}(f_3,f_1)f_1=f_2,\\
&\mathcal{A}(e_3,e_1)f_1=\mathcal{A}(e_3,f_1)e_1=f_2, \quad \mathcal{A}(f_3,e_1)f_1=\mathcal{A}(f_3,f_1)e_1=e_2.
\end{aligned}$$

Since $\rho=0$, $0\neq\mathcal{A}\in W^{\mathfrak{A}}_{+,12}$; this establishes Assertion (1).

We clear the previous notation and take:

$$\begin{aligned}
&\Theta_{+,11}{}^2=\varrho_1(x^1-\iota y^1), \qquad \Theta_{+,11}{}^1=\varrho_3(x^2-\iota y^2),\\
&\Theta_{+,12}{}^1=\Theta_{+,21}{}^1=\varrho_2(x^1-\iota y^1).
\end{aligned}$$

This implies:

$$\begin{aligned}
&\nabla_{e_1}e_1=\nabla_{f_1}f_1=\varrho_1(x^1e_2-y^1f_2)+\varrho_3(x^2e_1-y^2f_1),\\
&\nabla_{f_1}e_1=\nabla_{e_1}f_1=\varrho_1(x^1f_2-y^1e_2)+\varrho_3(x^2f_1-y^2e_1),\\
&\nabla_{e_1}e_2=\nabla_{f_1}f_2=\nabla_{e_2}e_1=\nabla_{f_2}f_1=\varrho_2(x^1e_1-y^1f_1),\\
&\nabla_{f_1}e_2=\nabla_{e_1}f_2=\nabla_{e_2}f_1=\nabla_{f_2}e_1=\varrho_2(x^1f_1-y^1e_1).
\end{aligned}$$

Set $\mathcal{A}:=\Xi\Theta_+\in\mathfrak{K}^{\mathfrak{A}}_{+,-}$. Then:

$$\begin{aligned}
&\mathcal{A}(e_1,f_1)e_1=2\varrho_1f_2, &&\mathcal{A}(e_1,f_1)f_1=2\varrho_1e_2,\\
&\mathcal{A}(e_1,f_1)e_2=2\varrho_2f_1, &&\mathcal{A}(e_1,f_1)f_2=2\varrho_2e_1,\\
&\mathcal{A}(e_1,e_2)e_1=\varrho_2e_1-\varrho_3e_1, &&\mathcal{A}(e_1,e_2)f_1=\varrho_2f_1-\varrho_3f_1,\\
&\mathcal{A}(e_1,f_2)e_1=\varrho_2f_1+\varrho_3f_1, &&\mathcal{A}(e_1,f_2)f_1=\varrho_2e_1+\varrho_3e_1,\\
&\mathcal{A}(f_1,f_2)e_1=-\varrho_2e_1+\varrho_3e_1, &&\mathcal{A}(f_1,f_2)f_1=-\varrho_2f_1+\varrho_3f_1,\\
&\mathcal{A}(f_1,e_2)e_1=-\varrho_2f_1-\varrho_3f_1, &&\mathcal{A}(f_1,e_2)f_1=-\varrho_2e_1-\varrho_3e_1.
\end{aligned}$$

We take $(\varrho_1,\varrho_2,\varrho_3)=(-\frac12,-\frac12,-\frac12)$ to create A_1 with:

$$\begin{aligned}
&A_1(e_1,f_1,e_1,f_2)=1, &&A_1(e_1,f_1,f_1,e_2)=-1,\\
&A_1(e_1,f_1,e_2,f_1)=1, &&A_1(e_1,f_1,f_2,e_1)=-1,\\
&A_1(e_1,e_2,e_1,e_1)=0, &&A(e_1,e_2,f_1,f_1)=0,\\
&A_1(e_1,f_2,e_1,f_1)=1, &&A_1(e_1,f_2,f_1,e_1)=-1,\\
&A_1(f_1,f_2,e_1,e_1)=0, &&A_1(f_1,f_2,f_1,f_1)=0,\\
&A_1(f_1,e_2,e_1,f_1)=-1, &&A_1(f_1,e_2,f_1,e_1)=1,\\
&\rho(A_1)(e_1,e_2)=1, &&\rho(A_1)(e_2,e_1)=1,\\
&\rho(A_1)(f_1,f_2)=-1, &&\rho(A_1)(f_2,f_1)=-1.
\end{aligned}$$

Interchanging the roles of the indices "1" and "2" then creates A_2 with:

$$\begin{aligned}
&A_2(e_2,f_2,e_2,f_1)=1, &&A_2(e_2,f_2,f_2,e_1)=-1,\\
&A_2(e_2,f_2,e_1,f_2)=1, &&A_2(e_2,f_2,f_1,e_2)=-1,
\end{aligned}$$

$$\begin{aligned}
&A_2(e_2,e_1,e_2,e_2)= \ \ 0, && A_2(e_2,e_1,f_2,f_2)= \ \ 0,\\
&A_2(e_2,f_1,e_2,f_2)= \ \ 1, && A_2(e_2,f_1,f_2,e_2)=-1,\\
&A_2(f_2,f_1,e_2,e_2)= \ \ 0, && A_2(f_2,f_1,f_2,f_2)= \ \ 0,\\
&A_2(f_2,e_1,e_2,f_2)=-1, && A_2(f_2,e_1,f_2,e_2)= \ \ 1,\\
&\rho(A_2)(e_2,e_1)= \ \ 1, && \rho(A_2)(e_1,e_2)= \ \ 1,\\
&\rho(A_2)(f_2,f_1)=-1, && \rho(A_2)(f_1,f_2)=-1.
\end{aligned}$$

These tensors are anti-symmetric in the last two indices so $\rho_{13}=-\rho$. We verify that $0\neq A_1-A_2\in W^{\mathfrak{A}}_{+,9}$. This proves Assertion (2).

Next, we take $(\varrho_1,\varrho_2,\varrho_3)=(\frac{1}{2},-\frac{1}{2},\frac{1}{2})$ to create A_3 with:

$$\begin{aligned}
&A_3(e_1,f_1,e_1,f_2)=-1, && A_3(e_1,f_1,f_1,e_2)= \ \ 1,\\
&A_3(e_1,f_1,e_2,f_1)= \ \ 1, && A_3(e_1,f_1,f_2,e_1)=-1,\\
&A_3(e_1,e_2,e_1,e_1)=-1, && A_3(e_1,e_2,f_1,f_1)= \ \ 1,\\
&A_3(e_1,f_2,e_1,f_1)= \ \ 0, && A_3(e_1,f_2,f_1,e_1)= \ \ 0,\\
&A_3(f_1,f_2,e_1,e_1)= \ \ 1, && A_3(f_1,f_2,f_1,f_1)=-1,\\
&A_3(f_1,e_2,e_1,f_1)= \ \ 0, && A_3(f_1,e_2,f_1,e_1)= \ \ 0,\\
&\rho(A_3)(e_1,e_2)= \ \ 1, && \rho(A_3)(f_1,f_2)=-1,\\
&\rho(A_3)(e_2,e_1)=-1, && \rho(A_3)(f_2,f_1)= \ \ 1.
\end{aligned}$$

Interchange the roles of the indices "1" and "2" to A_4 with:

$$\begin{aligned}
&A_4(e_2,f_2,e_2,f_1)=-1, && A_4(e_2,f_2,f_2,e_1)= \ \ 1,\\
&A_4(e_2,f_2,e_1,f_2)= \ \ 1, && A_4(e_2,f_2,f_1,e_2)=-1,\\
&A_4(e_2,e_1,e_2,e_2)=-1, && A_4(e_2,e_1,f_2,f_2)= \ \ 1,\\
&A_4(e_2,f_1,e_2,f_2)= \ \ 0, && A_4(e_2,f_1,f_2,e_2)= \ \ 0,\\
&A_4(f_2,f_1,e_2,e_2)= \ \ 1, && A_4(f_2,f_1,f_2,f_2)=-1,\\
&A_4(f_2,e_1,e_2,f_2)= \ \ 0, && A_4(f_2,e_1,f_2,e_2)= \ \ 0,\\
&\rho(A_4)(e_2,e_1)= \ \ 1, && \rho(A_4)(f_2,f_1)=-1,\\
&\rho(A_4)(e_1,e_2)=-1, && \rho(A_4)(f_1,f_2)= \ \ 1.
\end{aligned}$$

These two tensors are symmetric in the last two indices so $\rho_{13}=\rho$. We then have $0\neq A_3+A_4\in W^{\mathfrak{A}}_{+,10}$; Assertion (3) now follows.

We clear the previous notation and set ${\Theta_{+,11}}^2=x^3-\iota y^3$. Then:

$$\begin{aligned}
&\nabla_{e_1}e_1=\nabla_{f_1}f_1=x^3e_2-y^3f_2, && \nabla_{e_1}f_1=\nabla_{f_1}e_1=x^3f_2-y^3e_2,\\
&\mathcal{A}_5(e_3,e_1)e_1=\mathcal{A}_5(e_3,f_1)f_1=e_2, && \mathcal{A}_5(f_3,e_1)e_1=\mathcal{A}_5(f_3,f_1)f_1=-f_2,\\
&\mathcal{A}_5(e_3,e_1)f_1=\mathcal{A}_5(e_3,f_1)e_1=f_2, && \mathcal{A}_5(f_3,e_1)f_1=\mathcal{A}_5(f_3,f_1)e_1=-e_2,\\
&\rho_{13}(\mathcal{A}_5)=\rho(\mathcal{A}_5)=0.
\end{aligned}$$

We have $\mathcal{A}_5 \in \mathfrak{K}^{\mathfrak{A}}_{+,-} \cap \ker(\rho_{13}) \cap \ker(\rho) = W^{\mathfrak{A}}_{+,9} \oplus W^{\mathfrak{A}}_{+,10} \oplus W^{\mathfrak{A}}_{+,11}$. To prove Assertion (4), it suffices to show $\mathcal{A}_5$ has a non-zero component in $W^{\mathfrak{A}}_{+,11}$. Suppose to the contrary that $\mathcal{A}_5 \in W^{\mathfrak{A}}_{+,9} \oplus W^{\mathfrak{A}}_{+,10}$. We may then decompose $\mathcal{A} = \mathcal{A}_9 + \mathcal{A}_{10}$ where $\mathcal{A}_9 \in W^{\mathfrak{A}}_{+,9}$ and $\mathcal{A}_{10} \in W^{\mathfrak{A}}_{+,10}$. Then:

$$\begin{aligned} A_5(x,y,z,w) + A_5(x,y,w,z) &= A_9(x,y,z,w) + A_9(x,y,w,z) \\ &\quad + A_{10}(x,y,z,w) + A_{10}(x,y,w,z) \\ &= 2A_{10}(x,y,z,w). \end{aligned}$$

We obtain the desired contradiction by checking the Bianchi identity:

$$\begin{aligned} &A_{10}(f_3,f_1,e_2,e_1) + A_{10}(f_1,e_2,f_3,e_1) + A_{10}(e_2,f_3,f_1,e_1) \\ &= \tfrac{1}{2}\{A_5(f_3,f_1,e_2,e_1) + A_5(f_3,f_1,e_1,e_2) + A_5(f_1,e_2,f_3,e_1) \\ &\quad + A_5(f_1,e_2,e_1,f_3) + A_5(e_2,f_3,f_1,e_1) + A_5(e_2,f_3,e_1,f_1)\} \\ &= 0 - \tfrac{1}{2} + 0 + 0 + 0 + 0 \neq 0. \end{aligned}$$

This establishes Assertion (4). □

The argument given to establish Theorem 5.4.1 now establishes the following result from which the assertions of Theorem 1.5.2 and of Theorem 1.5.3 concerning the group $\mathcal{U}_+^\star$ in the para-complex Kähler setting will follow:

Theorem 5.5.1 *Let $m \geq 6$.*

(1) We have decompositions of the following spaces as modules for the group $\mathcal{U}_+^\star$:

$$\begin{aligned} \mathfrak{K}_{+,+} &= S_+^{2,\mathcal{U}_+} \oplus \Lambda_+^{2,\mathcal{U}_+} \oplus W^{\mathfrak{A}}_{+,12}, \\ \mathfrak{K}^{\mathfrak{A}}_{+,-} &= \mathbb{R} \oplus \chi \oplus 2 \cdot S_{0,-}^{2,\mathcal{U}_+} \oplus 2 \cdot \Lambda_{0,-}^{2,\mathcal{U}_+} \oplus W^{\mathfrak{A}}_{+,9} \oplus W^{\mathfrak{A}}_{+,10} \oplus W^{\mathfrak{A}}_{+,11}. \end{aligned}$$

(2) The modules $\{S_+^{2,\mathcal{U}_+}$, $\Lambda_+^{2,\mathcal{U}_+}$, $W^{\mathfrak{A}}_{+,12}$, $\mathbb{R}$, χ, $S_{0,-}^{2,\mathcal{U}_+}$, $\Lambda_{0,-}^{2,\mathcal{U}_+}$, $W^{\mathfrak{A}}_{+,9}$, $W^{\mathfrak{A}}_{+,10}$, $W^{\mathfrak{A}}_{+,11}\}$ are irreducible and inequivalent modules for the group $\mathcal{U}_+^\star$.

(3) Let $\mathcal{A} \in \mathfrak{K}^{\mathfrak{A}}_+$. We can choose linear functions $\{u_{ij}{}^k, v_{ij}{}^k\}$ on $\mathbb{R}^m$ with $u_{ij}{}^k = u_{ji}{}^k$ and $v_{ij}{}^k = v_{ji}{}^k$ defining an affine connection ∇ on $\mathbb{R}^m$ so

$$\begin{aligned} \nabla_{e_i} e_j &= \nabla_{f_i} f_j = u_{ij}{}^k e_k + v_{ij}{}^k f_k, \\ \nabla_{f_i} e_j &= \nabla_{e_i} f_j = v_{ij}{}^k e_k + u_{ij}{}^k f_k. \end{aligned}$$

Then ∇ is para-Kähler affine with $\mathcal{R}^\nabla(0) = \mathcal{A}$. Furthermore, if $\mathcal{A}$ belongs to $\mathfrak{K}^{\mathfrak{A}}_{+,+}$, then $\{u,v\}$ can be chosen to be para-holomorphic so that $\mathcal{R}(P) \in \mathfrak{K}^{\mathfrak{A}}_{+,+}$ for all $P \in \mathbb{R}^m$.

5.6 Structure of $\mathfrak{K}_\pm^{\mathfrak{A}}$ as a $\mathrm{GL}_\pm^\star$ Module

Section 5.6 is devoted to the proof of Theorem 1.5.1. In Theorem 1.3.1, we decomposed

$$\mathfrak{A} \approx \{\mathfrak{A} \cap \ker(\rho)\} \oplus \Lambda^2 \oplus S^2$$

as the direct sum of inequivalent irreducible general linear modules. The proof that we gave of Theorem 1.3.1 in Section 4.4 relied upon the orthogonal module decomposition of $\mathfrak{A}$ which was given in Theorem 1.4.1. Similarly, the proof of Theorem 1.5.1 given here of the decomposition of $\mathfrak{K}_\pm$ as a module for the group $\mathrm{GL}_\pm^\star$ will rely upon the decomposition of $\mathfrak{K}_\pm$ presented in Theorem 1.5.2.

Let $\langle\cdot,\cdot\rangle$ be an auxiliary inner product so that $(V,\langle\cdot,\cdot\rangle,J_\pm)$ is a para-Hermitian vector space (+) or a pseudo-Hermitian vector space (−). By Theorem 5.4.1 and Theorem 5.5.1,

$$\mathfrak{K}_{\pm,\pm}^{\mathfrak{A}} = S_\pm^{2,\mathcal{U}_\pm} \oplus \Lambda_\pm^{2,\mathcal{U}_\pm} \oplus \{\ker(\rho) \cap \mathfrak{K}_{\pm,\pm}\}$$

is a decomposition into irreducible and inequivalent modules for the group $\mathcal{U}_\pm^\star$; it is also a decomposition of $\mathfrak{K}_{-,-}^{\mathfrak{A}}$ into irreducible and inequivalent modules for the group $\mathcal{U}_-$. We also have a decomposition of the remaining summand $\mathfrak{K}_{\pm,\mp}^{\mathfrak{A}}$ into irreducible and inequivalent modules for the group $\mathcal{U}_\pm^\star$:

$$\mathfrak{K}_{\pm,\mp}^{\mathfrak{A}} = \mathbb{R} \oplus \chi \oplus 2\cdot S_{0,\mp}^{2,\mathcal{U}_\pm} \oplus 2\cdot \Lambda_{0,\mp}^{2,\mathcal{U}_\pm} \oplus W_{\pm,9}^{\mathfrak{A}} \oplus W_{\pm,10}^{\mathfrak{A}} \oplus W_{\pm,11}^{\mathfrak{A}}. \tag{5.6.a}$$

This leads to a module decomposition into inequivalent irreducibles of $\mathfrak{K}_{\pm,\mp}^{\mathfrak{A}} \cap \ker(\rho)$ in the form:

$$\mathfrak{K}_{\pm,\mp}^{\mathfrak{A}} \cap \ker(\rho) = S_{0,\mp}^{2,\mathcal{U}_\pm} \oplus \Lambda_{0,\mp}^{2,\mathcal{U}_\pm} \oplus W_{\pm,9}^{\mathfrak{A}} \oplus W_{\pm,10}^{\mathfrak{A}} \oplus W_{\pm,11}^{\mathfrak{A}}.$$

This is also a decomposition into irreducibles of $\mathfrak{K}_{-,+}^{\mathfrak{A}}$; we have $S_{0,+}^{2,\mathcal{U}_-} \approx \Lambda_{0,+}^{2,\mathcal{U}}$ as a module for the group $\mathcal{U}_-$ but otherwise the factors are inequivalent. We let $W_{\pm,7}^{\mathfrak{A}}$ and $W_{\pm,8}^{\mathfrak{A}}$ be the submodules of $\mathfrak{K}_{\pm,\mp}^{\mathfrak{A}} \cap \ker(\rho)$ abstractly isomorphic to $S_{0,\mp}^{2,\mathcal{U}_\pm}$ and $\Lambda_{0,\mp}^{2,\mathcal{U}_\pm}$, respectively:

$$S_{0,\mp}^{2,\mathcal{U}_\pm} \approx W_{\pm,7}^{\mathfrak{A}} \subset \mathfrak{K}_{\pm,\mp}^{\mathfrak{A}} \quad\text{and}\quad \Lambda_{0,\mp}^{2,\mathcal{U}_\pm} \approx W_{\pm,8}^{\mathfrak{A}} \subset \mathfrak{K}_{\pm,\mp}^{\mathfrak{A}}.$$

The tensor A^* played an important role in the proof of Lemma 5.4.2; it is given by setting:

$$A^*(x,y,z,w) := A(x,y,z,J_\pm w).$$

Lemma 5.6.1 *The map $T : A \to A^*$ satisfies:*

(1) $T^2 = \pm\mathrm{Id}$.

(2) T intertwines $\mathfrak{K}^{\mathfrak{A}}_{\pm,\mp} \cap \ker(\rho)$ *with* $\{\mathfrak{K}^{\mathfrak{A}}_{\pm,\mp} \cap \ker(\rho)\} \otimes \chi$ *as a module for the group* $\mathrm{GL}^\star_\pm$.

(3) T intertwines $W^{\mathfrak{A}}_{\pm,9}$ *with* $W^{\mathfrak{A}}_{\pm,10} \otimes \chi$, $W^{\mathfrak{A}}_{\pm,11}$ *with* $W^{\mathfrak{A}}_{\pm,11} \otimes \chi$, *and* $W^{\mathfrak{A}}_{\pm,7}$ *with* $W^{\mathfrak{A}}_{\pm,8} \otimes \chi$ *as a module for the group* $\mathcal{U}^\star_\pm$.

Proof. Assertion (1) is immediate. Let $A \in \mathfrak{K}^{\mathfrak{A}}_{\pm,\mp} \cap \ker(\rho)$. By expressing

$$\begin{aligned} A^*(x,y,z,w) &= A(x,y,z,J_\pm w) = \mp A(x,y,J_\pm z, J_\pm J_\pm w) \\ &= -A(x,y,J_\pm z, w), \end{aligned}$$

we see that $\rho(A^*)(y,z) = -\rho(A)(y,J_\pm z)$. It is immediate that A^* satisfies the Bianchi identity and that

$$\begin{aligned} &A^*(x,y,J_\pm z,J_\pm w) = A(x,y,J_\pm z,J_\pm J_\pm w) = \mp A(x,y,z,J_\pm w) \\ &\quad = \mp A^*(x,y,z,w), \\ &A^*(J_\pm x,J_\pm y,z,w) = A(J_\pm x,J_\pm y,z,J_\pm w) = \mp A(x,y,z,J_\pm w) \\ &\quad = \mp A^*(x,y,z,w). \end{aligned}$$

Assertion (2) now follows; the factor of χ arises as $J_\pm$ appears once in the definition of T. If $\psi \in \otimes^2 V^*$, we define $T\psi(x,y) := \psi(x, J_\pm y)$. It is then immediate that $\rho_{13}TA = T\rho_{13}A$ and thus T preserves $\ker(\rho_{13})$. We have:

$$\phi \in S^{2,\mathcal{U}_\pm}_\mp \quad \Rightarrow \quad T\phi \in \Lambda^{2,\mathcal{U}_\pm}_\mp \quad \text{and} \quad \psi \in \Lambda^{2,\mathcal{U}_\pm}_\mp \quad \Rightarrow \quad T\psi \in S^{2,\mathcal{U}_\pm}_\mp .$$

We see that T intertwines the representation $W^{\mathfrak{A}}_{\pm,9}$ with $W^{\mathfrak{A}}_{\pm,10} \otimes \chi$ by applying these relations to the last indices of the tensor. Since T is an (anti)-isometry, T intertwines $W^{\mathfrak{A}}_{\pm,11}$ with $W^{\mathfrak{A}}_{\pm,11} \otimes \chi$ since $W^{\mathfrak{A}}_{\pm,11}$ is the orthogonal complement of $W^{\mathfrak{A}}_{\pm,9} \oplus W^{\mathfrak{A}}_{\pm,10}$ in $\mathfrak{K}^{\mathfrak{A}}_{\pm,\mp} \cap \ker(\rho) \cap \ker(\rho_{13})$. We use the above display to conclude T intertwines $W^{\mathfrak{A}}_{\pm,7}$ with $W^{\mathfrak{A}}_{\pm,8} \otimes \chi$ since $W^{\mathfrak{A}}_{\pm,7} \oplus W^{\mathfrak{A}}_{\pm,8}$ is the orthogonal complement of $W^{\mathfrak{A}}_{\pm,9} \oplus W^{\mathfrak{A}}_{\pm,10} \oplus W^{\mathfrak{A}}_{\pm,11}$ in $\mathfrak{K}^{\mathfrak{A}}_{\pm,\mp} \cap \ker(\rho)$. □

We introduce some auxiliary notation:

Definition 5.6.1 Let $J_\pm$ be a (para)-complex structure on V. Set:

$$\begin{aligned} &(\sigma^{\mathfrak{A}}_{\pm,1}\phi_{\pm,1})(x,y)z := \phi_{\pm,1}(x,z)y - \phi_{\pm,1}(y,z)x \pm \phi_{\pm,1}(x,J_\pm z)J_\pm y \\ &\quad \mp\phi_{\pm,1}(y,J_\pm z)J_\pm x \pm 2\phi_{\pm,1}(x,J_\pm y)J_\pm z \qquad\qquad \text{for } \phi_{\pm,1} \in S^{2,\mathcal{U}_\pm}_\mp, \end{aligned}$$

$$(\sigma^{\mathfrak{A}}_{\pm,2}\phi_{\pm,2})(x,y)z := \phi_{\pm,2}(x,z)y - \phi_{\pm,2}(y,z)x \pm \phi_{\pm,2}(x,J_\pm z)J_\pm y$$
$$\mp\phi_{\pm,2}(y,J_\pm z)J_\pm x \qquad \text{for } \phi_{\pm,2}\in S^{2,\mathcal{U}_\pm}_\pm,$$
$$(\sigma^{\mathfrak{A}}_{\pm,3}\psi_{\pm,3})(x,y)z := \psi_{\pm,3}(x,z)y - \psi_{\pm,3}(y,z)x + 2\psi_{\pm,3}(x,y)z$$
$$\pm\psi_{\pm,3}(x,J_\pm z)J_\pm y \mp \psi_{\pm,3}(y,J_\pm z)J_\pm x \qquad \text{for } \psi_{\pm,3}\in \Lambda^{2,\mathcal{U}_\pm}_\mp,$$
$$(\sigma^{\mathfrak{A}}_{\pm,4}\psi_{\pm,4})(x,y)z := \psi_{\pm,4}(x,z)y - \psi_{\pm,4}(y,z)x + 2\psi_{\pm,4}(x,y)z$$
$$\pm\psi_{\pm,4}(x,J_\pm z)J_\pm y \mp \psi_{\pm,4}(y,J_\pm z)J_\pm x \pm 2\psi_{\pm,4}(x,J_\pm y)J_\pm z$$
$$\text{for } \psi_{\pm,4}\in \Lambda^{2,\mathcal{U}_\pm}_\pm.$$

Lemma 5.6.2

(1) If $\phi_{\pm,1}\in S^{2,\mathcal{U}_\pm}_\mp$, then $\sigma^{\mathfrak{A}}_{\pm,1}\phi_{\pm,1}\in\mathfrak{K}^{\mathfrak{A}}_{\pm,\mp}$ and $\rho\,\sigma^{\mathfrak{A}}_{\pm,1}\phi_{\pm,1} = -(m+2)\phi_{\pm,1}$.

(2) If $\phi_{\pm,2}\in S^{2,\mathcal{U}_\pm}_\pm$, then $\sigma^{\mathfrak{A}}_{\pm,2}\phi_{\pm,2}\in\mathfrak{K}^{\mathfrak{A}}_{\pm,\pm}$ and $\rho\,\sigma^{\mathfrak{A}}_{\pm,2}\phi_{\pm,2} = (2-m)\phi_{\pm,2}$.

(3) If $\psi_{\pm,3}\in \Lambda^{2,\mathcal{U}_\pm}_\mp$, then $\sigma^{\mathfrak{A}}_{\pm,3}\psi_{\pm,3}\in\mathfrak{K}^{\mathfrak{A}}_{\pm,\mp}$ and $\rho\,\sigma^{\mathfrak{A}}_{\pm,3}\psi_{\pm,3} = -(m+2)\psi_{\pm,3}$.

(4) If $\psi_{\pm,4}\in \Lambda^{2,\mathcal{U}_\pm}_\pm$, then $\sigma^{\mathfrak{A}}_{\pm,4}\psi_{\pm,4}\in\mathfrak{K}^{\mathfrak{A}}_{\pm,\pm}$ and $\rho\,\sigma^{\mathfrak{A}}_{\pm,4}\psi_{\pm,4} = -(2+m)\psi_{\pm,4}$.

(5) The maps $\sigma^{\mathfrak{A}}_{\pm,i}$ are $\mathrm{GL}^\star_\pm$ module morphisms splitting ρ. Thus we have module isomorphisms:

$$\mathfrak{K}_+ \approx \{\mathfrak{K}_+\cap\ker(\rho)\}\oplus S^{2,\mathcal{U}_\pm}_+ \oplus \Lambda^{2,\mathcal{U}_\pm}_+,$$
$$\mathfrak{K}_- \approx \{\mathfrak{K}_-\cap\ker(\rho)\}\oplus S^{2,\mathcal{U}_\pm}_- \oplus \Lambda^{2,\mathcal{U}_\pm}_-.$$

Proof. We begin with some basic parity observations:

$$\phi_{\pm,1}(x,J_\pm y) = \mp\phi_{\pm,1}(J_\pm x, J_\pm J_\pm y) = -\phi_{\pm,1}(J_\pm x, y),$$
$$\phi_{\pm,2}(x,J_\pm y) = \pm\phi_{\pm,2}(J_\pm x, J_\pm J_\pm y) = \phi_{\pm,2}(J_\pm x, y),$$
$$\psi_{\pm,3}(x,J_\pm y) = \mp\psi_{\pm,3}(J_\pm x, J_\pm J_\pm y) = -\psi_{\pm,3}(J_\pm x, y),$$
$$\psi_{\pm,4}(x,J_\pm y) = \pm\psi_{\pm,4}(J_\pm x, J_\pm J_\pm y) = \psi_{\pm,4}(J_\pm x, y).$$

It now follows that the tensors

$$\{\sigma^{\mathfrak{A}}_{\pm,1}\phi_{\pm,1}, \sigma^{\mathfrak{A}}_{\pm,2}\phi_{\pm,2}, \sigma^{\mathfrak{A}}_{\pm,3}\psi_{\pm,3}, \sigma^{\mathfrak{A}}_{\pm,4}\psi_{\pm,4}\}$$

are anti-symmetric in the first two arguments. We verify that the Bianchi identity is satisfied by these tensors and therefore that they belong to $\mathfrak{A}$ by computing:

$$
\begin{aligned}
&(\sigma^{\mathfrak{A}}_{\pm,1}\phi_{\pm,1})(x,y)z+(\sigma^{\mathfrak{A}}_{\pm,1}\phi_{\pm,1})(y,z)x+(\sigma^{\mathfrak{A}}_{\pm,1}\phi_{\pm,1})(z,x)y\\
&\quad= \phi_{\pm,1}(x,z)y-\phi_{\pm,1}(y,z)x\pm\phi_{\pm,1}(x,J_\pm z)J_\pm y\\
&\qquad+\phi_{\pm,1}(y,x)z-\phi_{\pm,1}(z,x)y\pm\phi_{\pm,1}(y,J_\pm x)J_\pm z\\
&\qquad+\phi_{\pm,1}(z,y)x-\phi_{\pm,1}(x,y)z\pm\phi_{\pm,1}(z,J_\pm y)J_\pm x\\
&\qquad\mp\phi_{\pm,1}(y,J_\pm z)J_\pm x\pm 2\phi_{\pm,1}(x,J_\pm y)J_\pm z\\
&\qquad\mp\phi_{\pm,1}(z,J_\pm x)J_\pm y\pm 2\phi_{\pm,1}(y,J_\pm z)J_\pm x\\
&\qquad\mp\phi_{\pm,1}(x,J_\pm y)J_\pm z\pm 2\phi_{\pm,1}(z,J_\pm x)J_\pm y\\
&\quad=0,
\end{aligned}
$$

$$
\begin{aligned}
&(\sigma^{\mathfrak{A}}_{\pm,2}\phi_{\pm,2})(x,y)z+(\sigma^{\mathfrak{A}}_{\pm,2}\phi_{\pm,2})(y,z)x+(\sigma^{\mathfrak{A}}_{\pm,2}\phi_{\pm,2})(z,x)y\\
&\quad= \phi_{\pm,2}(x,z)y-\phi_{\pm,2}(y,z)x\pm\phi_{\pm,2}(x,J_\pm z)J_\pm y\mp\phi_{\pm,2}(y,J_\pm z)J_\pm x\\
&\qquad+\phi_{\pm,2}(y,x)z-\phi_{\pm,2}(z,x)y\pm\phi_{\pm,2}(y,J_\pm x)J_\pm z\mp\phi_{\pm,2}(z,J_\pm x)J_\pm y\\
&\qquad+\phi_{\pm,2}(z,y)x-\phi_{\pm,2}(x,y)z\pm\phi_{\pm,2}(z,J_\pm y)J_\pm x\mp\phi_{\pm,2}(x,J_\pm y)J_\pm z\\
&\quad=0,
\end{aligned}
$$

$$
\begin{aligned}
&(\sigma^{\mathfrak{A}}_{\pm,3}\psi_{\pm,3})(x,y)z+(\sigma^{\mathfrak{A}}_{\pm,3}\psi_{\pm,3})(y,z)x+(\sigma^{\mathfrak{A}}_{\pm,3}\psi_{\pm,3})(z,x)y\\
&\quad= \psi_{\pm,3}(x,z)y-\psi_{\pm,3}(y,z)x+2\psi_{\pm,3}(x,y)z\\
&\qquad+\psi_{\pm,3}(y,x)z-\psi_{\pm,3}(z,x)y+2\psi_{\pm,3}(y,z)x\\
&\qquad+\psi_{\pm,3}(z,y)x-\psi_{\pm,3}(x,y)z+2\psi_{\pm,3}(z,x)y\\
&\qquad\pm\psi_{\pm,3}(x,J_\pm z)J_\pm y\mp\psi_{\pm,3}(y,J_\pm z)J_\pm x\\
&\qquad\pm\psi_{\pm,3}(y,J_\pm x)J_\pm z\mp\psi_{\pm,3}(z,J_\pm x)J_\pm y\\
&\qquad\pm\psi_{\pm,3}(z,J_\pm y)J_\pm x\mp\psi_{\pm,3}(x,J_\pm y)J_\pm z=0,
\end{aligned}
$$

$$
\begin{aligned}
&(\sigma^{\mathfrak{A}}_{\pm,4}\psi_{\pm,4})(x,y)z+(\sigma^{\mathfrak{A}}_{\pm,4}\psi_{\pm,4})(y,z)x+(\sigma^{\mathfrak{A}}_{\pm,4}\psi_{\pm,4})(z,x)y\\
&\quad= \psi_{\pm,4}(x,z)y-\psi_{\pm,4}(y,z)x+2\psi_{\pm,4}(x,y)z\\
&\qquad+\psi_{\pm,4}(y,x)z-\psi_{\pm,4}(z,x)y+2\psi_{\pm,4}(y,z)x\\
&\qquad+\psi_{\pm,4}(z,y)x-\psi_{\pm,4}(x,y)z+2\psi_{\pm,4}(z,x)y\\
&\qquad\pm\psi_{\pm,4}(x,J_\pm z)J_\pm y\mp\psi_{\pm,4}(y,J_\pm z)J_\pm x\pm 2\psi_{\pm,4}(x,J_\pm y)J_\pm z\\
&\qquad\pm\psi_{\pm,4}(y,J_\pm x)J_\pm z\mp\psi_{\pm,4}(z,J_\pm x)J_\pm y\pm 2\psi_{\pm,4}(y,J_\pm z)J_\pm x\\
&\qquad\pm\psi_{\pm,4}(z,J_\pm y)J_\pm x\mp\psi_{\pm,4}(x,J_\pm y)J_\pm z\pm 2\psi_{\pm,4}(z,J_\pm x)J_\pm y=0.
\end{aligned}
$$

We verify these endomorphisms commute with $J_\pm$ and thus belong to $\mathfrak{K}^{\mathfrak{A}}_\pm$ by comparing:

$$
\begin{aligned}
&(\sigma^{\mathfrak{A}}_{\pm,1}\phi_{\pm,1})(x,y)J_\pm z\\
&\quad= \phi_{\pm,1}(x,J_\pm z)y - \phi_{\pm,1}(y,J_\pm z)x \pm \phi_{\pm,1}(x,J_\pm J_\pm z)J_\pm y\\
&\qquad \mp\phi_{\pm,1}(y,J_\pm J_\pm z)J_\pm x \pm 2\phi_{\pm,1}(x,J_\pm y)J_\pm J_\pm z,\\
&J_\pm(\sigma^{\mathfrak{A}}_{\pm,1}\phi_{\pm,1})(x,y)z\\
&\quad= \phi_{\pm,1}(x,z)J_\pm y - \phi_{\pm,1}(y,z)J_\pm x \pm \phi_{\pm,1}(x,J_\pm z)J_\pm J_\pm y\\
&\qquad \mp\phi_{\pm,1}(y,J_\pm z)J_\pm J_\pm x \pm 2\phi_{\pm,1}(x,J_\pm y)J_\pm J_\pm z,
\end{aligned}
$$

$$
\begin{aligned}
&(\sigma^{\mathfrak{A}}_{\pm,2}\phi_{\pm,2})(x,y)J_\pm z = \phi_{\pm,2}(x,J_\pm z)y - \phi_{\pm,2}(y,J_\pm z)x\\
&\qquad \pm\phi_{\pm,2}(x,J_\pm J_\pm z)J_\pm y \mp \phi_{\pm,2}(y,J_\pm J_\pm z)J_\pm x,\\
&J_\pm(\sigma^{\mathfrak{A}}_{\pm,2}\phi_{\pm,2})(x,y)z = \phi_{\pm,2}(x,z)J_\pm y - \phi_{\pm,2}(y,z)J_\pm x\\
&\qquad \pm\phi_{\pm,2}(x,J_\pm z)J_\pm J_\pm y \mp \phi_{\pm,2}(y,J_\pm z)J_\pm J_\pm x,\\
&(\sigma^{\mathfrak{A}}_{\pm,3}\psi_{\pm,3})(x,y)J_\pm z = \psi_{\pm,3}(x,J_\pm z)y - \psi_{\pm,3}(y,J_\pm z)x\\
&\qquad +2\psi_{\pm,3}(x,y)J_\pm z \pm\psi_{\pm,3}(x,J_\pm J_\pm z)J_\pm y \mp \psi_{\pm,3}(y,J_\pm J_\pm z)J_\pm x,\\
&J_\pm(\sigma^{\mathfrak{A}}_{\pm,3}\psi_{\pm,3})(x,y)z = \psi_{\pm,3}(x,z)J_\pm y - \psi_{\pm,3}(y,z)J_\pm x\\
&\qquad +2\psi_{\pm,3}(x,y)J_\pm z \pm\psi_{\pm,3}(x,J_\pm z)J_\pm J_\pm y \mp \psi_{\pm,3}(y,J_\pm z)J_\pm J_\pm x,\\
&(\sigma^{\mathfrak{A}}_{\pm,4}\psi_{\pm,4})(x,y)J_\pm z = \psi_{\pm,4}(x,J_\pm z)y - \psi_{\pm,4}(y,J_\pm z)x\\
&\qquad +2\psi_{\pm,4}(x,y)J_\pm z \pm\psi_{\pm,4}(x,J_\pm J_\pm z)J_\pm y \mp \psi_{\pm,4}(y,J_\pm J_\pm z)J_\pm x\\
&\qquad \pm 2\psi_{\pm,4}(x,J_\pm y)J_\pm J_\pm z,\\
&J_\pm(\sigma^{\mathfrak{A}}_{\pm,4}\psi_{\pm,4})(x,y)z = \psi_{\pm,4}(x,z)J_\pm y - \psi_{\pm,4}(y,z)J_\pm x\\
&\qquad +2\psi_{\pm,4}(x,y)J_\pm z \pm\psi_{\pm,4}(x,J_\pm z)J_\pm J_\pm y \mp \psi_{\pm,4}(y,J_\pm z)J_\pm J_\pm x\\
&\qquad \pm 2\psi_{\pm,4}(x,J_\pm y)J_\pm J_\pm z.
\end{aligned}
$$

Let $\{e_i\}$ be a basis for V and let $\{e^i\}$ be the corresponding dual basis for V^*. We have $e^i(J_\pm e_i) = \mathrm{Tr}(J_\pm) = 0$. We examine the Ricci tensor:

$$
\begin{aligned}
&\rho(\sigma^{\mathfrak{A}}_{\pm,1}\phi_{\pm,1})(y,z)\\
&\quad= \phi_{\pm,1}(e_i,z)e^i(y) - \phi_{\pm,1}(y,z)e^i(e_i) \pm \phi_{\pm,1}(e_i,J_\pm z)e^i(J_\pm y)\\
&\qquad \mp\phi_{\pm,1}(y,J_\pm z)e^i(J_\pm e_i) \pm 2\phi_{\pm,1}(e_i,J_\pm y)e^i(J_\pm z)\\
&\quad= \phi_{\pm,1}(y,z) - m\phi_{\pm,1}(y,z) \pm \phi_{\pm,1}(J_\pm y,J_\pm z) \mp 0 \pm 2\phi_{\pm,1}(J_\pm z,J_\pm y)\\
&\quad= -(m+2)\phi_{\pm,1}(y,z),
\end{aligned}
$$

$$
\begin{aligned}
&\rho(\sigma^{\mathfrak{A}}_{\pm,2}\phi_{\pm,2})(y,z) = \phi_{\pm,2}(e_i,z)e^i(y) - \phi_{\pm,2}(y,z)e^i(e_i)\\
&\qquad \pm\phi_{\pm,2}(e_i,J_\pm z)e^i(J_\pm y) \mp \phi_{\pm,2}(y,J_\pm z)e^i(J_\pm e_i)\\
&\quad = \phi_{\pm,2}(y,z) - m\phi_{\pm,2}(y,z) \pm \phi_{\pm,2}(J_\pm y,J_\pm z) \mp 0 = (2-m)\phi_{\pm,2}(y,z),
\end{aligned}
$$

$$\begin{aligned}\rho(\sigma^{\mathfrak{A}}_{\pm,3}\psi_{\pm,3})(y,z) &= \psi_{\pm,3}(e_i,z)e^i(y) - \psi_{\pm,3}(y,z)e^i(e_i)\\&\quad+2\psi_{\pm,3}(e_i,y)e^i(z) \pm \psi_{\pm,3}(e_i,J_\pm z)e^i(J_\pm y) \mp \psi_{\pm,3}(y,J_\pm z)e^i(J_\pm e_i)\\&= \psi_{\pm,3}(y,z) - m\psi_{\pm,3}(y,z) + 2\psi_{\pm,3}(z,y) \pm \psi_{\pm,3}(J_\pm y, J_\pm z) \mp 0\\&= -(m+2)\psi_{\pm,3}(y,z),\\ \rho(\sigma^{\mathfrak{A}}_{\pm,4}\psi_{\pm,4})(y,z) &= \psi_{\pm,4}(e_i,z)e^i(y) - \psi_{\pm,4}(y,z)e^i(e_i)\\&\quad+2\psi_{\pm,4}(e_i,y)e^i(z) \pm \psi_{\pm,4}(e_i,J_\pm z)e^i(J_\pm y)\\&\quad\mp\psi_{\pm,4}(y,J_\pm z)e^i(J_\pm e_i) \pm 2\psi_{\pm,4}(e_i,J_\pm y)e^i(J_\pm z)\\&= \psi_{\pm,4}(y,z) - m\psi_{\pm,4}(y,z) + 2\psi_{\pm,4}(z,y) \pm \psi_{\pm,4}(J_\pm y, J_\pm z)\\&\quad\mp 0 \pm 2\psi_{\pm,4}(J_\pm z, J_\pm y) = (-2-m)\psi_{\pm,4}(y,z).\end{aligned}$$

The fact that these endomorphisms take values in the appropriate subspaces of $\mathfrak{K}^{\mathfrak{A}}_\pm$ now follows from Theorem 5.4.1 and from Theorem 5.5.1; the final Assertion now follows. □

We use Equation (5.6.a) to derive the decomposition:

$$\mathfrak{K}^{\mathfrak{A}}_{\pm,\mp} \cap \ker(\rho) = S^{2,\mathcal{U}_\pm}_{0,\mp} \oplus \Lambda^{2,\mathcal{U}_\pm}_{0,\mp} \oplus W^{\mathfrak{A}}_{\pm,9} \oplus W^{\mathfrak{A}}_{\pm,10} \oplus W^{\mathfrak{A}}_{\pm,11}.$$

The factors $S^{2,\mathcal{U}_\pm}_{0,\mp}$ and $\Lambda^{2,\mathcal{U}_\pm}_{0,\mp}$ are detected by ρ_{13}. This means that:

$$\rho_{13} : \mathfrak{K}^{\mathfrak{A}}_{\pm,\mp} \cap \ker(\rho) \to S^{2,\mathcal{U}_\pm}_{0,\mp} \oplus \Lambda^{2,\mathcal{U}_\pm}_{0,\mp} \to 0. \tag{5.6.b}$$

Let $\pi^{\mathfrak{A}}_{\pm,7}$ and $\pi^{\mathfrak{A}}_{\pm,8}$ be orthogonal projection on the submodules $W^{\mathfrak{A}}_{\pm,7}$ and $W^{\mathfrak{A}}_{\pm,8}$ of $\mathfrak{K}^{\mathfrak{A}}_{\pm,\mp}$ corresponding to $S^{2,\mathcal{U}_\pm}_{0,\mp}$ and $\Lambda^{2,\mathcal{U}_\pm}_{0,\mp}$, respectively. If ϕ belongs to $V^* \otimes V^*$, then set:

$$\begin{aligned}&\vartheta^{\mathfrak{A}}_\pm(\phi)(x,y,z,w) := \phi(x,w)\langle y,z\rangle - \phi(y,w)\langle x,z\rangle\\&\quad\mp\phi(x,J_\pm w)\langle y,J_\pm z\rangle \pm \phi(y,J_\pm w)\langle x,J_\pm z\rangle \pm 2\phi(z,J_\pm w)\langle x,J_\pm y\rangle.\end{aligned}$$

Lemma 5.6.3

(1) If $\phi \in S^{2,\mathcal{U}_\pm}_{0,\mp} \oplus \Lambda^{2,\mathcal{U}_\pm}_{0,\mp}$, then $\vartheta^{\mathfrak{A}}_\pm\phi \in \mathfrak{K}^{\mathfrak{A}}_{\pm,\mp}$.

(2) If $\phi_{\pm,1} \in S^{2,\mathcal{U}_\pm}_{0,\mp}$ and if $\psi_{\pm,3} \in \Lambda^{2,\mathcal{U}_\pm}_{0,\mp}$, then:

$$\begin{aligned}\rho\sigma^{\mathfrak{A}}_{\pm,1}\phi_{\pm,1} &= -(m+2)\phi_{\pm,1}, & \rho_{13}\sigma^{\mathfrak{A}}_{\pm,1}\phi_{\pm,1} &= 2\phi_{\pm,1},\\ \rho\vartheta^{\mathfrak{A}}_\pm\phi_{\pm,1} &= 2\phi_{\pm,1}, & \rho_{13}\vartheta^{\mathfrak{A}}_\pm\phi_{\pm,1} &= -(m+2)\phi_{\pm,1},\\ \rho\sigma^{\mathfrak{A}}_{\pm,3}\psi_{\pm,3} &= -(m+2)\psi_{\pm,3}, & \rho_{13}\sigma^{\mathfrak{A}}_{\pm,3}\psi_{\pm,3} &= -2\psi_{\pm,3},\\ \rho\vartheta^{\mathfrak{A}}_\pm\psi_{\pm,3} &= -2\psi_{\pm,3}, & \rho_{13}\vartheta^{\mathfrak{A}}_\pm\psi_{\pm,3} &= -(m+2)\psi_{\pm,3}.\end{aligned}$$

(3) $\pi^{\mathfrak{A}}_{\pm,7} = -\frac{1}{m(m+4)}\{2\sigma^{\mathfrak{A}}_{\pm,1} + (m+2)\vartheta^{\mathfrak{A}}_\pm\}\rho_{13,s}$.

(4) $\pi^{\mathfrak{A}}_{\pm,8} = -\frac{1}{m(m+4)}\{-2\sigma^{\mathfrak{A}}_{\pm,3} + (m+2)\vartheta^{\mathfrak{A}}_{\pm}\}\rho_{13,a}$.

Proof. It is immediate from the definition that $\vartheta^{\mathfrak{A}}_{\pm}(\phi)$ is anti-symmetric in the first two arguments. Note that

$$\phi(x, J_\pm y) = \mp\phi(J_\pm x, J_\pm J_\pm y) = -\phi(J_\pm x, y).$$

We verify that $\vartheta^{\mathfrak{A}}_{\pm}\phi$ satisfies the Bianchi identity by computing:

$$\begin{aligned}
&\vartheta^{\mathfrak{A}}_{\pm}(\phi)(x,y,z,w) + \vartheta^{\mathfrak{A}}_{\pm}(\phi)(y,z,x,w) + \vartheta^{\mathfrak{A}}_{\pm}(\phi)(z,x,y,w)\\
&\quad = \mp\phi(x,J_\pm w)\langle y,J_\pm z\rangle \pm \phi(y,J_\pm w)\langle x,J_\pm z\rangle \pm 2\phi(z,J_\pm w)\langle x,J_\pm y\rangle\\
&\quad\quad \mp\phi(y,J_\pm w)\langle z,J_\pm x\rangle \pm \phi(z,J_\pm w)\langle y,J_\pm x\rangle \pm 2\phi(x,J_\pm w)\langle y,J_\pm z\rangle\\
&\quad\quad \mp\phi(z,J_\pm w)\langle x,J_\pm y\rangle \pm \phi(x,J_\pm w)\langle z,J_\pm y\rangle \pm 2\phi(y,J_\pm w)\langle z,J_\pm x\rangle\\
&\quad\quad +\phi(x,w)\langle y,z\rangle - \phi(y,w)\langle x,z\rangle\\
&\quad\quad +\phi(y,w)\langle z,x\rangle - \phi(z,w)\langle y,x\rangle\\
&\quad\quad +\phi(z,w)\langle x,y\rangle - \phi(x,w)\langle z,y\rangle = 0.
\end{aligned}$$

We will show that $\vartheta^{\mathfrak{A}}_{\pm}\phi \in \mathfrak{K}^{\mathfrak{A}}_{\pm,\mp}$ by demonstrating that:

$$\vartheta^{\mathfrak{A}}_{\pm}\phi(x,y,z,w) = \mp\vartheta^{\mathfrak{A}}_{\pm}\phi(x,y,J_\pm z,J_\pm w) = \mp\vartheta^{\mathfrak{A}}_{\pm}\phi(J_\pm x,J_\pm y,z,w).$$

We compare:

$$\begin{aligned}
&\vartheta^{\mathfrak{A}}_{\pm}(\phi)(x,y,z,w) = \phi(x,w)\langle y,z\rangle - \phi(y,w)\langle x,z\rangle\\
&\quad\quad \mp\phi(x,J_\pm w)\langle y,J_\pm z\rangle \pm \phi(y,J_\pm w)\langle x,J_\pm z\rangle \pm 2\phi(z,J_\pm w)\langle x,J_\pm y\rangle,\\
&\vartheta^{\mathfrak{A}}_{\pm}(\phi)(x,y,J_\pm z,J_\pm w) = \phi(x,J_\pm w)\langle y,J_\pm z\rangle - \phi(y,J_\pm w)\langle x,J_\pm z\rangle\\
&\quad\quad \mp\phi(x,J_\pm J_\pm w)\langle y,J_\pm J_\pm z\rangle \pm \phi(y,J_\pm J_\pm w)\langle x,J_\pm J_\pm z\rangle\\
&\quad\quad \pm 2\phi(J_\pm z,J_\pm J_\pm w)\langle x,J_\pm y\rangle,\\
&\vartheta^{\mathfrak{A}}_{\pm}(\phi)(J_\pm x,J_\pm y,z,w) = \phi(J_\pm x,w)\langle J_\pm y,z\rangle - \phi(J_\pm y,w)\langle J_\pm x,z\rangle\\
&\quad\quad \mp\phi(J_\pm x,J_\pm w)\langle J_\pm y,J_\pm z\rangle \pm \phi(J_\pm y,J_\pm w)\langle J_\pm x,J_\pm z\rangle\\
&\quad\quad \pm 2\phi(z,J_\pm w)\langle J_\pm x,J_\pm J_\pm y\rangle.
\end{aligned}$$

We use Lemma 5.6.2 to determine $\rho\sigma^{\mathfrak{A}}_{\pm,1}$ and $\rho\sigma^{\mathfrak{A}}_{\pm,3}$. Since $\phi \perp \langle\cdot,\cdot\rangle$ and $\phi \perp \Omega$, $\varepsilon^{il}\phi(e_i,e_l) = \varepsilon^{il}\phi(e_i,J_\pm e_l) = 0$. We compute $\rho\vartheta^{\mathfrak{A}}_{\pm}$:

$$\begin{aligned}
&\rho\vartheta^{\mathfrak{A}}_{\pm}(\phi)(y,z) = \varepsilon^{il}\phi(e_i,e_l)\langle y,z\rangle - \varepsilon^{il}\phi(y,e_l)\langle e_i,z\rangle\\
&\quad\quad \mp\varepsilon^{il}\phi(e_i,J_\pm e_l)\langle y,J_\pm z\rangle \pm \varepsilon^{il}\phi(y,J_\pm e_l)\langle e_i,J_\pm z\rangle\\
&\quad\quad \pm 2\varepsilon^{il}\phi(z,J_\pm e_l)\langle e_i,J_\pm y\rangle\\
&\quad = 0 - \phi(y,z) \mp 0 \pm \phi(y,J_\pm J_\pm z) \pm 2\phi(z,J_\pm J_\pm y)\\
&\quad = -\phi(y,z) + \phi(y,z) + 2\phi(z,y) = 2\phi(z,y).
\end{aligned}$$

We examine ρ_{13}:

$$\begin{aligned}
&\rho_{13}\vartheta_\pm^{\mathfrak{A}}(\phi)(y,w) = \varepsilon^{ik}\phi(e_i,w)\langle y,e_k\rangle - \varepsilon^{ik}\phi(y,w)\langle e_i,e_k\rangle \\
&\quad \mp\varepsilon^{ik}\phi(e_i,J_\pm w)\langle y,J_\pm e_k\rangle \pm \varepsilon^{ik}\phi(y,J_\pm w)\langle e_i,J_\pm e_k\rangle \\
&\quad \pm 2\varepsilon^{ik}\phi(e_k,J_\pm w)\langle e_i,J_\pm y\rangle \\
&\quad = \phi(y,w) - m\phi(y,w) \pm \phi(J_\pm y,J_\pm w) \pm 0 \pm 2\phi(J_\pm y,J_\pm w) \\
&\quad = -(m+2)\phi(y,w), \\
&\rho_{13}(\sigma_{\pm,1}^{\mathfrak{A}}\phi_{\pm,1})(y,w) = \varepsilon^{ik}\phi_{\pm,1}(e_i,e_k)\langle y,w\rangle - \varepsilon^{ik}\phi_{\pm,1}(y,e_k)\langle e_i,w\rangle \\
&\quad \pm\varepsilon^{ik}\phi_{\pm,1}(e_i,J_\pm e_k)\langle J_\pm y,w\rangle \mp \varepsilon^{ik}\phi_{\pm,1}(y,J_\pm e_k)\langle J_\pm e_i,w\rangle \\
&\quad \pm 2\varepsilon^{ik}\phi_{\pm,1}(e_i,J_\pm y)\langle J_\pm e_k,w\rangle \\
&\quad = 0 - \phi_{\pm,1}(y,w) \pm 0 \pm \phi_{\pm,1}(y,J_\pm J_\pm w) \mp 2\phi_{\pm,1}(J_\pm w,J_\pm y) \\
&\quad = 2\phi_{\pm,1}(y,w), \\
&\rho(\sigma_{\pm,3}^{\mathfrak{A}}\psi_{\pm,3})(y,w) = \varepsilon^{ik}\psi_{\pm,3}(e_i,e_k)\langle y,w\rangle - \varepsilon^{ik}\psi_{\pm,3}(y,e_k)\langle e_i,w\rangle \\
&\quad +2\varepsilon^{ik}\psi_{\pm,3}(e_i,y)\langle e_k,w\rangle \pm \varepsilon^{ik}\psi_{\pm,3}(e_i,J_\pm e_k)\langle J_\pm y,w\rangle \\
&\quad \mp\varepsilon^{ik}\psi_{\pm,3}(y,J_\pm e_k)\langle J_\pm e_i,w\rangle \\
&\quad = 0 - \psi_{\pm,3}(y,w) + 2\psi_{\pm,3}(w,y) \pm 0 \pm \psi_{\pm,3}(y,J_\pm J_\pm w) \\
&\quad = -2\psi_{\pm,3}(y,w).
\end{aligned}$$

We introduce temporary notation and set:

$$\begin{aligned}
\tilde\pi_{\pm,7}^{\mathfrak{A}} &:= -\tfrac{1}{m^2+4m}\{2\sigma_{\pm,1}^{\mathfrak{A}} + (m+2)\vartheta_\pm^{\mathfrak{A}}\}, \\
\tilde\pi_{\pm,8}^{\mathfrak{A}} &:= -\tfrac{1}{m^2+4m}\{-2\sigma_{\pm,3}^{\mathfrak{A}} + (m+2)\vartheta_\pm^{\mathfrak{A}}\}.
\end{aligned}$$

We show that $\tilde\pi_{\pm,7}^{\mathfrak{A}}$ and $\tilde\pi_{\pm,8}^{\mathfrak{A}}$ split the action of ρ_{13} described in Equation (5.6.b) and complete the proof by using Assertion (2) to see:

$$\begin{aligned}
\rho\tilde\pi_{\pm,7}^{\mathfrak{A}}\phi_{\pm,1} &= -\tfrac{1}{m^2+4m}\{-(m+2)2 + 2(m+2)\}\phi_{\pm,1} = 0, \\
\rho_{13}\tilde\pi_{\pm,7}^{\mathfrak{A}}\phi_{\pm,1} &= -\tfrac{1}{m^2+4m}\{4-(m+2)^2\}\phi_{\pm,1} = \phi_{\pm,1}, \\
\rho\tilde\pi_{\pm,8}^{\mathfrak{A}}\psi_{\pm,3} &= -\tfrac{1}{m^2+4m}\{(m+2)2 - 2(m+2)\}\psi_{\pm,3} = 0, \\
\rho_{13}\tilde\pi_{\pm,8}^{\mathfrak{A}}\psi_{\pm,3} &= -\tfrac{1}{m^2+4m}\{4-(m+2)^2\}\phi_{\pm,3} = \psi_{\pm,3}.
\end{aligned}$$ □

We now examine the remaining orthogonal projections.

Lemma 5.6.4 *Let $\pi_{\pm,i}$ for $i = 9, 10, 11$ be orthogonal projection on the modules for the group $\mathcal{U}_\pm^\star$ $W_{\pm,i}^{\mathfrak{A}}$. Let $A \in \mathfrak{K}_{\pm,\mp}^{\mathfrak{A}} \cap \ker(\rho)$.*

(1) If $\rho_{13}(A) \in \Lambda_{0,\mp}^{2,\mathcal{U}_\pm}$, then $\pi_{\pm,9}^{\mathfrak{A}}(A)(x,y,z,w)$

$$= \tfrac{1}{4}\{A(x,y,z,w) + A(y,x,w,z) + A(z,w,x,y) + A(w,z,y,x)\}.$$

(2) If $\rho_{13}(A) \in S_{0,\mp}^{2,\mathcal{U}_\pm}$*, then* $\pi_{\pm,10}(A)(x,y,z,w) = \pm\frac{1}{4}\{A(x,y,z,J_\pm J_\pm w)$
$+A(y,x,J_\pm w,J_\pm z) + A(z,J_\pm w,x,J_\pm y) + A(J_\pm w,z,y,J_\pm x)\}$.

(3) $\pi_{\pm,11}(A) = \operatorname{Id} - \pi_{\pm,9} - \pi_{\pm,10}$.

Proof. Clearly $\pi^{\mathfrak{A}}_{\pm,9}(A)$ is anti-symmetric in (x,y). We verify that $\pi^{\mathfrak{A}}_{\pm,9}(A)$ satisfies the Bianchi identity and show $\pi^{\mathfrak{A}}_{\pm,9}(A) \in \mathfrak{A}$ by computing:

$$\begin{aligned}
&\pi^{\mathfrak{A}}_{\pm,9}(A)(x,y,z,w) + \pi^{\mathfrak{A}}_{\pm,9}(A)(y,z,x,w) + \pi^{\mathfrak{A}}_{\pm,9}(A)(z,x,y,w)\\
&\quad= \tfrac{1}{4}\{A(x,y,z,w) + A(y,x,w,z) + A(z,w,x,y) + A(w,z,y,x)\}\\
&\qquad+\tfrac{1}{4}\{A(y,z,x,w) + A(z,y,w,x) + A(x,w,y,z) + A(w,x,z,y)\}\\
&\qquad+\tfrac{1}{4}\{A(z,x,y,w) + A(x,z,w,y) + A(y,w,z,x) + A(w,y,x,z)\}\\
&\quad= \tfrac{1}{4}\{A(w,z,y,x) + A(z,y,w,x) + A(y,w,z,x)\}\\
&\qquad+\tfrac{1}{4}\{A(z,w,x,y) + A(w,x,z,y) + A(x,z,w,y)\}\\
&\qquad+\tfrac{1}{4}\{A(y,x,w,z) + A(x,w,y,z) + A(w,y,x,z)\}\\
&\qquad+\tfrac{1}{4}\{A(x,y,z,w) + A(y,z,x,w) + A(z,x,y,w)\}\\
&\quad= 0.
\end{aligned}$$

We show $\pi^{\mathfrak{A}}_{\pm,9}(A) \in \mathfrak{K}^{\mathfrak{A}}_{\pm,\mp}$ by comparing:

$$\begin{aligned}
&\pi^{\mathfrak{A}}_{\pm,9}(A)(x,y,z,w) = \tfrac{1}{4}\{A(x,y,z,w) + A(y,x,w,z)\\
&\qquad+A(z,w,x,y) + A(w,z,y,x)\},\\
&\pi^{\mathfrak{A}}_{\pm,9}(A)(x,y,J_\pm z,J_\pm w) = \tfrac{1}{4}\{A(x,y,J_\pm z,J_\pm w) + A(y,x,J_\pm w,J_\pm z)\\
&\qquad+A(J_\pm z,J_\pm w,x,y) + A(J_\pm w,J_\pm z,y,x)\},\\
&\pi^{\mathfrak{A}}_{\pm,9}(A)(J_\pm x,J_\pm y,z,w) = \tfrac{1}{4}\{A(J_\pm x,J_\pm y,z,w) + A(J_\pm y,J_\pm x,w,z)\\
&\qquad+A(z,w,J_\pm x,J_\pm y) + A(w,z,J_\pm y,J_\pm x)\}.
\end{aligned}$$

It is immediate that $\pi^{\mathfrak{A}}_{\pm,9}(A)$ is anti-symmetric in the last two indices; thus $\rho(\pi^{\mathfrak{A}}_{\pm,9}(A)) = -\rho_{13}(\pi^{\mathfrak{A}}_{\pm,9}(A))$. We show $\pi^{\mathfrak{A}}_{\pm,9}(A) \in \ker(\rho)$ and therefore that $\pi^{\mathfrak{A}}_{\pm,9}(A)$ takes values in $W_{\pm,9}$ by computing:

$$\begin{aligned}
&\rho(\pi^{\mathfrak{A}}_{\pm,9}(A))(y,z) = \tfrac{1}{4}\varepsilon^{il}A(e_i,y,z,e_l) + \tfrac{1}{4}\varepsilon^{il}A(y,e_i,e_l,z)\\
&\qquad+\tfrac{1}{4}\varepsilon^{il}A(z,e_l,e_i,y) + \tfrac{1}{4}\varepsilon^{il}A(e_l,z,y,e_i)\\
&\quad= \tfrac{1}{4}\{\rho(y,z) - \rho_{13}(y,z) - \rho_{13}(z,y) + \rho(z,y)\} = 0.
\end{aligned}$$

Suppose A is anti-symmetric in (z,w). Then $A \in \mathfrak{R}$ by Lemma 4.1.1; it is then immediate that $\pi^{\mathfrak{A}}_{\pm,9}(A)(x,y,z,w) = A(x,y,z,w)$. This completes the proof of Assertion (1).

Since the map T of Lemma 5.6.1 is an (anti)-isometry, Lemma 5.6.1 shows that $\pm T\pi_{\pm,9}T = \pi_{\pm,10}$; Assertion (2) now follows from Assertion

(1). Assertion (3) is immediate from Assertions (1) and (2) and from Theorem 5.4.1 and Theorem 5.5.1. □

The analysis performed previously shows that

$$\{\ker(\rho) \cap \mathfrak{K}^{\mathfrak{A}}_{\pm,\pm}, \ker(\rho) \cap \mathfrak{K}^{\mathfrak{A}}_{\pm,\mp}, S_{+}^{2,\mathcal{U}_\pm}, S_{-}^{2,\mathcal{U}_\pm}, \Lambda_{+}^{2,\mathcal{U}_\pm}, \Lambda_{-}^{2,\mathcal{U}_\pm}\}$$

are inequivalent modules when the structure group is $\mathcal{U}_\pm^\star$; $S_{+}^{2,\mathcal{U}_-} \approx \Lambda_{+}^{2,\mathcal{U}_-}$ as a module when the structure group is $\mathcal{U}_-$. Thus these are inequivalent $\mathrm{GL}_\pm^\star$ modules. Furthermore the following modules are irreducible:

$$\{S_{+}^{2,\mathcal{U}_\pm}, S_{-}^{2,\mathcal{U}_\pm}, \Lambda_{+}^{2,\mathcal{U}_\pm}, \Lambda_{-}^{2,\mathcal{U}_\pm}\}.$$

The decomposition $\mathfrak{K}^{\mathfrak{A}}_{\pm} = \mathfrak{K}^{\mathfrak{A}}_{\pm,+} \oplus \mathfrak{K}^{\mathfrak{A}}_{\pm,-}$ is a decomposition of modules with structure group $\mathrm{GL}_\pm^\star$. By Lemma 5.6.2, we have the following decompositions as modules with structure group $\mathrm{GL}_\pm^\star$:

$$\begin{aligned}
\mathfrak{K}^{\mathfrak{A}}_{\pm,\pm} &= \{\ker(\rho) \cap \mathfrak{K}_{\pm,\pm}\} \oplus \sigma^{\mathfrak{A}}_{\pm,2} S_{\pm}^{2,\mathcal{U}_\pm} \oplus \sigma^{\mathfrak{A}}_{\pm,4} \Lambda_{\pm}^{2,\mathcal{U}_\pm}, \\
\mathfrak{K}^{\mathfrak{A}}_{\pm,\mp} &= \{\ker(\rho) \cap \mathfrak{K}_{\pm,\mp}\} \oplus \sigma^{\mathfrak{A}}_{\pm,1} S_{\mp}^{2,\mathcal{U}_\pm} \oplus \sigma^{\mathfrak{A}}_{\pm,3} \Lambda_{\mp}^{2,\mathcal{U}_\pm}.
\end{aligned}$$

By Theorem 5.4.1 and Theorem 5.5.1,

$$\{\ker(\rho) \cap \mathfrak{K}^{\mathfrak{A}}_{\pm,\pm}\}$$

is an irreducible $\mathcal{U}_\pm^\star$ and module for the group $\mathcal{U}_-$ and hence necessarily an irreducible module with respect to the structure group $\mathrm{GL}_\pm^\star$. Thus to complete the proof of Theorem 1.5.1 for the group $\mathrm{GL}_\pm^\star$, we need only show:

Theorem 5.6.1 *$\ker(\rho) \cap \mathfrak{K}^{\mathfrak{A}}_{\pm}$ is an irreducible module with structure group $\mathrm{GL}_\pm^\star$.*

Proof. We extend arguments given in [Brozos-Vázquez, Gilkey, and Nikčević (2011b)] from the positive definite setting to the indefinite and the para-Hermitian settings. We suppose to the contrary that ξ is a non-trivial proper submodule of $\ker(\rho) \cap \mathfrak{K}^{\mathfrak{A}}_{\pm}$ with respect to the structure group $\mathrm{GL}_\pm^\star$. We introduce an auxiliary para-Hermitian inner product (+) or an auxiliary pseudo-Hermitian inner product (−) $\langle \cdot, \cdot \rangle$. Necessarily, then ξ is a non-trivial proper submodule for the group $\mathcal{U}_\pm^\star$ so we may apply Theorem 5.4.1 and Theorem 5.5.1 to see that there is a set of indices $I \subset \{7, 8, 9, 10, 11\}$ so that:

$$\xi = \oplus_{i \in I} W^{\mathfrak{A}}_{\pm,i}.$$

We choose an orthonormal basis $\{e_1, f_1, ..., e_{\bar{m}}, f_{\bar{m}}\}$ for V so $J_\pm e_i = f_i$ and $J_\pm f_i = \pm e_i$. All elements of $\otimes^4 V$ considered in the proof of Theorem 5.6.1 will be anti-symmetric in the first two indices.

Case I: Suppose that $W_{\pm,9} \subset \xi$. Let $A_{\pm,9}$ be determined by the relations:

$$\begin{array}{ll}
A_{\pm,9}(e_1,f_1,e_1,f_2) = -1, & A_{\pm,9}(e_1,f_1,f_1,e_2) = 1, \\
A_{\pm,9}(e_1,f_1,e_2,f_1) = -1, & A_{\pm,9}(e_1,f_1,f_2,e_1) = 1, \\
A_{\pm,9}(e_1,f_2,e_1,f_1) = -1, & A_{\pm,9}(e_1,f_2,f_1,e_1) = 1, \\
A_{\pm,9}(e_1,f_2,e_2,f_2) = 1, & A_{\pm,9}(e_1,f_2,f_2,e_2) = -1, \\
A_{\pm,9}(f_1,e_2,e_1,f_1) = 1, & A_{\pm,9}(f_1,e_2,f_1,e_1) = -1, \\
A_{\pm,9}(f_1,e_2,e_2,f_2) = -1, & A_{\pm,9}(f_1,e_2,f_2,e_2) = 1, \\
A_{\pm,9}(e_2,f_2,e_1,f_2) = 1, & A_{\pm,9}(e_2,f_2,f_1,e_2) = -1, \\
A_{\pm,9}(e_2,f_2,e_2,f_1) = 1, & A_{\pm,9}(e_2,f_2,f_2,e_1) = -1.
\end{array}$$

It is then immediate by inspection that $A_{\pm,9} \in W_{\pm,9}$. We adopt the notation established in the proof of Lemma 4.4.2. Let

$$g_{1,\varepsilon}(e_i) := \left\{ \begin{array}{ll} \varepsilon e_1 & \text{if } i = 1 \\ e_i & \text{if } i \neq 1 \end{array} \right\}, \quad g_{1,\varepsilon}(e^i) := \left\{ \begin{array}{ll} \varepsilon^{-1} e^1 & \text{if } i = 1 \\ e^i & \text{if } i \neq 1 \end{array} \right\},$$
$$g_{1,\varepsilon}(f_i) := \left\{ \begin{array}{ll} \varepsilon f_1 & \text{if } i = 1 \\ f_i & \text{if } i \neq 1 \end{array} \right\}, \quad g_{1,\varepsilon}(f^i) := \left\{ \begin{array}{ll} \varepsilon^{-1} f^1 & \text{if } i = 1 \\ f^i & \text{if } i \neq 1 \end{array} \right\}.$$

Since ξ is a linear subspace of finite dimension, it is closed. Consequently

$$B_{\pm,1} := \lim_{\varepsilon \to 0} \varepsilon g_{1,\varepsilon}^* A_{\pm,9} \in \xi.$$

The non-zero components of $B_{\pm,1}$ and ρ_{13} are determined by:

$$\begin{array}{ll}
B_{\pm,1}(e_2,f_2,e_2,f_1) = 1, & B_{\pm,1}(e_2,f_2,f_2,e_1) = -1, \\
\rho_{13}(B_{\pm,1})(e_2,e_1) = \mp 1, & \rho_{13}(B_{\pm,1})(f_2,f_1) = 1.
\end{array} \tag{5.6.c}$$

By interchanging the roles of $\{e_1, f_1\}$ and $\{e_2, f_2\}$ we can create an element $B_{\pm,2} \in \xi$ with

$$\begin{array}{ll}
B_{\pm,2}(e_1,f_1,e_1,f_2) = 1, & B_{\pm,2}(e_1,f_1,f_1,e_2) = -1, \\
\rho_{13}(B_{\pm,2})(e_1,e_2) = \mp 1, & \rho_{13}(B_{\pm,2})(f_1,f_2) = 1.
\end{array}$$

Thus $B_{\pm,1} + B_{\pm,2}$ has a non-zero component in $W_{\pm,7}^{\mathfrak{A}}$ and $B_{\pm,1} - B_{\pm,2}$ has a non-zero component in $W_{\pm,8}^{\mathfrak{A}}$. This shows that:

$$W_{\pm,9}^{\mathfrak{A}} \subset \xi \quad \Rightarrow \quad W_{\pm,7}^{\mathfrak{A}} \oplus W_{\pm,8}^{\mathfrak{A}} \subset \xi.$$

Let $B^*_{\pm,i} := TB_{\pm,i}$. Instead of studying $\pi^{\mathfrak{A}}_{\pm,10}(B_{\pm,1}+B_{\pm,2})$, we examine $\pi^{\mathfrak{A}}_{\pm,9}(B^*_{\pm,1}+B^*_{\pm,2})$. We have

$$\begin{aligned}
(B^*_{\pm,1}+B^*_{\pm,2})(e_1,f_1,e_1,e_2)&=1, & (B^*_{\pm,1}+B^*_{\pm,2})(e_1,f_1,f_1,f_2)&=\mp1,\\
(B^*_{\pm,1}+B^*_{\pm,2})(e_2,f_2,e_2,e_1)&=1, & (B^*_{\pm,1}+B^*_{\pm,2})(e_2,f_2,f_2,f_1)&=\mp1,\\
\rho_{13}(B^*_{\pm,1}+B^*_{\pm,2})(f_1,e_2)&=1, & \rho_{13}(B^*_{\pm,1}+B^*_{\pm,2})(e_2,f_1)&=-1,\\
\rho_{13}(B^*_{\pm,1}+B^*_{\pm,2})(f_2,e_1)&=1, & \rho_{13}(B^*_{\pm,1}+B^*_{\pm,2})(e_1,f_2)&=-1.
\end{aligned}$$

Since $\rho_{13}(B^*_{\pm,1}+B^*_{\pm,2})$ is anti-symmetric, we have by Lemma 5.6.4 that:

$$\pi^{\mathfrak{A}}_{\pm,9}(B^*_{\pm,1}+B^*_{\pm,2})(e_1,f_1,e_1,e_2)=\tfrac14.$$

Consequently $\pi^{\mathfrak{A}}_{\pm,10}(B_{\pm,1}+B_{\pm,2})\neq 0$. This implies:

$$W^{\mathfrak{A}}_{\pm,9}\subset\xi \quad\Rightarrow\quad W^{\mathfrak{A}}_{\pm,10}\subset\xi.$$

Suppose $m\geq 6$. Set

$$\begin{aligned}
g_{2,\varepsilon}(e_i)&:=\left\{\begin{array}{ll} e_3-\varepsilon e_1 & \text{if } i=3\\ e_i & \text{if } i\neq 3\end{array}\right\}, & g_{2,\varepsilon}(e^i)&:=\left\{\begin{array}{ll} e^1+\varepsilon e^3 & \text{if } i=1\\ e^i & \text{if } i\neq 1\end{array}\right\},\\
g_{2,\varepsilon}(f_i)&:=\left\{\begin{array}{ll} f_3-\varepsilon f_1 & \text{if } i=3\\ f_i & \text{if } i\neq 3\end{array}\right\}, & g_{2,\varepsilon}(f^i)&:=\left\{\begin{array}{ll} f^1+\varepsilon f^3 & \text{if } i=1\\ f^i & \text{if } i\neq 1\end{array}\right\}.
\end{aligned}$$

Let $B_{\pm,3}:=\partial_\varepsilon\{g^*_{2,\varepsilon}A_{\pm,9}\}|_{\varepsilon=0}$. We then have:

$$\begin{aligned}
B_{\pm,3}(e_1,f_1,e_2,f_3)&=-1, & B_{\pm,3}(e_1,f_1,f_2,e_3)&=1,\\
B_{\pm,3}(e_1,f_2,e_1,f_3)&=-1, & B_{\pm,3}(e_1,f_2,f_1,e_3)&=1,\\
B_{\pm,3}(f_1,e_2,e_1,f_3)&=1, & B_{\pm,3}(f_1,e_2,f_1,e_3)&=-1,\\
B_{\pm,3}(e_2,f_2,e_2,f_3)&=1, & B_{\pm,3}(e_2,f_2,f_2,e_3)&=-1.
\end{aligned}$$

Note that $B_{\pm,3}\in\ker(\rho_{13})$. Let $\eta=e_3$ or $\eta=f_3$. Then

$$B_{\pm,3}(\eta,*,*,*)=B_{\pm,3}(*,\eta,*,*)=B_{\pm,3}(*,*,\eta,*)=0.$$

Since $|B^{\mathfrak{A}}_{\pm,3}(*,*,*,*)|\leq 1$, we have

$$\left|\pi^{\mathfrak{A}}_{\pm,9}B_{\pm,3}(e_1,f_1,e_2,\eta)\right|\leq\tfrac14 \quad\text{and}\quad \left|\pi^{\mathfrak{A}}_{\pm,10}B_{\pm,3}(e_1,f_1,e_2,\eta)\right|\leq\tfrac14.$$

The component of $B_{\pm,3}$ in W_{11} is non-trivial because

$$\{\pi^{\mathfrak{A}}_{\pm,9}B_{\pm,3}+\pi^{\mathfrak{A}}_{\pm,10}+\pi^{\mathfrak{A}}_{\pm,11}B_{\pm,3}\}(e_1,f_1,e_2,e_3)=1.$$

Consequently $W^{\mathfrak{A}}_{\pm,11}\subset\xi$. We summarize our conclusions:

$$W^{\mathfrak{A}}_{\pm,9}\subset\xi \quad\Rightarrow\quad \xi=\mathfrak{K}^{\mathfrak{A}}_{\pm,\mp}\cap\ker(\rho).$$

Case II: Suppose that $W_{\pm,10} \subset \xi$. We use Lemma 5.6.1 to interchange the roles of $W^{\mathfrak{A}}_{\pm,9}$ and $W^{\mathfrak{A}}_{\pm,10}$ and then apply Case I to see:

$$W^{\mathfrak{A}}_{\pm,10} \subset \xi \quad \Rightarrow \quad \xi = \mathfrak{K}^{\mathfrak{A}}_{\pm,\mp} \cap \ker(\rho).$$

Case III: Suppose that $m \geq 6$ and that $W^{\mathfrak{A}}_{\pm,11} \subset \xi$. Set:

$$\begin{array}{ll}
A_{\pm,11}(e_1,e_2,e_1,e_3) = \mp 1, & A_{\pm,11}(e_1,e_2,f_1,f_3) = 1, \\
A_{\pm,11}(e_1,f_2,e_1,f_3) = -1, & A_{\pm,11}(e_1,f_2,f_1,e_3) = 1, \\
A_{\pm,11}(e_1,e_3,e_1,e_2) = \pm 1, & A_{\pm,11}(e_1,e_3,f_1,f_2) = -1, \\
A_{\pm,11}(e_1,f_3,e_1,f_2) = 1, & A_{\pm,11}(e_1,f_3,f_1,e_2) = -1, \\
A_{\pm,11}(f_1,e_2,e_1,f_3) = 1, & A_{\pm,11}(f_1,e_2,f_1,e_3) = -1, \\
A_{\pm,11}(f_1,f_2,e_1,e_3) = 1, & A_{\pm,11}(f_1,f_2,f_1,f_3) = \mp 1, \\
A_{\pm,11}(f_1,e_3,e_1,f_2) = -1, & A_{\pm,11}(f_1,e_3,f_1,e_2) = 1, \\
A_{\pm,11}(f_1,f_3,e_1,e_2) = -1, & A_{\pm,11}(f_1,f_3,f_1,f_2) = \pm 1.
\end{array}$$

We verify by inspection that $A_{\pm,11} \in \mathfrak{K}^{\mathfrak{A}}_{\pm,\mp} \cap \ker(\rho) \cap \ker(\rho_{13})$. We study:

$$\begin{aligned}
\pi^{\mathfrak{A}}_{\pm,9}(A_{\pm,11})(x,y,z,w) = \tfrac{1}{4}\{&A_{\pm,11}(x,y,z,w) + A_{\pm,11}(y,x,w,z) \\
&+A_{\pm,11}(z,w,x,y) + A_{\pm,11}(w,z,y,x)\}.
\end{aligned}$$

Set $U := \{e_2, f_2, e_3, f_3\}$. For this to be non-zero, either $x \in U$ or $y \in U$ and either $z \in U$ or $w \in U$. If x and z belong to U, then

$$\begin{aligned}
&A_{\pm,11}(x,y,z,w) = -A_{\pm,11}(z,w,x,y), \quad \text{and} \\
&A_{\pm,11}(y,x,w,z) = A_{\pm,11}(w,z,y,x) = 0.
\end{aligned}$$

Thus $\pi^{\mathfrak{A}}_{\pm,9}A_{\pm,11}(x,y,z,w) = 0$ in this special case. Since $\pi^{\mathfrak{A}}_{\pm,9}A_{\pm,11}$ is anti-symmetric in the first two indices and in the last two indices, we see that $\pi^{\mathfrak{A}}_{\pm,9}A_{\pm,11} = 0$ in the remaining cases. To examine $\pi^{\mathfrak{A}}_{\pm,10}$, we consider the dual tensor:

$$\begin{array}{ll}
A^*_{\pm,11}(e_1,e_2,e_1,f_3) = -1, & A^*_{\pm,11}(e_1,e_2,f_1,e_3) = 1, \\
A^*_{\pm,11}(e_1,f_2,e_1,e_3) = -1, & A^*_{\pm,11}(e_1,f_2,f_1,f_3) = \pm 1, \\
A^*_{\pm,11}(e_1,e_3,e_1,f_2) = 1, & A^*_{\pm,11}(e_1,e_3,f_1,e_2) = -1, \\
A^*_{\pm,11}(e_1,f_3,e_1,e_2) = 1, & A^*_{\pm,11}(e_1,f_3,f_1,f_2) = \mp 1, \\
A^*_{\pm,11}(f_1,e_2,e_1,e_3) = 1, & A^*_{\pm,11}(f_1,e_2,f_1,f_3) = \mp 1, \\
A^*_{\pm,11}(f_1,f_2,e_1,f_3) = \pm 1, & A^*_{\pm,11}(f_1,f_2,f_1,e_3) = \mp 1, \\
A^*_{\pm,11}(f_1,e_3,e_1,e_2) = -1, & A^*_{\pm,11}(f_1,e_3,f_1,f_2) = \pm 1, \\
A^*_{\pm,11}(f_1,f_3,e_1,f_2) = \mp 1, & A^*_{\pm,11}(f_1,f_3,f_1,e_2) = \pm 1.
\end{array}$$

Once again $x \in U$ and $z \in U$ implies

$$A^*_{\pm,11}(x,y,z,w) + A^*_{\pm,11}(z,w,x,y) = 0,$$
$$A^*_{\pm,11}(y,x,w,z) = A^*_{\pm,11}(w,z,y,x) = 0.$$

The argument given above to show $\pi^{\mathfrak{A}}_{\pm,9}A_{\pm,11} = 0$ then shows $\pi^{\mathfrak{A}}_{\pm,9}A^*_{\pm,11} = 0$ and hence $\pi^{\mathfrak{A}}_{\pm,10}A_{\pm,11} = 0$. Consequently since $\rho(A_{\pm,11}) = \rho_{13}(A_{\pm,11}) = 0$, we may conclude that $A_{\pm,11} \in W^{\mathfrak{A}}_{\pm,11}$. Set:

$$g_{3,\varepsilon}(e_i) := \left\{\begin{matrix} \varepsilon e_3 \text{ if } i=3 \\ e_i \quad \text{if } i \neq 3 \end{matrix}\right\}, \quad g_{3,\varepsilon}(e^i) := \left\{\begin{matrix} \varepsilon^{-1}e^3 \text{ if } i=3 \\ e^i \quad\quad \text{if } i \neq 3 \end{matrix}\right\},$$
$$g_{3,\varepsilon}(f_i) := \left\{\begin{matrix} \varepsilon f_3 \text{ if } i=3 \\ f_i \quad \text{if } i \neq 3 \end{matrix}\right\}, \quad g_{3,\varepsilon}(f^i) := \left\{\begin{matrix} \varepsilon^{-1}f^3 \text{ if } i=3 \\ f^i \quad\quad \text{if } i \neq 3 \end{matrix}\right\}.$$

We set $B_{\pm,4} := \lim_{\varepsilon\to 0} \varepsilon g^*_{3,\varepsilon} A_{\pm,11} \in \xi$. We see that the non-zero components of $B_{\pm,4}$ are determined by:

$$\begin{array}{ll} B_{\pm,4}(e_1,e_2,e_1,e_3) = \mp 1, & B_{\pm,4}(e_1,e_2,f_1,f_3) = 1, \\ B_{\pm,4}(e_1,f_2,e_1,f_3) = -1, & B_{\pm,4}(e_1,f_2,f_1,e_3) = 1, \\ B_{\pm,4}(f_1,e_2,e_1,f_3) = 1, & B_{\pm,4}(f_1,e_2,f_1,e_3) = -1, \\ B_{\pm,4}(f_1,f_2,e_1,e_3) = 1, & B_{\pm,4}(f_1,f_2,f_1,f_3) = \mp 1. \end{array}$$

We verify that $\rho(B_{\pm,4}) = \rho_{13}(B_{\pm,4}) = 0$. We use Lemma 5.6.4 to see:

$$\pi^{\mathfrak{A}}_{\pm,9}(B_{\pm,4})(e_1,e_2,e_1,e_3) = \tfrac14 B_{\pm,4}(e_1,e_2,e_1,e_3) = \mp\tfrac14,$$
$$\pi^{\mathfrak{A}}_{\pm,10}(B_{\pm,4})(e_1,e_2,e_1,e_3) = \pm\tfrac14 \pi^{\mathfrak{A}}_{\pm,9}(B^*_{\pm,4})(e_1,e_2,e_1,f_3)$$
$$= \tfrac14 B_{\pm,4}(e_1,e_2,e_1,e_3) = \mp\tfrac14.$$

We may conclude:

$$\text{if } m \geq 6 \text{ then } W^{\mathfrak{A}}_{\pm,11} \subset \xi \;\Rightarrow\; W^{\mathfrak{A}}_{\pm,9} \oplus W^{\mathfrak{A}}_{\pm,10} \subset \xi$$
$$\Rightarrow\; W^{\mathfrak{A}}_{\pm,7} \oplus W^{\mathfrak{A}}_{\pm,8} \subset \xi \;\Rightarrow\; \xi = \mathfrak{K}^{\mathfrak{A}}_{\pm,\mp} \cap \ker(\rho).$$

Case IV: Suppose that $W^{\mathfrak{A}}_{\pm,7} \subset \xi$. Let

$$\phi := e^1 \otimes e^2 + e^2 \otimes e^1 \mp f^1 \otimes f^2 \mp f^2 \otimes f^1 \in S^{2,\mathcal{U}_\pm}_{0,\mp}.$$

We use Lemma 5.6.3 to find $A_{\pm,7} \in W^{\mathfrak{A}}_{\pm,7}$ so that $\rho_{13}A_{\pm,7} = \phi$. We shall not compute all the terms in $A_{\pm,7}$ as this would be a bit of a bother and shall content ourselves with determining:

$$c_1 := A_{\pm,7}(e_2,f_2,e_2,e_1) \quad \text{and} \quad c_2 := A_{\pm,7}(e_2,f_2,e_2,f_1).$$

We observe:

$$\begin{aligned}
&\langle\sigma^{\mathfrak{A}}_{\pm,1}\phi(e_2,f_2)e_2,e_1\rangle=0,\\
&\langle\sigma^{\mathfrak{A}}_{\pm,1}\phi(e_2,f_2)e_2,f_1\rangle=0,\\
&\vartheta^{\mathfrak{A}}_{\pm}(\phi)(e_2,f_2,e_2,e_1):=\phi(e_2,e_1)\langle f_2,e_2\rangle-\phi(f_2,e_1)\langle e_2,e_2\rangle\\
&\qquad\mp\phi(e_2,J_\pm e_1)\langle f_2,J_\pm e_2\rangle\pm\phi(f_2,J_\pm e_1)\langle e_2,J_\pm e_2\rangle\\
&\qquad\pm2\phi(e_2,J_\pm e_1)\langle e_2,J_\pm f_2\rangle=0,\\
&\vartheta^{\mathfrak{A}}_{\pm}(\phi)(e_2,f_2,e_2,f_1):=\phi(e_2,f_1)\langle f_2,e_2\rangle-\phi(f_2,f_1)\langle e_2,e_2\rangle\\
&\qquad\mp\phi(e_2,J_\pm f_1)\langle f_2,J_\pm e_2\rangle\pm\phi(f_2,J_\pm f_1)\langle e_2,J_\pm e_2\rangle\\
&\qquad\pm2\phi(e_2,J_\pm f_1)\langle e_2,J_\pm f_2\rangle=0\pm1\pm1\pm0\pm2\neq0.
\end{aligned}$$

Thus by Lemma 5.6.3, $c_1=0$ and $c_2\neq0$. Let $\Phi\in\mathcal{U}_\pm$ be reflection in the subspace perpendicular to the plane determined by $\{e_1,f_1\}$. This means that:

$$\Phi e_i:=\left\{\begin{array}{r}-e_1\text{ if }i=1\\ e_i\text{ if }i>1\end{array}\right\},\qquad \Phi f_i:=\left\{\begin{array}{r}-f_1\text{ if }i=1\\ f_i\text{ if }i>1\end{array}\right\}.$$

Since $\Phi^*\phi=-\phi$, we have $\Phi^*A_{\pm,7}=-A_{\pm,7}$. If $A_{\pm,7}(x_1,x_2,x_3,x_4)\neq0$, then the number of times that x_i is either f_1 or e_1 must be odd. Similarly the number of times that $x_i\in\{e_2,f_2\}$ is odd. Define $g_{4,\varepsilon_1,\varepsilon_2}\in\mathrm{GL}_\pm$ by setting:

$$g_{4,\varepsilon_1,\varepsilon_2}e_i=\left\{\begin{array}{ll}\varepsilon_1e_1&\text{if }i=1\\ \varepsilon_2e_2&\text{if }i=2\\ e_i&\text{if }i\geq3\end{array}\right\},\quad g_{4,\varepsilon_1,\varepsilon_2}e^i=\left\{\begin{array}{ll}\varepsilon_1^{-1}e^1&\text{if }i=1\\ \varepsilon_2^{-1}e^2&\text{if }i=2\\ e^i&\text{if }i\geq3\end{array}\right\},$$

$$g_{4,\varepsilon_1,\varepsilon_2}f_i=\left\{\begin{array}{ll}\varepsilon_1f_1&\text{if }i=1\\ \varepsilon_2f_2&\text{if }i=2\\ f_i&\text{if }i\geq3\end{array}\right\},\quad g_{4,\varepsilon_1,\varepsilon_2}f^i=\left\{\begin{array}{ll}\varepsilon_1^{-1}f^1&\text{if }i=1\\ \varepsilon_2^{-1}f^2&\text{if }i=2\\ f^i&\text{if }i\geq3\end{array}\right\}.$$

We may expand $g^*_{4,\varepsilon_1,\varepsilon_2}A_{\pm,7}$ as a finite Laurent polynomial in $\{\varepsilon_1,\varepsilon_2\}$. Since $g^*_{4,\varepsilon_1,\varepsilon_2}A_{\pm,7}\in\xi$, all the coefficient curvature tensors also belong to ξ. Let $B_{\pm,5}\in\xi$ be the coefficient of $\varepsilon_1^{-1}\varepsilon_2^3$ in $g^*_{4,\varepsilon_1,\varepsilon_2}A_{\pm,7}$;

$$B_{\pm,5}=\left\{\tfrac16\varepsilon_1\partial^3_{\varepsilon_2}g^*_{4,\varepsilon_1,\varepsilon_2}A_{\pm,7}\right\}\Big|_{\varepsilon_1=0,\varepsilon_2=0}.$$

The only (possibly) non-zero components of $B_{\pm,5}$ are given by:

$$\begin{aligned}
&B_{\pm,5}(e_2,f_2,e_2,e_1)=A_{\pm,7}(e_2,f_2,e_2,e_1)=0,\\
&B_{\pm,5}(e_2,f_2,e_2,f_1)=A_{\pm,7}(e_2,f_2,e_2,f_1)=c_2,\\
&B_{\pm,5}(e_2,f_2,f_2,e_1)=-B_{\pm,5}(e_2,f_2,e_2,f_1)=-c_2,\\
&B_{\pm,5}(e_2,f_2,f_2,f_1)=\mp B_{\pm,5}(e_2,f_2,e_2,e_1)=0.
\end{aligned}$$

The only non-zero components of $\rho_{13}(B_{\pm,5})$ are given by:

$$\rho_{13}(B_{\pm,5})(e_2,e_1) = \mp c_2 \quad \text{and} \quad \rho_{13}(B_{\pm,5})(f_2,f_1) = c_2. \tag{5.6.d}$$

Interchanging the roles of the indices "1" and "2" is an isometry preserving ϕ; this creates a tensor $B_{\pm,6} \in \xi$ so that

$$\begin{aligned} &B_{\pm,6}(e_1,f_1,f_1,e_2) = -c_2, \quad B_{\pm,6}(e_1,f_1,e_1,f_2) = c_2, \\ &\rho_{13}(B_{\pm,6})(e_1,e_2) = \mp c_2, \quad \rho_{13}(B_{\pm,6})(f_1,f_2) = c_2. \end{aligned}$$

In particular $B_{\pm,5} - B_{\pm,6}$ has an anti-symmetric Ricci tensor so we may use Lemma 5.6.4 to compute

$$\pi^{\mathfrak{A}}_{\pm,9}(B_{\pm,5} - B_{\pm,6})(e_2,f_2,e_2,f_1) = \tfrac{1}{4}c_2 \neq 0.$$

This implies $W_{\pm,9} \subset \xi$ and hence by Case I,

$$W^{\mathfrak{A}}_{\pm,7} \subset \xi \quad \Rightarrow \quad \xi = \mathfrak{K}^{\mathfrak{A}}_{\pm,\mp} \cap \ker(\rho).$$

Case V: Suppose that $W^{\mathfrak{A}}_{\pm,8} \subset \xi$. We use the duality operator, Lemma 5.6.1, and Case IV to see that

$$W^{\mathfrak{A}}_{\pm,8} \subset \xi \quad \Rightarrow \quad \xi = \mathfrak{K}^{\mathfrak{A}}_{\pm,\mp} \cap \ker(\rho).$$

The proof Theorem 5.6.1, which deals with the group $\mathrm{GL}^\star_\pm$, now follows from the discussion given above. □

We now turn to the consideration of $\mathfrak{K}^{\mathfrak{A}}_- \cap \ker(\rho)$ as a module for the group GL_-. The analysis is just a bit different since $W^{\mathfrak{A}}_{-,7} \approx W^{\mathfrak{A}}_{-,8}$ and $W^{\mathfrak{A}}_{-,9} \approx W^{\mathfrak{A}}_{-,10}$ as modules for the group $\mathcal{U}_-$. We suppose $m \geq 6$ to simplify the discussion. We begin our study with the following technical observation:

Lemma 5.6.5

(1) If W is a non-trivial proper $\mathcal{U}_-$ invariant subspace of $W^{\mathfrak{A}}_{-,7} \oplus W^{\mathfrak{A}}_{-,8}$, then there exists $(a_1,a_2) \neq 0$ so that $W = \{a_1A + a_2TA\}_{A \in W^{\mathfrak{A}}_{-,7}}$.

(2) If W is a non-trivial proper $\mathcal{U}_-$ invariant subspace of $W^{\mathfrak{A}}_{-,9} \oplus W^{\mathfrak{A}}_{-,10}$, then there exists $(b_1,b_2) \neq 0$ so that $W = \{b_1A + b_2TA\}_{A \in W^{\mathfrak{A}}_{-,9}}$.

Proof. By Lemma 5.1.1, $\dim\{\mathcal{I}_2^{\mathcal{U}_-}(\mathfrak{K}^{\mathfrak{A}}_-)\} \leq 20$. On the other hand, in the decomposition of $\mathfrak{K}^{\mathfrak{A}}_-$, which is discussed in Theorem 5.4.1, the four modules $\{S^{2,\mathcal{U}_-}_-, \Lambda^{2,\mathcal{U}_-}_-, W^{\mathfrak{A}}_{-,11}, W^{\mathfrak{A}}_{-,12}\}$ each appear with multiplicity one, the two modules $\{\mathbb{R}, W^{\mathfrak{A}}_{-,9}\}$ each appear with multiplicity two, and the single module $S^{2,\mathcal{U}_-}_{0,+}$ appears with multiplicity four. Applying Lemma 2.2.2 yields the inequality:

$$\dim\left\{\mathcal{I}_2^{\mathcal{U}_-}(\mathfrak{K}_-)\right\} \geq 4\cdot 1 + 2\cdot 3 + 1\cdot 10 = 20.$$

Thus equality holds and Lemma 2.2.1 shows

$$\dim\left\{\mathrm{Hom}_{\mathcal{U}_-}(W^{\mathfrak{A}}_{-,7}, W^{\mathfrak{A}}_{-,7})\right\} = \dim\left\{\mathrm{Hom}_{\mathcal{U}_-}(W^{\mathfrak{A}}_{-,9}, W^{\mathfrak{A}}_{-,9})\right\} = 1.$$

The desired conclusions now follow from Lemma 2.2.2. □

We now establish the second main result of Section 5.6 and establish Theorem 1.5.1 for the group GL_-:

Theorem 5.6.2 $\ker(\rho)\cap\mathfrak{K}^{\mathfrak{A}}_-$ *is an irreducible module for the structure group* GL_-.

Proof. Let ξ be a non-trivial submodule of $\mathfrak{K}^{\mathfrak{A}}_-\cap\ker(\rho)$. The operators g_ε used in the analysis of Cases I-V given in the proof of Theorem 5.6.1 above all belonged to GL_-; thus the analysis of these Cases pertains in the present setting. If $m \geq 6$ and if $W^{\mathfrak{A}}_{-,11}\subset\xi$, the analysis of Case III shows $\xi = \mathfrak{K}^{\mathfrak{A}}_-\cap\ker(\rho)$. There are two remaining cases, which we examine seriatum:

Case VI: Suppose that ξ contains a submodule for the group $\mathcal{U}_-$ isomorphic to $W^{\mathfrak{A}}_{-,9}$. We adopt the notation of Case I above to define $A_{-,9}\in W^{\mathfrak{A}}_{-,9}$. We apply Lemma 5.6.5 to see that $a_1A_{-,9} + a_2TA_{-,9}\in\xi$ for suitably chosen $(a_1,a_2)\neq(0,0)$. Let $B_{-,1}$ be as in Case I. As T is a morphism in the category of modules for the group GL_-, we may define

$$C_{-,1} = a_1B_{-,1} + a_2TB_{-,1}\in\xi.$$

We use Equation (5.6.c) to see that the only (possibly) non-zero components of $\rho_{13}(C_{-,1})$ are given by:

$$\begin{aligned}\rho_{13}(C_{-,1})(e_2,e_1) &= \ \ a_1, & \rho_{13}(C_{-,1})(f_2,f_1) &= a_1,\\ \rho_{13}(C_{-,1})(e_2,f_1) &= -a_2, & \rho_{13}(C_{-,1})(f_2,e_1) &= a_2.\end{aligned}\tag{5.6.e}$$

Interchanging the basis $\{e_1,f_1,e_2,f_2\}$ by $\{f_1,-e_1,e_2,f_2\}$ (or, equivalently, by applying a suitable element of $\mathcal{U}_-$) yields a tensor $C_{-,2}$ in ξ so that the only (possibly) non-zero entries in ρ_{13} are given by:

$$\begin{aligned}\rho_{13}(C_{-,2})(e_2,f_1) &= a_1, & \rho_{13}(C_{-,2})(f_2,e_1) &= -a_1,\\ \rho_{13}(C_{-,2})(e_2,e_1) &= a_2, & \rho_{13}(C_{-,2})(f_2,f_1) &= \ \ a_2.\end{aligned}$$

Setting $C_{-,3} := a_1C_{-,1} + a_2C_{-,2}$ yields:

$$\begin{aligned} \rho_{13}(C_{-,3})(e_2,e_1) &= a_1^2 + a_2^2, & \rho_{13}(C_{-,3})(f_2,f_1) &= a_1^2 + a_2^2, \\ \rho_{13}(C_{-,3})(e_2,f_1) &= 0, & \rho_{13}(C_{-,3})(f_2,e_1) &= 0. \end{aligned}$$

Interchanging the roles of the indices "1" and "2" then yields a tensor $C_{-,4}$

$$\begin{aligned} \rho_{13}(C_{-,4})(e_1,e_2) &= a_1^2 + a_2^2, & \rho_{13}(C_{-,3})(f_1,f_2) &= a_1^2 + a_2^2, \\ \rho_{13}(C_{-,4})(e_1,f_2) &= 0, & \rho_{13}(C_{-,3})(f_1,e_2) &= 0. \end{aligned}$$

Since $\rho_{13}(C_{-,3}+C_{-,4})$ is symmetric and non-trivial, this generates a tensor with a non-zero component in $W_{-,7}^{\mathfrak{A}}$ and a trivial component in $W_{-,8}^{\mathfrak{A}}$. The analysis of Case IV now pertains to show $\xi = \mathfrak{K}_-^{\mathfrak{A}} \cap \ker(\rho)$.

Case VII: Suppose that ξ contains a submodule for the group $\mathcal{U}_-$ isomorphic to $W_{-,7}^{\mathfrak{A}}$. Let $A_{-,7} \in W_{-,7}^{\mathfrak{A}}$ be as constructed in Case IV. We apply Lemma 5.6.5 to see that $b_1A_{-,7} + b_2TA_{-,7} \in \xi$ for a suitably chosen pair $(b_1, b_2) \neq (0,0)$. Let $B_{-,5}$ be as in Case IV. Then we have that $C_{-,5} := c_2^{-1}\{b_1B_{-,5} + b_2TB_{-,5}\}$ belongs ξ. By Equation (5.6.d), the only (possibly) non-zero components of $\rho_{13}(C_{-,5})$ are given by:

$$\begin{aligned} \rho_{13}(C_{-,5})(e_2,e_1) &= \ \ b_1, & \rho_{13}(C_{-,5})(f_2,f_1) &= b_1, \\ \rho_{13}(C_{-,5})(e_2,f_1) &= -b_2, & \rho_{13}(C_{-,5})(f_2,e_1) &= b_2. \end{aligned}$$

This tensor has the same Ricci tensor as that of $C_{-,1}$ given in Equation (5.6.e) above if we set $b_1 = a_1$ and $b_2 = a_2$; the argument now proceeds exactly the same as in Case VI. □

Chapter 6

Riemannian Geometry

In Chapter 6, we turn our attention to Riemannian geometry. The metric g now plays a central role. Throughout Chapter 6, we shall let V be a real vector space of dimension m equipped with a non-degenerate symmetric bilinear form $\langle\cdot,\cdot\rangle$ of signature (p,q). We shall let $\mathfrak{R} \subset \otimes^4 V^*$ be the set of tensors A so that we have the following identities for all x, y, z, w in V:

$$A(x,y,z,w) = -A(y,x,z,w) = A(z,w,x,y),$$
$$A(x,y,z,w) + A(y,z,x,w) + A(z,x,y,w) = 0.$$

If $A \in \mathfrak{R}$, then the triple $(V,\langle\cdot,\cdot\rangle,A)$ is said to be a *pseudo-Riemannian curvature model.* We let R be the curvature tensor of the Levi-Civita connection of a pseudo-Riemannian manifold (M,g); $R_P \in \mathfrak{R}(T_PM,g_P)$ for any $P \in M$.

In Section 6.1, we show that any curvature model is geometrically realizable. We also prove Theorem 1.6.1 and establish the equality

$$\dim\{\mathfrak{R}\} = \tfrac{1}{12}(m^4 - m^2).$$

This equality was used previously. In Section 6.2, the Weyl conformal curvature tensor is introduced and shown to be a conformal invariant. The Cauchy–Kovalevskaya Theorem, which will play a central role in our treatment of manifolds of constant scalar curvature, is the subject of Section 6.3. In Section 6.4, we examine the scalar curvature and establish various results concerning realizations by metrics of constant scalar curvature. In Section 6.5, we derive the fundamental curvature relation in Weyl geometry given in Theorem 1.7.1. We also prove Theorem 1.7.2 giving the curvature decomposition of [Higa (1993)] and [Higa (1994)] in Weyl geometry. We then use results of [Gilkey, Nikčević, and Simon (2011)] to prove Theorem 1.7.3; this examines certain realization questions in Weyl geometry.

6.1 The Riemann Curvature Tensor

We begin our study with:

Lemma 6.1.1 *Every pseudo-Riemannian curvature model is geometrically realizable.*

Proof. Let $(V, \langle\cdot,\cdot\rangle, A)$ be a pseudo-Riemannian curvature model. Let M be a small neighborhood of $0 \in V$, let $P = 0$, let $(x^1, \ldots, x^m)$ be the system of local coordinates on V induced by a basis $\{e_i\}$ for V, and let

$$g_{ik} := \varepsilon_{ik} - \tfrac{1}{3} A_{ijlk} x^j x^l.$$

Clearly $g_{ik} = g_{ki}$. As $g_{ik}(0) = \varepsilon_{ik}$ is non-singular, g is a pseudo-Riemannian metric on some neighborhood of the origin. Let

$$g_{ij/k} := \partial_{x_k} g_{ij} \quad \text{and} \quad g_{ij/kl} := \partial_{x_k}\partial_{x_l} g_{ij}.$$

The Christoffel symbols of the first kind are given by:

$$\Gamma_{ijk} := g(\nabla_{\partial_{x_i}} \partial_{x_j}, \partial_{x_k}) = \tfrac{1}{2}(g_{jk/i} + g_{ik/j} - g_{ij/k}).$$

As $g = \varepsilon + O(|x|^2)$ and $\Gamma = O(|x|)$, we may compute:

$$\begin{aligned} R_{ijkl} &= \{\partial_{x_i}\Gamma_{jkl} - \partial_{x_j}\Gamma_{ikl}\} + O(|x|^2) \\ &= \tfrac{1}{2}\{g_{jl/ik} + g_{ik/jl} - g_{jk/il} - g_{il/jk}\} + O(|x|^2) \\ &= \tfrac{1}{6}\{-A_{jikl} - A_{jkil} - A_{ijlk} - A_{iljk} \\ &\qquad + A_{jilk} + A_{jlik} + A_{ijkl} + A_{ikjl}\} + O(|x|^2) \\ &= \tfrac{1}{6}\{4A_{ijkl} - 2A_{iljk} - 2A_{iklj}\} + O(|x|^2) \\ &= A_{ijkl} + O(|x|^2). \end{aligned}$$

□

Proof of Theorem 1.6.1. Let $(V, \langle\cdot,\cdot\rangle)$ be an inner product space. Let $\phi \in S^2$ and $\psi \in \Lambda^2$. We recall the notation established in Definition 1.6.1 and set:

$$\begin{aligned} A_\phi(x,y,z,w) &:= \phi(x,w)\phi(y,z) - \phi(x,z)\phi(y,w), \\ A_\psi(x,y,z,w) &:= \psi(x,w)\psi(y,z) - \psi(x,z)\psi(y,w) - 2\psi(x,y)\psi(z,w). \end{aligned}$$

To prove Theorem 1.6.1, we must show:

$$\begin{aligned} &\dim\{\mathfrak{R}\} = \tfrac{1}{12} m^2(m^2-1), \quad \text{and} \\ &\mathfrak{R} = \operatorname{Span}_{\phi\in S^2}\{A_\phi\} = \operatorname{Span}_{\psi\in\Lambda^2}\{A_\psi\}. \end{aligned}$$

The original proof [Fiedler (2003)] of this result involved the use of representation theory and Young diagrams. We shall give here a quite different proof. This proof was originally presented in [Gilkey (2001)] as Theorem 1.8.2. We also note that a third proof was given in [Díaz-Ramos et al. (2004)].

We use an analysis similar to that used in the proof of Lemma 4.1.3 to see that $A_\phi \in \mathfrak{A}$ and that $A_\psi \in \mathfrak{A}$. Since

$$A_\phi(x,y,z,w) = A_\phi(z,w,x,y), \quad \text{and}$$
$$A_\psi(x,y,z,w) = A_\psi(z,w,x,y),$$

it follows $A_\phi \in \mathfrak{R}$ and $A_\psi \in \mathfrak{R}$.

We drop the Einstein convention and do not sum over repeated indices. For the remainder of the proof, we introduce temporary notation:

$$\mathfrak{S} := \operatorname{Span}_{\phi\in S^2} A_\phi \quad \text{and} \quad \mathfrak{L} := \operatorname{Span}_{\psi\in\Lambda^2} A_\psi .$$

Let $\{e_i\}$ be a basis for V and let $\{e^i\}$ be the corresponding dual basis for V^*. Then $\{e^i \otimes e^j \otimes e^k \otimes e^l\}$ is a basis for $\otimes^4 V^*$. We define a tensor satisfying the first symmetries of Equation (1.6.a), but which does not in general satisfy the first Bianchi identity, by symmetrizing over the action of $\mathbb{Z}_2 \oplus \mathbb{Z}_2 \oplus \mathbb{Z}_2$ to define:

$$\begin{aligned}T_{ijkl} := {} & e^i \otimes e^j \otimes e^k \otimes e^l + e^k \otimes e^l \otimes e^i \otimes e^j \\ & - e^j \otimes e^i \otimes e^k \otimes e^l - e^k \otimes e^l \otimes e^j \otimes e^i \\ & - e^i \otimes e^j \otimes e^l \otimes e^k - e^l \otimes e^k \otimes e^i \otimes e^j \\ & + e^j \otimes e^i \otimes e^l \otimes e^k + e^l \otimes e^k \otimes e^j \otimes e^i .\end{aligned}$$

Let $A \in \mathfrak{R}$. Then A is a linear combination of the tensors T_{ijkl} so we may decompose A in the form:

$$A = \sum_{i,j \text{ distinct}} c_{ijji} T_{ijji} \tag{6.1.a}$$
$$+ \sum_{i,j,k \text{ distinct}} c_{ijki} T_{ijki} \tag{6.1.b}$$
$$+ \sum_{i,j,k,l \text{ distinct}} c_{ijkl} T_{ijkl}, \tag{6.1.c}$$

where the coefficients may be chosen to satisfy the relations:

$$c_{ijji} = c_{jiij}, \quad c_{ijki} = c_{ikji}, \quad c_{ijkl} = c_{klij} = -c_{jikl}.$$

Since A satisfies the first Bianchi identity, we have that:

$$c_{ijkl} + c_{iklj} + c_{iljk} = 0.$$

We shall complete the proof by studying the tensors appearing in Displays (6.1.a), (6.1.b), and (6.1.c) separately.

To study the tensors in Equation (6.1.a), we let i and j be distinct indices. Let

$$\phi = e^i \otimes e^j + e^j \otimes e^i, \quad \psi = e^i \otimes e^j - e^j \otimes e^i.$$

We then have $A_\phi = \lambda T_{ijji}$ and $A_\psi = \mu T_{ijji}$ for some universal constants λ and μ that are independent of i, j, and m. Evaluating on (e_i, e_j, e_j, e_i) yields

$$A_\phi(e_i, e_j, e_j, e_i) = -1, \quad A_\psi(e_i, e_j, e_j, e_i) = 3, \quad T_{ijji}(e_i, e_j, e_j, e_i) = 2.$$

Consequently

$$T_{ijji} = -2A_\phi = \tfrac{2}{3} A_\psi \in \mathfrak{S} \cap \mathfrak{L}.$$

Next, we study the tensors in Display (6.1.b) by taking $\{i, j, k\}$ to be distinct indices. We clear the previous notation and define:

$$\phi = e^i \otimes e^j + e^j \otimes e^i + e^i \otimes e^k + e^k \otimes e^i,$$
$$\psi = e^i \otimes e^j - e^j \otimes e^i + e^i \otimes e^k - e^k \otimes e^i.$$

We express A_ϕ and A_ψ in the form

$$A_\phi = \lambda_1 T_{ijki} + \lambda_2 T_{ijji} + \lambda_3 T_{ikki},$$
$$A_\psi = \mu_1 T_{ijki} + \mu_2 T_{ijji} + \mu_3 T_{ikki}$$

for suitably chosen universal constants $\{\lambda_1, \lambda_2, \lambda_3, \mu_1, \mu_2, \mu_3\}$. Since we already know that $T_{ijji} \in \mathfrak{S} \cap \mathfrak{L}$ and $T_{ikki} \in \mathfrak{S} \cap \mathfrak{L}$, it suffices to prove $\lambda_1 \neq 0$ and $\mu_1 \neq 0$. We verify that this is true by evaluating on (e_i, e_j, e_k, e_i):

$$A_\phi(e_i, e_j, e_k, e_i) = -1, \quad A_\psi(e_i, e_j, e_k, e_i) = 3,$$
$$T_{ijji}(e_i, e_j, e_k, e_i) = 0, \ T_{ikki}(e_i, e_j, e_k, e_i) = 0,$$
$$T_{ijki}(e_i, e_j, e_k, e_i) = 1.$$

Finally, let $\{i, j, k, l\}$ be distinct indices. We clear the previous notation and set

$$\phi = e^i \otimes e^k + e^k \otimes e^i + e^j \otimes e^l + e^l \otimes e^j,$$
$$\psi = e^i \otimes e^k - e^k \otimes e^i + e^j \otimes e^l - e^l \otimes e^j.$$

We use the first Bianchi identity to see that:

$$\begin{aligned}A_\phi &= \lambda_1 T_{ijkl} + \lambda_2 T_{iklj} + \lambda_3 T_{iljk} + \lambda_4 T_{ijji} + \lambda_5 T_{kllk},\\ A_\psi &= \mu_1 T_{ijkl} + \mu_2 T_{iklj} + \mu_3 T_{iljk} + \mu_4 T_{ijji} + \mu_5 T_{kllk}\end{aligned}$$

for suitably chosen universal constants. We compute:

$$\begin{aligned}&A_\phi(e_i,e_j,e_k,e_l) = -1, \quad A_\phi(e_i,e_k,e_l,e_j) = 0, \quad A_\phi(e_i,e_l,e_j,e_k) = 1,\\ &A_\psi(e_i,e_j,e_k,e_l) = -1, \quad A_\psi(e_i,e_k,e_l,e_j) = 2, \quad A_\psi(e_i,e_l,e_j,e_k) = -1.\end{aligned}$$

Since $T_{ijkl}(e_i,e_j,e_k,e_l) = 1$, the constants satisfy the relations:

$$\begin{aligned}\lambda_1 &= -1, \quad \lambda_2 = 0, \quad \lambda_3 = 1,\\ \mu_1 &= -1, \quad \mu_2 = 2, \quad \mu_3 = -1.\end{aligned}$$

This shows that:

$$\begin{aligned}T_{ijkl} - T_{iljk} &= -A_\phi + \star T_{ikkii} + \star T_{jllj} \in \mathfrak{S},\\ -T_{ijkl} + 2T_{iklj} - T_{iljk} &= A_\psi + \star T_{ikkii} + \star T_{jllj} \in \mathfrak{L}.\end{aligned}$$

We cyclically permute the indices $\{j,k,l\}$ to see

$$T_{iklj} - T_{ijkl} \in \mathfrak{S} \quad \text{and} \quad -T_{iklj} + 2T_{iljk} - T_{ijkl} \in \mathfrak{L}.$$

We subtract to see $T_{iklj} - T_{iljk} \in \mathfrak{L}$. We cyclically permute the indices $\{j,k,l\}$ to see

$$T_{ijkl} - T_{iklj} \in \mathfrak{S} \cap \mathfrak{L} \text{ and } T_{iljk} - T_{ijkl} \in \mathfrak{S} \cap \mathfrak{L}.$$

Let $A \in \mathfrak{R}$. As A satisfies the first Bianchi identity, $c_{ijkl} + c_{iklj} + c_{iljk} = 0$ and thus the terms in Display (6.1.c) are expressible in terms of the tensors $T_{ijkl} - T_{iljk}$ and $T_{ijkl} - T_{iklj}$ which belong to $\mathfrak{S} \cap \mathfrak{L}$. This completes the proof of Assertion (1) in Theorem 1.6.1.

We now turn to Assertion (2) and compute $\dim\{\mathfrak{R}\}$. We have shown that the tensors T_{ijji}, T_{ijki}, and $T_{ijkl} - T_{iklj}$ span $\mathfrak{R}$. There are $\frac{1}{2}m(m-1)$ possible pairs of indices $1 \le i < j \le m$ and thus $\frac{1}{2}m(m-1)$ of the tensors T_{ijji} are required. If $m \ge 3$, there are m ways to choose the index i and $\frac{1}{2}(m-1)(m-2)$ ways to choose the indices $1 \le j < k \le m$ with $i \ne j$ and $i \ne k$. Thus there are $\frac{1}{2}m(m-1)(m-2)$ of the tensors T_{ijki} needed. If $m \ge 4$, there are $\frac{1}{24}m(m-1)(m-2)(m-3)$ ways to choose four distinct indices

$$1 \le i < j < k < l \le m.$$

Given such a choice, there are two distinct tensors $\{T_{ijkl}-T_{iklj}, T_{ijkl}-T_{iljk}\}$ needed. The evaluations discussed above on

$$\{(e_i,e_j,e_j,e_i),(e_i,e_j,e_k,e_i),(e_i,e_j,e_k,e_l),(e_i,e_k,e_l,e_j)\}$$

show the tensors $T_{ijkl}-T_{iklj}$ and $T_{ijkl}-T_{iljk}$ are linearly independent. Thus we conclude that:

$$\begin{aligned}\dim\mathfrak{R} &= \tfrac{1}{2}m(m-1)+\tfrac{1}{2}m(m-1)(m-2)\\ &\quad+\tfrac{2}{24}m(m-1)(m-2)(m-3)\\ &= \tfrac{1}{12}m(m-1)\{6+6(m-2)+(m^2-5m+6)\}\\ &= \tfrac{1}{12}m(m-1)(m^2+m)=\tfrac{1}{12}m^2(m^2-1).\end{aligned}\qquad\square$$

6.2 The Weyl Conformal Curvature Tensor

Let θ be a symmetric tensor of rank two. We adopt the notation of Equation (4.1.e) and define:

$$\begin{aligned}\sigma(\theta)(x,y,z,w):= &\ \tfrac{1}{m-2}\{\theta(x,w)\langle y,z\rangle+\langle x,w\rangle\theta(y,z)\}\\ &-\tfrac{1}{m-2}\{\theta(x,z)\langle y,w\rangle+\langle x,z\rangle\theta(y,w)\}\\ &-\tfrac{\tau}{(m-1)(m-2)}\{\langle x,w\rangle\langle y,z\rangle-\langle x,z\rangle\langle y,w\rangle\}.\end{aligned}$$

We showed in the proof of Theorem 4.1.1 that σ splits the canonical projection defined by the Ricci tensor from $\mathfrak{R}$ to S^2 and defines the decomposition

$$\mathfrak{R}=\{\ker(\rho)\cap\mathfrak{R}\}\oplus S_0^2\oplus\mathbb{R}.$$

The *Weyl conformal curvature tensor* is defined as the image of π_W (orthogonal projection on $\ker(\rho)$):

$$\pi_W A := A-\sigma(\rho(A)). \qquad (6.2.a)$$

If $\phi\in C^\infty(M)$, let $H(\phi,g)_{ij}:=\phi_{;ij}$ be the *Hessian*. This gives the components of the second covariant derivative of ϕ with respect to the Levi-Civita connection of g; it is a symmetric tensor of rank two. If θ is a symmetric tensor of rank two, let

$$\kappa(\theta,g)_{ijkl}:=g_{jl}\theta_{ik}+g_{ik}\theta_{jl}-g_{il}\theta_{jk}-g_{jk}\theta_{il}.$$

We use the curvature decomposition of Theorem 4.1.1 given by [Singer and Thorpe (1969)] to show that the Weyl conformal curvature tensor is a conformal invariant:

Lemma 6.2.1 *Let $m \geq 4$. Let $g_\varepsilon := e^{2\varepsilon\phi} g$, let R_ε be the curvature of g_ε, and let W_ε be the Weyl conformal curvature tensor of g_ε. Then:*

(1) $\partial_\varepsilon R_\varepsilon = 2\phi R_\varepsilon + \kappa(H(\phi, g_\varepsilon), g_\varepsilon)$.
(2) $W_\varepsilon = e^{2\varepsilon\phi} W_0$.

Proof. By replacing g by $\tilde{g} := e^{2\varepsilon_0\phi} g$, we see that to prove Assertion (1) in general, it suffices to prove Assertion (1) when $\varepsilon = 0$. This means that we must show:

$$\partial_\varepsilon R_\varepsilon|_{\varepsilon=0} = 2\phi R + \kappa(H(\phi, g), g).$$

Fix a point P of M. We apply Lemma 3.1.1 to choose local coordinates so $g = g(P) + O(|x|^2)$. Then:

$$\begin{aligned} \Gamma_{ijk}(\varepsilon) = \tfrac{1}{2}(1 + 2\varepsilon\phi)&(g_{jk/i} + g_{ik/j} - g_{ij/k}) \\ &+\varepsilon(g_{jk}\phi_{/i} + g_{ik}\phi_{/j} - g_{ij}\phi_{/k}) + O(\varepsilon^2). \end{aligned}$$

Since $dg = O(|x|)$ and since $d\phi$ is multiplied by ε, the quadratic terms in the curvature arising from the Christoffel symbols play no role in the computation at P. Furthermore $H(\phi, g)_{ij} = \phi_{/ij} + O(|x|)$. Consequently

$$\begin{aligned} R_{ijkl}(\varepsilon, P) &= (1 + 2\varepsilon\phi) R_{ijkl}(P) \\ &\quad +\varepsilon(g_{jl}\phi_{/ik} + g_{ik}\phi_{/jl} - g_{il}\phi_{/jk} - g_{jk}\phi_{/il})(P) + O(\varepsilon^2), \\ \partial_\varepsilon R|_{\varepsilon=0}(P) &= 2\phi R(P) + \kappa(H(\phi, g), g)(P). \end{aligned}$$

Assertion (1) now follows since P was arbitrary.

The curvature decomposition of Theorem 4.1.1 shows there is an isomorphism of orthogonal modules:

$$\mathfrak{R} \approx W_6^{\mathcal{O}} \oplus \mathbb{R} \oplus S_0^2.$$

These are inequivalent irreducible orthogonal modules. Since $\kappa : S^2 \to \mathfrak{R}$, κ takes values in the module isomorphic to $\mathbb{R} \oplus S_0^2$ and thus $\pi_W \kappa(H(\phi, g)) = 0$. One applies π_W to the identity of Assertion (1) to see:

$$\partial_\varepsilon \pi_W(R_\varepsilon) = 2\phi \pi_W(R_\varepsilon).$$

Integrating this equation yields Assertion (2). Furthermore, $\pi_W(R)$ also can be seen to be a conformal invariant. □

6.3 The Cauchy–Kovalevskaya Theorem

In Section 6.3, we state the versions of the *Cauchy–Kovalevskaya Theorem* that we shall need; we refer to [Evans (1998)] pages 221–233 for the proof. Introduce coordinates $x = (x^1, \ldots, x^m)$ on $\mathbb{R}^m$ and let $\partial_{x_i} := \frac{\partial}{\partial x_i}$. Set

$$x = (y, x^m) \quad \text{where} \quad y = (x^1, \ldots, x^{m-1}) \in \mathbb{R}^{m-1}.$$

Let W be an auxiliary real vector space; in what follows, we will take $W = \mathbb{R}$ to consider a single scalar equation and $W = \mathbb{R}^2$ to consider a pair of scalar equations. Let

$$u := (u_0, u_1, \ldots, u_m) \in W \otimes \mathbb{R}^{m+1}.$$

We begin by discussing *quasi-linear partial differential equations of order two.* Let $\psi(x, u)$ be the germ of a real analytic function taking values in W and let $\psi^{ij}(x, u) = \psi^{ji}(x, u)$ be the germs of real analytic functions taking values in $\text{End}(W)$. Given the germ of a real analytic function U mapping $\mathbb{R}^m$ to W, the associated *Cauchy data* is given by setting

$$u(x) := (u_0(x), \ldots, u_m(x))$$

where

$$u_0(x) := U(x), \quad u_1(x) := \partial_1 U(x), \quad \ldots, \quad u_m(x) := \partial_m U(x).$$

Theorem 6.3.1 [**Cauchy–Kovalevskaya**] *If* $\det \psi^{mm}(0) \neq 0$, *there exists* $\varepsilon > 0$ *and there exists a unique real analytic* U *defined for* $|x| < \varepsilon$ *satisfying the following equations:*

$$\psi^{ij}(x, u(x))\partial_{x_i}\partial_{x_j}U(x) + \psi(x, u(x)) = 0,$$
$$U(y, 0) = 0, \quad \text{and} \quad \partial_m U(y, 0) = 0.$$

Next, we turn our attention to *scalar quasi-linear partial differential equations of order four.* Let $\Phi = \Phi(u)$ be real analytic. We consider the jets of order three of Φ:

$$\xi := \{\Phi,\ \partial_{j_1}\Phi,\ \partial_{j_1}\partial_{j_2}\Phi,\ \partial_{j_1}\partial_{j_2}\partial_{j_3}\Phi\}.$$

We consider a quasi-linear partial differential equation in Φ which is of order four where the coefficients $\psi^{i_1 i_2 i_3 i_4}$ and ψ are real analytic functions of ξ.

Theorem 6.3.2 *If $\psi^{mmmm}(0) \neq 0$, there exists $\varepsilon > 0$ and there exists a unique real analytic Φ defined for $|x| < \varepsilon$ satisfying*

$$\psi^{i_1 i_2 i_3 i_4}(\xi)\partial_{i_1}\partial_{i_2}\partial_{i_3}\partial_{i_4}\Phi + \psi(x,\xi) = 0 \tag{6.3.a}$$

with:

$$\begin{aligned} \Phi(y,0) &= 0, & \partial_m\Phi(y,0) &= 0, \\ \partial_m\partial_m\Phi(y,0) &= 0, & \partial_m\partial_m\partial_m\Phi(y,0) &= 0. \end{aligned} \tag{6.3.b}$$

6.4 Geometric Realizations of Riemann Curvature Tensors

In Section 6.4 we use results of [Brozos-Vázquez et al. (2009)] to establish Theorem 1.6.3; these are joint works with H. Kang and G. Weingart.

We say that a pseudo-Riemannian curvature model $(V, \langle\cdot,\cdot\rangle, A)$ is *conformally flat* if the associated Weyl conformal curvature tensor π_W vanishes. Similarly, we say that a pseudo-Riemannian manifold (M, g) is conformally flat if $\pi_W(R_g)$ vanishes at each point of M. We begin with a technical observation.

Lemma 6.4.1 *Every conformally flat pseudo-Riemannian curvature model is geometrically realizable by a conformally flat Riemannian manifold.*

Proof. We choose an orthonormal basis for V and let $\varepsilon_{ij} := \langle e_i, e_j\rangle$; $\varepsilon_{ij} = 0$ for $i \neq j$ and $\varepsilon_{ij} = \pm 1$ if $i = j$. Let $\phi(x) = \Phi_{ij}x^i x^j$ for $\Phi \in S^2$ be a quadratic polynomial function. We consider the metric

$$g = (1 + 2\phi)\varepsilon = \varepsilon + O(|x|^2).$$

This is non-degenerate at the origin and hence non-degenerate in a neighborhood of the origin. Since $\phi = O(|x|^2)$, we may apply Lemma 3.3.2 to see the non-zero components of the curvature are given by:

$$R_{ijki} = \left\{ \begin{array}{rl} -\varepsilon_{ii}\phi_{jk} & \text{if } j \neq k \\ -\varepsilon_{ii}\phi_{jj} - \varepsilon_{jj}\phi_{ii} & \text{if } j = k \end{array} \right\} \quad \text{do not sum over } i.$$

The map $\phi \to \rho(R_\phi)(0)$ defines a map $\Xi : S^2 \to S^2$. Taking $\Phi = \varepsilon$ yields an element with $\tau \neq 0$. Taking $\phi \in S_0^2$ yields a non-zero element in S_0^2. By Theorem 2.4.1, $S^2 = \mathbb{R} \cdot \varepsilon \oplus S_0^2$ is the orthogonal direct sum of inequivalent irreducible orthogonal modules. It now follows that Ξ is surjective. If A is conformally flat, then A is completely determined by its Ricci tensor. Theorem 1.6.3 now follows. □

If (M, g) is a pseudo-Riemannian manifold and if ϕ is a smooth function so that $1 + 2\phi$ never vanishes, we can consider the conformal variation

$$g_\phi = (1 + 2\phi)g.$$

The metrics constructed to prove Lemma 6.1.1 and Lemma 6.4.1 were quadratic polynomials and hence real analytic. Theorem 1.6.3 will follow from these two results and from the following result:

Lemma 6.4.2 *Let P be a point of a real analytic pseudo-Riemannian manifold (M, g) of dimension $m \geq 3$. There exists the germ of a real analytic function ϕ defined near P so that ϕ vanishes to third order at P and so that $(1 + 2\phi)g$ has constant scalar curvature.*

Proof. Let R be the curvature tensor and let τ be the scalar curvature of g, respectively. Let $x = (x^1, \ldots, x^m)$ be a system of local real analytic coordinates on M centered at P and let $y = (x^1, \ldots, x^{m-1})$ be the first $m-1$ coordinates. We set $\varepsilon_{ij} := g(\partial_{x_i}, \partial_{x_j})(0)$. By making a linear change of coordinates, we may suppose that $\{\partial_{x_i}\}$ is an orthonormal frame at P, or, in other words, that

$$\varepsilon_{ij} = \begin{Bmatrix} 0 & \text{if } i \neq j \\ \pm 1 & \text{if } i = j \end{Bmatrix}.$$

Let ϕ be a real analytic function with $\phi(0) = 0$. We set $\phi_i := \partial_{x_i}\phi$ and $\phi_{ij} := \partial_{x_i}\partial_{x_j}\phi$. We assume

$$\phi(y, 0) = 0 \quad \text{and} \quad \phi_m(y, 0) = 0.$$

We consider the conformal variation $h := (1 + 2\phi)g$. Since $\phi(0) = 0$, h is non-singular on some neighborhood of 0. Let R_h be the curvature tensor of h and let τ_h be the scalar curvature of h. We work modulo terms $\psi(x, \phi, \phi_1, \ldots, \phi_m)$ where $\psi(0) = 0$ to define an equivalence relation $\equiv$. Then

$$\begin{aligned} R_{h,ijkl} &\equiv R_{ijkl} + g_{jl}\phi_{ik} - g_{il}\phi_{jk} - g_{jk}\phi_{il} + g_{ik}\phi_{jl}, \\ \tau_h - \tau_g(0) &\equiv h^{il}h^{jk}\{g_{jl}\phi_{ik} - g_{il}\phi_{jk} - g_{jk}\phi_{il} + g_{ik}\phi_{jl}\}. \end{aligned}$$

We set $h^{jk} = \varepsilon^{jk}$ and compute

$$\begin{aligned} &\varepsilon^{il}\varepsilon^{jk}\{\varepsilon_{jl}\phi_{ik} - \varepsilon_{il}\phi_{jk} - \varepsilon_{jk}\phi_{il} + \varepsilon_{ik}\phi_{jl}\} \\ &\quad \equiv \varepsilon^{ik}\phi_{ik} - m\varepsilon^{jk}\phi_{jk} - m\varepsilon^{il}\phi_{il} + \varepsilon^{jl}\phi_{jl}. \end{aligned}$$

The coefficient of ϕ_{mm} is thus seen to be $(2-2m)\varepsilon^{mm} \neq 0$. Consequently, Theorem 6.3.1 is applicable and we may choose ϕ to solve the equations:

$$\tau_h - \tau_g(0) = 0, \quad \phi(y,0) = 0, \quad \partial_m \phi(y,0) = 0.$$

The first and second order jets of ϕ vanish at the origin. And the only possibly non-zero second order jet of ϕ at the origin is ϕ_{mm}. The relation $\psi^{ij}\phi_{ij} \equiv 0$ implies $\psi^{mm}\phi_{mm}(0) = 0$. Thus all the second order jets of ϕ vanish at the origin. □

6.5 Weyl Geometry II

Let (M,g) be a pseudo-Riemannian manifold. If $\phi = \phi_i e^i$ is a 1-form, we use the metric to lower an index and to define the corresponding vector field $\xi := \phi_i e_i$; this is characterized by the identity $\phi(x) = g(x,\xi)$ for any tangent vector field x. The following is classic – see, for example, the discussion in [Ganchev and Ivanov (1994)], [Hayden (1932)], [Pedersen and Swann (1991)], and [Pedersen and Tod (1993)]. It summarizes the fundamental curvature conditions in Weyl geometry:

Theorem 6.5.1 *Let ∇ be an affine connection on a pseudo-Riemannian manifold (M,g). Let ∇^g be the Levi-Civita connection of g.*

(1) The following assertions are equivalent:

(a) $\nabla g = -2\phi \otimes g$.

(b) $\nabla_x y = \nabla^g_x y + \phi(x)y + \phi(y)x - g(x,y)\xi$.

(2) If either of the conditions in (1) are satisfied, then (M,g,∇) is said to be a Weyl manifold *and we have:*

(a) $R(x,y,z,w) + R(x,y,w,z) = \frac{2}{m}\{\rho(R)(y,x) - \rho(R)(x,y)\}g(z,w)$.

(b) $\rho_a(R) = -md\phi$.

Proof. Let $\Gamma_{ij}{}^k$ be the Christoffel symbols of the connection ∇ and let $\Gamma^g_{ij}{}^k$ be the Christoffel symbols of the Levi-Civita connection relative to some orthonormal frame $\{e_i\}$. We may express $\Gamma_{ijk} = \Gamma^g_{ijk} + \theta_{ijk}$. We have

$$g_{jk;i} := e_i g(e_j,e_k) - g(\nabla_{e_i}e_j,e_k) - g(e_j,\nabla_{e_i}e_k) = -\theta_{ijk} - \theta_{ikj}.$$

Thus the assertion that $\nabla g = -2\phi \otimes g$ is equivalent to the identity:

$$2\phi_i \varepsilon_{jk} = \theta_{ijk} + \theta_{ikj}. \tag{6.5.a}$$

Similarly, the assertion $\nabla_x y = \nabla_x^g y + \phi(x)y + \phi(y)x - g(x,y)\xi$ is equivalent to the identity

$$\theta_{ijk} = \phi_i\varepsilon_{jk} + \phi_j\varepsilon_{ik} - \phi_k\varepsilon_{ij}. \tag{6.5.b}$$

Thus to establish that Assertions (1a) and (1b) are equivalent, we must show that Equation (6.5.a) and Equation (6.5.b) are equivalent algebraically if we assume the symmetry $\theta_{ijk} = \theta_{jik}$ (which is the condition that the torsion tensor of ∇ vanishes). Suppose first that Equation (6.5.a) is satisfied. We compute:

$$\begin{aligned}\theta_{ijk} &= -\theta_{ikj} + 2\phi_i\varepsilon_{jk} = -\theta_{kij} + 2\phi_i\varepsilon_{jk} = \theta_{kji} + 2\phi_i\varepsilon_{jk} - 2\phi_k\varepsilon_{ij}\\ &= \theta_{jki} + 2\phi_i\varepsilon_{jk} - 2\phi_k\varepsilon_{ij} = -\theta_{jik} + 2\phi_i\varepsilon_{jk} - 2\phi_k\varepsilon_{ij} + 2\phi_j\varepsilon_{ik}.\end{aligned}$$

Equation (6.5.b) now follows. Conversely, suppose that Equation (6.5.b) is satisfied. We complete the proof of Assertion (1) by checking:

$$\begin{aligned}\theta_{ijk} + \theta_{ikj} &= \phi_i\varepsilon_{jk} + \phi_j\varepsilon_{ik} - \phi_k\varepsilon_{ij} + \phi_i\varepsilon_{kj} + \phi_k\varepsilon_{ij} - \phi_j\varepsilon_{ik}\\ &= 2\phi_i\varepsilon_{jk}.\end{aligned}$$

Suppose that Assertion (1b) is satisfied. We have

$$R_{ijkl} = \partial_{x_i}\Gamma_{jkl} - \partial_{x_j}\Gamma_{ikl} + \varepsilon^{ns}\Gamma_{inl}\Gamma_{jks} - \varepsilon^{ns}\Gamma_{jnl}\Gamma_{iks}.$$

Fix a point P of M. Choose local coordinates on M centered at P so that $g_{ij} = \varepsilon_{ij} + O(|x|^2)$. We then will have

$$\begin{aligned}R_{ijkl}(0) &= \{R^g_{ijkl} + \partial_{x_i}\theta_{jkl} - \partial_{x_j}\theta_{ikl} + \varepsilon^{ns}\theta_{inl}\theta_{jks} - \varepsilon^{ns}\theta_{jnl}\theta_{iks}\}(0),\\ \rho_{jk}(0) &= \rho^g_{jk}(0) + \varepsilon^{il}\{\partial_{x_i}\theta_{jkl} - \partial_{x_j}\theta_{ikl} + \varepsilon^{ns}\theta_{inl}\theta_{jks} - \varepsilon^{ns}\theta_{jnl}\theta_{iks}\}(0).\end{aligned}$$

The quadratic terms in θ and ρ^g_{jk} are symmetric in $\{j,k\}$. We have

$$\begin{aligned}\varepsilon^{il}\theta_{ikl} &= \varepsilon^{il}\{\phi_i\varepsilon_{kl} + \phi_k\varepsilon_{il} - \phi_l\varepsilon_{ik}\} = m\phi_k,\\ (\rho_{jk} - \rho_{kj})(0) &= m\{\partial_{x_k}\phi_j - \partial_{x_j}\phi_k\}(0).\end{aligned}$$

This verifies Assertion (2b). Furthermore, since $R^g_{ijkl} + R^g_{ijlk} = 0$ and $\theta_{ijk} + \theta_{ikj} = 2\phi_i\varepsilon_{jk}$, we verify Assertion (2a) by computing:

$$\begin{aligned}\{R_{ijkl} + R_{ijlk}\}(0) &= \{\partial_{x_i}(\theta_{jkl} + \theta_{jlk}) - \partial_{x_j}(\theta_{ikl} + \theta_{ilk})\\ &\quad + \varepsilon^{ns}(\theta_{inl}\theta_{jks} + \theta_{ink}\theta_{jls} - \theta_{jnl}\theta_{iks} - \theta_{jnk}\theta_{ils})\}(0)\\ &= 2\varepsilon_{kl}(\partial_{x_i}\phi_j - \partial_{x_j}\phi_i)(0) + \varepsilon^{ns}\{(-\theta_{iln} + 2\phi_i\varepsilon_{ln})\theta_{jks}\\ &\quad + (-\theta_{ikn} + 2\phi_i\varepsilon_{kn})\theta_{jls} - \theta_{jsl}\theta_{ikn} - \theta_{jsk}\theta_{iln}\}(0)\\ &= 2\varepsilon_{kl}(\partial_{x_i}\phi_j - \partial_{x_j}\phi_i)(0) + 2\varepsilon^{ns}\{\phi_i\varepsilon_{ln}\theta_{jks} - \theta_{iln}\phi_j\varepsilon_{sk}\end{aligned}$$

$$
\begin{aligned}
&\quad +\phi_i\varepsilon_{kn}\theta_{jls} - \theta_{ikn}\phi_j\varepsilon_{sl}\}(0)\\
&= 2\varepsilon_{kl}(\partial_{x_i}\phi_j - \partial_{x_j}\phi_i)(0)\\
&\quad +2\{\phi_i\theta_{jkl} - \theta_{ilk}\phi_j + \phi_i\theta_{jlk} - \theta_{ikl}\phi_j\}(0)\\
&= 2\varepsilon_{kl}(\partial_{x_i}\phi_j - \partial_{x_j}\phi_i)(0) + 2\varepsilon_{kl}\{2\phi_i\phi_j - 2\phi_j\phi_i\}(0)\\
&= 2\varepsilon_{kl}(\partial_{x_i}\phi_j - \partial_{x_j}\phi_i)(0) = \tfrac{2}{m}\{\rho_{ji} - \rho_{ij}\}\varepsilon_{kl}. \qquad \square
\end{aligned}
$$

Weyl geometry is a conformal theory. In other words:

Theorem 6.5.2 *Let (M, g, ∇) be a Weyl manifold and let $f \in C^\infty(M)$. Let $g_1 := e^{2f}g$ be a conformally equivalent metric. Then (M, g_1, ∇) also is a Weyl manifold where $\phi_1 = \phi - df$.*

Proof. We compute

$$\nabla g_1 = \nabla(e^{2f}g) = e^{2f}\nabla g + 2df \otimes e^{2f}g = (-2\phi + 2df) \otimes g_1. \qquad \square$$

We say that $A \in \mathfrak{A}$ is a *Weyl curvature operator* if it satisfies the fundamental curvature relation derived in Theorem 6.5.1:

$$A(x,y,z,w) + A(x,y,w,z) = \tfrac{2}{m}\{\rho(A)(y,x) - \rho(A)(x,y)\}\langle z,w\rangle. \tag{6.5.c}$$

Let $\mathfrak{W}$ be the set of all such tensors. We adopt the notation of Definition 4.1.1 and set $\sigma^{\mathfrak{W}} := \sigma_4 - \sigma_5$; if $\psi \in \Lambda^2$, then:

$$\{\sigma^{\mathfrak{W}}\psi\}_{ijkl} := 2\psi_{ij}\varepsilon_{kl} + \psi_{ik}\varepsilon_{jl} - \psi_{jk}\varepsilon_{il} - \psi_{il}\varepsilon_{jk} + \psi_{jl}\varepsilon_{ik}. \tag{6.5.d}$$

Theorem 1.7.2 will follow from:

Theorem 6.5.3 *There is an orthogonal direct sum module decomposition $\mathfrak{W} = \mathfrak{R} \oplus \mathfrak{P}$ where $\mathfrak{P} = \sigma^{\mathfrak{W}}\Lambda^2$.*

Proof. In the proof of Theorem 4.1.1, we defined

$$\pi_{\Lambda\otimes S}(A)(x,y,z,w) := \tfrac{1}{2}\{A(x,y,z,w) + A(x,y,w,z)\}$$

and showed we had a short exact sequence

$$0 \to \mathfrak{R} \to \mathfrak{A} \to \Lambda^2 \otimes S^2 \to 0.$$

We may decompose

$$S^2 = \mathbb{R}\cdot\langle\cdot,\cdot\rangle \oplus S_0^2.$$

Let $A \in \mathfrak{W}$. Equation (6.5.c) shows that

$$\sigma_{\Lambda\otimes S}A = -\frac{2}{m}\rho_a \otimes \langle\cdot,\cdot\rangle$$

and consequently $\pi_{\Lambda\otimes S}(A)$ takes values in $\Lambda^2\otimes\mathbb{R}\cdot\langle\cdot,\cdot\rangle$. Thus the orthogonal module decomposition of $\mathfrak{W}$ has one of the following two forms:

$$\mathfrak{W}=\mathfrak{R}\quad\text{or}\quad\mathfrak{W}\approx\mathfrak{R}\oplus\Lambda^2.$$

We argue that the second possibility pertains. If $\psi\in\Lambda^2$, let $R=\sigma^{\mathfrak{W}}\psi$. This belongs to the submodule of $\mathfrak{A}$ abstractly isomorphic to $\Lambda^2\oplus\Lambda^2$. We show that $R\in\mathfrak{W}$ and complete the proof by computing:

$$\begin{aligned}\rho_{jk}&=\varepsilon^{il}\{2\psi_{ij}\varepsilon_{kl}+\psi_{ik}\varepsilon_{jl}-\psi_{jk}\varepsilon_{il}-\psi_{il}\varepsilon_{jk}+\psi_{jl}\varepsilon_{ik}\}\\&=2\psi_{kj}+\psi_{jk}-m\psi_{jk}+0+\psi_{jk}=-m\psi_{jk},\\R_{ijkl}+R_{ijlk}&=4\psi_{ij}\varepsilon_{kl}=\tfrac{2}{m}(\rho_{ji}-\rho_{ij})\varepsilon_{kl}.\end{aligned}$$ □

Theorem 1.7.3 will follow from:

Theorem 6.5.4 *Let $\mathcal{A}\in\mathfrak{W}(V,\langle\cdot,\cdot\rangle)$. Then there is a Weyl structure (g,∇) on a neighborhood of the origin in V so*

$$g_0=\langle\cdot,\cdot\rangle\quad\textit{and}\quad\mathcal{R}_0=\mathcal{A}.$$

Proof. If $\Theta\in S^2\otimes S^2$ and if $\Phi\in\otimes^2V^*$, define the germ g of a pseudo-Riemannian metric on V and a 1-form ϕ by setting:

$$g_{ij}=\varepsilon_{ij}+\sum_{k,l}\Theta_{ijkl}x^kx^l\quad\text{and}\quad\phi:=\sum_{ij}\Phi_{ij}x^idx^j.$$

Let ξ be the corresponding dual vector field. We form the connection:

$$\nabla_xy:=\nabla^g_xy+\phi(x)y+\phi(y)x-g(x,y)\xi.$$

We may then apply Theorem 6.5.1 to conclude (g,∇,ϕ) forms a Weyl structure. Since $g_{ij}=\varepsilon_{ij}+O(|x|^2)$ and $\phi(0)=0$, only the second derivatives of the metric and the first derivatives of ϕ play a role. We compute:

$$\Gamma_{jkl}=\Gamma^g_{jkl}+x^i\{\Phi_{ij}\varepsilon_{kl}+\Phi_{ik}\varepsilon_{jl}-\varepsilon_{jk}\Phi_{il}\}.$$

We use Lemma 3.3.2 to compute R^g; this shows that:

$$\begin{aligned}R_{ijkl}(0)=&\tfrac12\{\Theta_{ikjl}+\Theta_{jlik}-\Theta_{iljk}-\Theta_{jkil}\}\\&+(\Phi_{ij}-\Phi_{ji})\varepsilon_{kl}+\Phi_{ik}\varepsilon_{jl}-\Phi_{il}\varepsilon_{jk}-\Phi_{jk}\varepsilon_{il}+\Phi_{jl}\varepsilon_{ik}.\end{aligned}$$

This yields an $\mathcal{O}$ equivariant map

$$\mathcal{E}:\{S^2\otimes S^2\}\oplus\{V^*\otimes V^*\}\to\mathfrak{W}.$$

If we take $\Phi = 0$, then we are essentially studying the question of the geometrical realizability of any Riemannian algebraic curvature tensor; Theorem 1.6.3 then shows that

$$\mathcal{E} : S^2 \otimes S^2 \to \mathfrak{R} \to 0.$$

Thus the only question is whether or not $\mathcal{E}(V^* \otimes V^*) \subset \mathfrak{R}$. We may decompose

$$V^* \otimes V^* = \mathbb{R} \oplus S_0^2 \oplus \Lambda^2$$

as the sum of three irreducible orthogonal modules where S_0^2 are the symmetric tensor of rank two having trace 0. We take $\Phi \in \Lambda^2$ with non-zero components:

$$\Phi_{12} = -\Phi_{21} = 1.$$

We then have

$$R_{1211}(0) = \varepsilon_{11}\{1 - (-1) + 0 + 0 + 1 - 1\} \neq 0.$$

Thus $\mathcal{E}$ induces a non-zero map from

$$\Lambda^2 \to \mathfrak{P} \oplus \mathbb{R} \oplus S_0^2 \oplus W_6^{\mathcal{O}}.$$

The modules $\{\mathfrak{P}, \mathbb{R}, S_0^2, W_6^{\mathcal{O}}\}$ are inequivalent. Since Λ^2 and $\mathfrak{P}$ are irreducible orthogonal modules,

$$\mathcal{E} : \Lambda^2 \approx \mathfrak{P}.$$

This completes the proof. □

We now turn to the proof of Theorem 1.7.4; it is an interesting illustration of the extent to which the geometric category is determined by the algebraic setting. We restate Theorem 1.7.4 for the convenience of the reader:

Theorem 6.5.5 *Let $\mathcal{W} = (M, g, \nabla)$ be a Weyl manifold with $H^1(M;\mathbb{R}) = 0$. The following assertions are equivalent. If any is satisfied, we say that $\mathcal{W}$ is* trivial.

(1) $d\phi = 0$.
(2) $\nabla = \nabla^{\tilde{g}}$ *for some $\tilde{g}$ in the conformal class defined by g.*
(3) $\nabla = \nabla^{\tilde{g}}$ *for some semi-Riemannian metric $\tilde{g}$.*
(4) $R_P(\nabla) \in \mathfrak{R}$ *for every $P \in M$.*
(5) ∇ *is Ricci symmetric.*

Proof. Suppose that $d\phi = 0$. Since

$$H^1(M;\mathbb{R}) = 0,$$

we can express $\phi = df$ for some function f. Then ∇ is the Levi-Civita connection for the conformally equivalent metric $\tilde{g} := e^{2f}g$ by Theorem 6.5.2. Thus Assertion (1) implies Assertion (2). Clearly Assertion (2) implies Assertion (3). Since the curvature tensor of the Levi-Civita connection belongs to $\mathfrak{R}$, Assertion (3) implies Assertion (4). If Assertion (4) holds, then ρ is symmetric and Assertion (5) holds. Suppose that Assertion (5) holds so ρ is symmetric. We apply Theorem 6.5.1 (2b) to see $d\phi = -\frac{1}{n}\rho_a = 0$ and thus Assertion (5) implies Assertion (1). □

Chapter 7

Complex Riemannian Geometry

In Chapter 7, we shall study a para-Hermitian vector space $(V, \langle\cdot,\cdot\rangle, J_+)$ or a pseudo-Hermitian vector space $(V, \langle\cdot,\cdot\rangle, J_-)$ in the algebraic setting, and we shall study a para-Hermitian manifold (M, g, J_+) or a pseudo-Hermitian manifold (M, g, J_-) in the geometric setting. We shall always work in dimension $m \geq 4$.

In Section 7.1, we decompose $\mathfrak{R}$ as a module for the group $\mathcal{U}_-$ and $\mathcal{U}_\pm^\star$. This leads to the proof of Theorem 1.8.1. The original decomposition of [Tricerri and Vanhecke (1981)] was in the positive definite setting. The subsequent extension to the pseudo-Hermitian setting was given by [Brozos-Vázquez et al. (2009a)]; the decomposition in the para-Hermitian setting is new.[1] A corresponding decomposition of the space of (para)-Kähler curvature tensors is established in Theorem 1.10.3.

In Section 7.2, we investigate the submodules of $\mathfrak{R}$ arising from the Ricci tensors in the para-Hermitian setting and in the pseudo-Hermitian setting to derive certain results of [Tricerri and Vanhecke (1981)] that will play a central role in our discussion of (para)-Kähler geometry in Section 7.5. In Section 7.3, we use the results of Section 7.1 to prove Theorem 1.9.2 and Theorem 1.9.3. In Section 7.4, we establish Theorem 1.8.3. In Section 7.5, we prove Theorem 1.10.2.

In Section 7.6 we study (para)-complex Weyl geometry and work in dimension $m \geq 6$. We first prove Theorem 1.11.1; Theorem 1.11.1 shows that any para-Hermitian or pseudo-Hermitian Kähler–Weyl structure is trivial. The curvature decomposition of Theorem 1.11.3 giving the structure of $\mathfrak{W}$

[1]The proof of Theorem 1.8.1 in the para-Hermitian setting given originally in [Brozos-Vázquez et al. (2009a)] was incorrect as was the proof of Theorem 1.10.3 in the para-Kähler setting [Brozos-Vázquez, Gilkey, and Merino (2010)]. The decompositions stated there were for the group $\mathcal{U}_+$ and are false in that context; one must work with structure group $\mathcal{U}_+^\star$ instead.

as a module for the groups $\mathcal{U}_-$ and $\mathcal{U}_\pm^\star$ follows from Theorem 1.7.2 and from Theorem 1.8.1. This curvature decomposition plays a crucial role in the proof of Theorem 1.11.4. Theorem 1.11.2 then follows directly from Theorem 1.11.4. We refer to [Gilkey and Nikčević (2011a)] for a treatment of the setting in dimension $m = 4$ as it is very different.

In Section 7.7 we once again change focus and discuss the covariant derivative of the Kähler form. We establish the geometrical realization results of Theorem 1.12.1 and Theorem 1.12.2; these results let us pass from the algebraic to the geometric setting. We prove Theorem 1.12.3 decomposing the space $\mathfrak{H}_\pm$ of algebraic covariant derivative (para)-Kähler tensors as a module for the group $\mathcal{U}_\pm^\star$ and as a module for the group $\mathcal{U}_-$. We conclude Section 7.7 by verifying Theorem 1.12.6 showing there is a pseudo-Hermitian manifold with $\nabla\Omega_- \in \xi$ for any submodule ξ of $\mathfrak{H}_-$ in any signature $(2\bar{p}, 2\bar{q})$ where $2\bar{p} + 2\bar{q} \geq 10$.

One can pass formally from the pseudo-Hermitian to the para-Hermitian context by setting $J_+ = \sqrt{-1}J_-$; this observation motivates certain changes of sign.

7.1 The Decomposition of $\mathfrak{R}$ as Modules over $\mathcal{U}_\pm^\star$

In Section 7.1, we will decompose $\mathfrak{R}$ as a module with structure groups $\mathrm{GL}_\pm^\star$ and $\mathcal{U}_\pm^\star$ to prove Theorem 1.8.1. The decomposition of $\mathfrak{R}$ as a module with structure group $\mathcal{U}_-$ in the Riemannian setting is due to [Tricerri and Vanhecke (1981)].

We recall the notation of Equation (1.2.a), Equation (1.8.a), and Equation (1.8.b). We also recall the notation of Definition 1.8.1. These conventions may be restated for the convenience of the reader as follows:

$$\begin{aligned}
&\mathrm{GL}_\pm := \{T \in \mathrm{GL} : TJ_\pm = J_\pm T\},\\
&\mathrm{GL}_\pm^\star := \{T \in \mathrm{GL} : TJ_\pm = J_\pm T \text{ or } TJ_\pm = -J_\pm T\},\\
&\mathcal{U}_\pm := \mathcal{O} \cap \mathrm{GL}_\pm, \qquad \mathcal{U}_\pm^\star := \mathcal{O} \cap \mathrm{GL}_\pm^\star,\\
&\rho_{J_\pm}(x,y) := \varepsilon^{il} A(e_i, x, J_\pm y, J_\pm e_l),\\
&\tau_{J_\pm} := \varepsilon^{il}\varepsilon^{jk} A(e_i, e_j, J_\pm e_k, J_\pm e_l),\\
&\mathfrak{R}_+^{\mathcal{U}_\pm} := \{A \in \mathfrak{R} : A(J_\pm x, J_\pm y, J_\pm z, J_\pm w) = A(x,y,z,w)\},\\
&\mathfrak{R}_-^{\mathcal{U}_\pm} := \{A \in \mathfrak{R} : A(J_\pm x, J_\pm y, J_\pm z, J_\pm w) = -A(x,y,z,w)\},\\
&\mathfrak{K}_\pm^{\mathfrak{R}} := \{A \in \mathfrak{R} : A(x,y,z,w) = \mp A(J_\pm x, J_\pm y, z, w)\},
\end{aligned}$$

$$\begin{aligned}
\mathcal{G}_\pm(T)(x,y,z,w) &:= T(x,y,z,w) + T(J_\pm x, J_\pm y, J_\pm z, J_\pm w)\\
&\pm T(J_\pm x, J_\pm y, z, w) \pm T(J_\pm x, y, J_\pm z, w) \pm T(J_\pm x, y, z, J_\pm w)\\
&\pm T(x, J_\pm y, J_\pm z, w) \pm T(x, J_\pm y, z, J_\pm w) \pm T(x, y, J_\pm z, J_\pm w),\\
\mathfrak{G}_\pm &:= \mathfrak{R} \cap \ker(\mathcal{G}_\pm), \qquad W^{\mathfrak{R}}_{\pm,3} := \mathfrak{K}^{\mathfrak{R}}_\pm \cap \ker(\rho),\\
W^{\mathfrak{R}}_{\pm,6} &:= \{\mathfrak{K}^{\mathfrak{R}}_\pm\}^\perp \cap \mathfrak{G}_\pm \cap \mathfrak{R}^{\mathcal{U}_\pm}_+ \cap \ker(\rho \oplus \rho_{J_\pm}),\\
W^{\mathfrak{R}}_{\pm,7} &:= \{A \in \mathfrak{R} : A(J_\pm x, y, z, w) = A(x, y, J_\pm z, w)\},\\
W^{\mathfrak{R}}_{\pm,10} &:= \mathfrak{R}^{\mathcal{U}_\pm}_- \cap \ker(\rho \oplus \rho_{J_\pm}).
\end{aligned}$$

We begin by examining the space of quadratic invariants. We adopt the notation of Definition 2.3.2 and set:

$$\begin{aligned}
&\psi^{\mathfrak{R}}_1 := \mathcal{I}\{(1,2,2,1)(3,4,4,3)\}, \quad \psi^{\mathcal{U}_\pm}_4 := \mathcal{I}\{(1,2,J_\pm 2,J_\pm 1)(3,4,4,3)\},\\
&\psi^{\mathfrak{R}}_2 := \mathcal{I}\{(1,2,3,1)(4,2,3,4)\}, \quad \psi^{\mathcal{U}_\pm}_5 := \mathcal{I}\{(1,2,3,1)(4,J_\pm 2,J_\pm 3,4)\},\\
&\psi^{\mathfrak{R}}_3 := \mathcal{I}\{(1,2,3,4)(1,2,3,4)\}, \quad \psi^{\mathcal{U}_\pm}_6 := \mathcal{I}\{(1,2,3,1)(J_\pm 4,J_\pm 2,3,4)\},\\
&\psi^{\mathcal{U}_\pm}_7 := \mathcal{I}\{(J_\pm 1,2,3,1)(J_\pm 4,2,3,4)\},\\
&\psi^{\mathcal{U}_\pm}_8 := \mathcal{I}\{(1,2,3,4)(1,2,J_\pm 3,J_\pm 4)\},\\
&\psi^{\mathcal{U}_\pm}_9 := \mathcal{I}\{(1,2,3,4)(1,J_\pm 2,3,J_\pm 4)\},\\
&\psi^{\mathcal{U}_\pm}_{10} := \mathcal{I}\{(1,2,J_\pm 2,J_\pm 1)(3,4,J_\pm 4,J_\pm 3)\},\\
&\psi^{\mathcal{U}_\pm}_{11} := \mathcal{I}\{(1,2,3,J_\pm 1)(4,J_\pm 2,J_\pm 3,J_\pm 4)\},\\
&\psi^{\mathcal{U}_\pm}_{12} := \mathcal{I}\{(1,2,3,4)(J_\pm 1,J_\pm 2,J_\pm 3,J_\pm 4)\}.
\end{aligned}$$

Lemma 7.1.1 *Let $(V, \langle\cdot,\cdot\rangle, J_\pm)$ be a para-Hermitian $(+)$ or a pseudo-Hermitian $(-)$ vector space.*

(1) The invariants $\{\psi^{\mathfrak{R}}_1, \psi^{\mathfrak{R}}_2, \psi^{\mathfrak{R}}_3, \psi^{\mathcal{U}_\pm}_4, \dots, \psi^{\mathcal{U}_\pm}_{12}\}$ span $\mathcal{I}^{\mathcal{U}_-}_2(\mathfrak{R})$ and $\mathcal{I}^{\mathcal{U}^\star_\pm}_2(\mathfrak{R})$.
(2) We have $\dim\{\mathcal{I}^{\mathcal{U}_-}_2(\mathfrak{R})\} \le 12$ and $\dim\{\mathcal{I}^{\mathcal{U}^\star_\pm}_2(\mathfrak{R})\} \le 12$.

Remark 7.1.1 It follows from the proof we shall give of Theorem 1.8.1 that equality holds in Lemma 7.1.1 if $m \ge 8$; thus the invariants of Assertion (1) form a basis for the associated spaces of invariants in this setting.

Proof. We argue as follows and at some length:

(1) General remarks:
   (a) We use the Bianchi identity to conclude that it is unnecessary to consider strings $(\nu, J_\pm\nu, *, *)(*, *, *, *)$ for any ν.
   (b) If $S = \mathcal{I}\{(1,2,3,4)(*,*,*,*)\}$, we can use the Bianchi identity on $(*,*,*,*)$ to replace the invariant under consideration by sums of invariants where the indices "1" and "$J_\pm 1$" do not touch the indices "4" and "$J_\pm 4$" in the remaining variable.

(c) If we replace the basis $\{e^1_{i_1}\}$ by $\{J_+e^1_{i_1}\}$, we must change the sign since $J_+^*\langle\cdot,\cdot\rangle = -\langle\cdot,\cdot\rangle$. Thus $\mathcal{I}\{(-,1,-,J_\pm 1,-)\}$ $= \mp\mathcal{I}\{(-,J_\pm 1,-,J_\pm J_\pm 1,-)\} = -\mathcal{I}\{(-,J_\pm 1,-,1,-)\}$ so interchanging a "1" index with a "$J_\pm 1$" index changes the sign. Thus $\mathcal{I}\{(\nu,\mu,\mu,J_\pm\nu)(*,*,*,*)\} = -\mathcal{I}\{(J_\pm\nu,\mu,\mu,\nu)(*,*,*,*)\}$ so this invariant vanishes. We use "$-$" as a placeholder above as we do not know in exactly what location the index "1" appears.

(d) We will stratify the invariants by the number of times $J_\pm$ appears as a decoration; this gives rise to five basic cases. Within a given case, we consider the three subcases where the indices decouple, where two indices appear with multiplicity one in each monomial, and where all four indices appear with multiplicity one in each monomial. Each subcase may be divided into various possibilities.

(2) No indices are decorated by $J_\pm$. Such invariants belong to $\mathcal{I}_2^{\mathcal{O}}(\mathfrak{R})$.

(a) The indices decouple in the two monomials. We use the usual $\mathbb{Z}_2$ symmetries to see this gives rise to a single invariant:
$\psi_1^{\mathfrak{R}} := \mathcal{I}\{(1,2,2,1)(3,4,4,3)\} = \tau^2$.

(b) Two indices appear with multiplicity one in each monomial:
$\psi_2^{\mathfrak{R}} := \mathcal{I}\{(1,2,3,1)(4,2,3,4)\} = |\rho|^2$.

(c) Each index appears in each monomial:

i. $\psi_3^{\mathfrak{R}} := \mathcal{I}\{(1,2,3,4)(1,2,3,4)\} = |R|^2$.

ii. $\Psi_1 := \mathcal{I}\{(1,2,3,4)(1,3,2,4)\}$.

iii. $\Psi_2 := \mathcal{I}\{(1,2,3,4)(1,4,2,3)\}$
$= \mathcal{I}\{(1,3,4,2)(1,2,3,4)\} = -\Psi_1$,
$\Psi_2 = -\mathcal{I}\{(1,2,3,4)(1,2,3,4)\} - \mathcal{I}\{(1,2,3,4)(1,3,4,2)\}$
$= -\psi_3^{\mathfrak{R}} + \Psi_1$.

iv. $\Psi_1 = -\Psi_2 = \frac{1}{2}\psi_3^{\mathfrak{R}}$.

(3) One index is decorated by $J_\pm$. There are several possibilities all of which yield the zero invariant:

(a) The indices decouple in the two variables:
$\mathcal{I}\{(1,2,2,1)(3,4,4,J_\pm 3)\}$ – this invariant is zero by (1c).

(b) Two indices appear with multiplicity one in each monomial.

i. $\mathcal{I}\{(1,2,3,1)(4,2,3,J_\pm 4)\} = -\mathcal{I}\{(1,2,3,1)(J_\pm 4,2,3,4)\}$
$= -\mathcal{I}\{(1,3,2,1)(J_\pm 4,3,2,4)\} = -\mathcal{I}\{(1,2,3,1)(4,2,3,J_\pm 4)\}$ so this invariant is zero.

ii. $\mathcal{I}\{(1,2,3,1)(4,J_\pm 2,3,4)\} = -\mathcal{I}\{(1,J_\pm 2,3,1)(4,2,3,4)\}$
$= -\mathcal{I}\{(4,J_\pm 2,3,4)(1,2,3,1)\}$ so this invariant is zero.

(c) Each index appears in each monomial. We suppose 4 is the decorated index. We need only consider the invariants by (1b):

i. $\mathcal{I}\{(1,2,3,4)(1,2,3,J_\pm 4)\} = -\mathcal{I}\{(1,2,3,J_\pm 4)(1,2,3,4)\}$ so this invariant is zero.

ii. $\mathcal{I}\{(1,2,3,4)(1,3,2,J_\pm 4)\} = -\mathcal{I}\{(1,2,3,J_\pm 4)(1,3,2,4)\}$ $= -\mathcal{I}\{(1,3,2,J_\pm 4)(1,2,3,4)\}$ so this invariant is zero.

(4) Two indices are decorated by $J_\pm$.

(a) The indices decouple in the two variables.

i. $\psi_4^{\mathcal{U}_\pm} := \mathcal{I}\{(1,2,J_\pm 2,J_\pm 1)(3,4,4,3)\}$.

ii. $\mathcal{I}\{(1,2,J_\pm 2,1)(3,4,J_\pm 4,3)\}$ – this invariant is zero by (1c).

(b) Two indices appear with multiplicity one in each monomial:

i. $\psi_5^{\mathcal{U}_\pm} := \mathcal{I}\{(1,2,3,1)(4,J_\pm 2,J_\pm 3,4)\}$.

ii. $\psi_6^{\mathcal{U}_\pm} := \mathcal{I}\{(1,2,3,1)(J_\pm 4,J_\pm 2,3,4)\}$.

iii. $\psi_7^{\mathcal{U}_\pm} := \mathcal{I}\{(J_\pm 1,2,3,1)(J_\pm 4,2,3,4)\}$.

(c) Each index appears in each monomial. We move the $J_\pm$ terms across to assume one term is $(1,2,3,4)$. We assume "4" is one of the two decorated indices. By (1b), we assume the indices "1" and "$J_\pm 1$" do not touch the index "$J_\pm 4$" in the other variable. The possibilities then become:

i. $\psi_8^{\mathcal{U}_\pm} := \mathcal{I}\{(1,2,3,4)(1,2,J_\pm 3,J_\pm 4)\}$
$= -\mathcal{I}\{(1,2,3,4)(2,J_\pm 3,1,J_\pm 4)\} - \mathcal{I}\{(1,2,3,4)(J_\pm 3,1,2,J_\pm 4)\}$
$= \mathcal{I}\{(2,1,3,4)(2,J_\pm 3,1,J_\pm 4)\} + \mathcal{I}\{(1,2,3,4)(1,J_\pm 3,2,J_\pm 4)\}$
$= \mathcal{I}\{(1,2,3,4)(1,J_\pm 3,2,J_\pm 4)\} + \mathcal{I}\{(1,2,3,4)(1,J_\pm 3,2,J_\pm 4)\}$
$= 2\Psi_4$. See below for the definition of Ψ_4.

ii. $\psi_9^{\mathcal{U}_\pm} := \mathcal{I}\{(1,2,3,4)(1,J_\pm 2,3,J_\pm 4)\}$.

iii. $\mathcal{I}\{(1,2,3,4)(J_\pm 1,2,3,J_\pm 4)\} = \mathcal{I}\{(2,1,3,4)(2,J_\pm 1,3,J_\pm 4)\}$ $= \mathcal{I}\{(1,2,3,4)(1,J_\pm 2,3,J_\pm 4)\} = \psi_9^{\mathcal{U}_\pm}$.

iv. $\Psi_3 := \mathcal{I}\{(1,2,3,4)(1,3,J_\pm 2,J_\pm 4)\}$.

v. $\Psi_4 := \mathcal{I}\{(1,2,3,4)(1,J_\pm 3,2,J_\pm 4)\}$
$= \mathcal{I}\{(1,2,J_\pm 3,J_\pm 4)(1,3,2,4)\} = \mathcal{I}\{(1,3,J_\pm 2,J_\pm 4)(1,2,3,4)\}$
$= \Psi_3$.

vi. $\Psi_5 := \mathcal{I}\{(1,2,3,4)(J_\pm 1,3,2,J_\pm 4)\}$
$= -\mathcal{I}\{(1,2,3,4)(3,2,J_\pm 1,J_\pm 4)\} - \mathcal{I}\{(1,2,3,4)(2,J_\pm 1,3,J_\pm 4)\}$
$= -\mathcal{I}\{(2,1,3,4)(2,3,J_\pm 1,J_\pm 4)\} + \mathcal{I}\{(2,1,3,4)(2,J_\pm 1,3,J_\pm 4)\}$
$= -\mathcal{I}\{(1,2,3,4)(1,3,J_\pm 2,J_\pm 4)\} + \mathcal{I}\{(1,2,3,4)(1,J_\pm 2,3,J_\pm 4)\}$
$= -\Psi_3 + \psi_9^{\mathcal{U}_\pm}$.

vii. $\Psi_3 = \Psi_4 = \frac{1}{2}\psi_8^{\mathcal{U}_\pm}$, $\Psi_5 = -\frac{1}{2}\psi_8^{\mathcal{U}_\pm} + \psi_9^{\mathcal{U}_\pm}$.

(5) Three indices are decorated by $J_\pm$. There are several possibilities all of which yield the zero invariant:

(a) The indices decouple in the two variables:
$\mathcal{I}\{(1,2,J_\pm 2,J_\pm 1)(3,4,J_\pm 4,3)\}$. This vanishes by (1c).

(b) Two indices appear with multiplicity one in each monomial.

i. $\mathcal{I}\{(1,2,3,1)(4,J_\pm 2,J_\pm 3,J_\pm 4)\}$
$= -\mathcal{I}\{(1,2,3,1)(J_\pm 4,J_\pm 2,J_\pm 3,4)\}$
$= -\mathcal{I}\{(1,3,2,1)(J_\pm 4,J_\pm 3,J_\pm 2,4)\}$
$= -\mathcal{I}\{(1,2,3,1)(4,J_\pm 2,J_\pm 3,J_\pm 4)\}$ so this vanishes.

ii. $\mathcal{I}\{(J_\pm 1,J_\pm 2,3,1)(J_\pm 4,2,3,4)\}$
$= -\mathcal{I}\{(J_\pm 1,2,3,1)(J_\pm 4,J_\pm 2,3,4)\}$
$= -\mathcal{I}\{(J_\pm 4,2,3,4)(J_\pm 1,J_\pm 2,3,1)\}$ so this vanishes.

(c) Each index appears in each variable. One variable is assumed to be $(1,2,3,4)$. We assume four is the undecorated variable. By (1b), we assume that $J_\pm 1$ does not touch 4 in the other variable.

i. $\mathcal{I}\{(1,2,3,4)(J_\pm 1,J_\pm 2,J_\pm 3,4)\}$
$= -\mathcal{I}\{(J_\pm 1,J_\pm 2,J_\pm 3,4)(1,2,3,4)\}$ so this vanishes.

ii. $\mathcal{I}\{(1,2,3,4)(J_\pm 1,J_\pm 3,J_\pm 2,4)\}$
$= -\mathcal{I}\{(J_\pm 1,J_\pm 2,J_\pm 3,4)(1,3,2,4)\}$
$= -\mathcal{I}\{(J_\pm 1,J_\pm 3,J_\pm 2,4)(1,2,3,4)\}$ so this vanishes.

(6) Each index is decorated.

(a) The indices decouple in the two variables.

i. $\psi_{10}^{\mathcal{U}_\pm} := \mathcal{I}\{(1,2,J_\pm 2,J_\pm 1)(3,4,J_\pm 4,J_\pm 3)\}$.

(b) Two indices appear with multiplicity one in each monomial.

i. $\psi_{11}^{\mathcal{U}_\pm} := \mathcal{I}\{(1,2,3,J_\pm 1)(4,J_\pm 2,J_\pm 3,J_\pm 4)\}$.

(c) Each index appears in each monomial.

i. $\psi_{12}^{\mathcal{U}_\pm} := \mathcal{I}\{(1,2,3,4)(J_\pm 1,J_\pm 2,J_\pm 3,J_\pm 4)\}$.

ii. $\Psi_6 := \mathcal{I}\{(1,2,3,4)(J_\pm 1,J_\pm 3,J_\pm 2,J_\pm 4)\}$.

iii. $\Psi_7 := \mathcal{I}\{(1,2,3,4)(J_\pm 1,J_\pm 4,J_\pm 2,J_\pm 3)\}$
$= \mathcal{I}\{(1,3,4,2)(J_\pm 1,J_\pm 2,J_\pm 3,J_\pm 4)\}$
$= \mathcal{I}\{(J_\pm 1,J_\pm 2,J_\pm 3,J_\pm 4)(1,3,4,2)\}$
$= \mathcal{I}\{(1,2,3,4)(J_\pm 1,J_\pm 3,J_\pm 4,J_\pm 2)\}$
$= -\mathcal{I}\{(1,2,3,4)(J_\pm 1,J_\pm 3,J_\pm 2,J_\pm 4)\} = -\Psi_6$
$= -\mathcal{I}\{(1,2,3,4)(J_\pm 1,J_\pm 2,J_\pm 3,J_\pm 4)\}$
$-\mathcal{I}\{(1,2,3,4)(J_\pm 1,J_\pm 3,J_\pm 4,J_\pm 2)\} = -\psi_{12}^{\mathcal{U}_\pm} + \Psi_6$.

iv. $\Psi_7 = -\Psi_6 = -\frac{1}{2}\psi_{12}^{\mathcal{U}_\pm}$.

We have exhausted the cases to construct twelve invariants spanning $\mathcal{I}_2^{\mathcal{U}_-}(\mathfrak{R})$. As $J_\pm$ appears an even number of times, these are also $\mathcal{U}_\pm^\star$ invariants. □

There is a basic parity constraint satisfied by the tensors ρ and $\rho_{J_\pm}$:

Lemma 7.1.2

(1) $\rho : \mathfrak{R}_+^{\mathcal{U}_\pm} \to S_\mp^{2,\mathcal{U}_\pm}$ *and* $\rho : \mathfrak{R}_-^{\mathcal{U}_\pm} \to S_\pm^{2,\mathcal{U}_\pm}$.

(2) $\rho_{J_\pm} : \mathfrak{R}_+^{\mathcal{U}_\pm} \to S_\mp^{2,\mathcal{U}_\pm}$ *and* $\rho_{J_\pm} : \mathfrak{R}_-^{\mathcal{U}_\pm} \to \Lambda_\pm^{2,\mathcal{U}_\pm}$.

Proof. Let $A_{\pm,\delta} \in \mathfrak{R}_\delta^\pm$. As $J_\pm^* \varepsilon = \mp\varepsilon$ and as ρ is a symmetric tensor of rank two, we prove Assertion (1) by computing:

$$\begin{aligned}\rho(A_{\pm,\delta})(x,y) &= \varepsilon^{ij} A_\pm(e_i,x,y,e_j) = \delta\varepsilon^{ij} A_{\pm,\delta}(J_\pm e_i, J_\pm x, J_\pm y, J_\pm e_j)\\ &= \delta(J_\pm^*\varepsilon)^{ij} A_{\pm,\delta}(e_i, J_\pm x, J_\pm y, e_j) = \mp\delta\rho(A_{\pm,\delta})(J_\pm x, J_\pm y).\end{aligned}$$

Next we study $\rho_{J_\pm}$:

$$\begin{aligned}&\rho_{J_\pm}(A_{\pm,\delta})(J_\pm x, J_\pm y) = \varepsilon^{ij} A_{\pm,\delta}(e_i, J_\pm x, J_\pm J_\pm y, J_\pm e_j)\\ &\quad = (J_\pm^*\varepsilon)^{ij} A_{\pm,\delta}(J_\pm e_i, J_\pm x, J_\pm J_\pm y, J_\pm J_\pm e_j) = \mp\delta\rho_{J_\pm}(A_{\pm,\delta})(x,y),\\ &\rho_{J_\pm}(A_{\pm,\delta})(y,x) = \varepsilon^{ij} A_{\pm,\delta}(e_i, y, J_\pm x, J_\pm e_j)\\ &\quad = \delta\varepsilon^{ij} A_{\pm,\delta}(J_\pm e_i, J_\pm y, J_\pm J_\pm x, J_\pm J_\pm e_j)\\ &\quad = \delta\varepsilon^{ij} A_{\pm,\delta}(e_j, x, J_\pm y, J_\pm e_i) = \delta\rho_{J_\pm}(A_{\pm,\delta})(x,y).\end{aligned}$$ □

We now come to a useful method for constructing examples.

Definition 7.1.1 Let $m = 2\bar{m}$, let $\{e_1, \dots, e_m\}$ be the usual basis for $\mathbb{R}^m$, and let $u = (u^1, \dots, u^m)$ be the associated coordinate system. Define a (para)-complex structure $J_\pm$ on $\mathbb{R}^m$ by setting:

$$J_\pm : \partial_{u_i} \to \partial_{u_{i+\bar{m}}}, \quad J_\pm : \partial_{u_{i+\bar{m}}} \to \pm\partial_{u_i} \quad \text{for} \quad 1 \le i \le \bar{m}. \tag{7.1.a}$$

Let $\varepsilon_{\pm,ij}$ denote the components of a $J_\pm$ (skew)-invariant non-degenerate inner product on $\mathbb{R}^m$ and let $\Theta_\pm \in S_\mp^{2,\mathcal{U}_\pm} \otimes S^2$. Define:

$$g_{\pm,ij} := \varepsilon_{\pm,ij} - 2\Theta_{\pm,ijkl} u^k u^l.$$

Let $\Xi_\pm(\Theta_\pm)$ denote the curvature of the associated Levi-Civita connection at the origin. By Lemma 3.3.2,

$$\Xi_\pm(\Theta_\pm)_{ijkl} = \Theta_{\pm,iljk} + \Theta_{\pm,jkil} - \Theta_{\pm,ikjl} - \Theta_{\pm,jlik}. \tag{7.1.b}$$

Furthermore, by Theorem 1.9.1

$$\Xi_\pm : S_\mp^{2,\mathcal{U}_\pm} \otimes S^2 \to \mathfrak{G}_\pm.$$

We now study the modules $\mathfrak{R}_-^{\mathcal{U}_\pm}$ in further detail:

Lemma 7.1.3

(1) Ξ_+ *defines a* $\mathcal{U}_+^\star$ *equivariant map* $\Xi_+ : S_-^{2,\mathcal{U}_+} \otimes S_\pm^{2,\mathcal{U}_+} \to \mathfrak{R}_\mp^{\mathcal{U}_+}$.

(2) Ξ_- *defines a* $\mathcal{U}_-^\star$ *equivariant map* $\Xi_- : S_+^{2,\mathcal{U}_-} \otimes S_\pm^{2,\mathcal{U}_-} \to \mathfrak{R}_\pm^{\mathcal{U}_-}$.

(3) We have $(\rho \oplus \rho_{J_+})\Xi_+ : S_-^{2,\mathcal{U}_+} \otimes S_+^{2,\mathcal{U}_+} \to S_+^{2,\mathcal{U}_+} \oplus \Lambda_+^{2,\mathcal{U}_+} \to 0$.

(4) We have $(\rho \oplus \rho_{J_-})\Xi_- : S_+^{2,\mathcal{U}_-} \otimes S_-^{2,\mathcal{U}_-} \to S_-^{2,\mathcal{U}_-} \oplus \Lambda_-^{2,\mathcal{U}_-} \to 0$.

(5) If $m \geq 6$, *then* $W_{\pm,10}^{\mathfrak{R}} \cap \operatorname{Range}(\Xi_\pm) \neq \{0\}$.

Proof. Assertions (1) and (2) follow from the discussion given above and from Equation (7.1.b). To simplify the notation, we let $x^i := u^i$ and $y^i := u^{i+\bar{m}}$ for $1 \leq i \leq \bar{m}$. We then have

$$J_\pm \partial_{x_i} = \partial_{y_i} \quad \text{and} \quad J_\pm \partial_{y_i} = \pm \partial_{x_i} \quad \text{for} \quad 1 \leq i \leq \bar{m}, \tag{7.1.c}$$
$$\varepsilon_\pm(\partial_{x_i}, \partial_{x_i}) = \mp \varepsilon_\pm(\partial_{y_i}, \partial_{y_i}).$$

We use the construction of Definition 7.1.1. Consider the metrics:

$$g_\pm := \varepsilon_\pm - \{\varrho_1((x^2)^2 \pm (y^2)^2) + 2\varrho_2(x^1y^2 + x^2y^1)\}$$
$$\times (dx^1 \otimes dx^1 \mp dy^1 \otimes dy^1).$$

The perturbation $\Theta_\pm$ defining $g_\pm$ is:

$$\Theta_\pm = (dx^1 \otimes dx^1 \mp dy^1 \otimes dy^1) \times \{\varrho_1(dx^2 \otimes dx^2 \pm dy^2 \otimes dy^2)$$
$$+\varrho_2(dx^1 \otimes dy^2 + dy^2 \otimes dx^1 + dx^2 \otimes dy^1 + dy^1 \otimes dx^2)\}.$$

Since $\Theta_+ \in S_-^{2,\mathcal{U}_+} \otimes S_+^{2,\mathcal{U}_+}$ and $\Theta_- \in S_+^{2,\mathcal{U}_-} \otimes S_-^{2,\mathcal{U}_-}$, we apply Equation (7.1.b) to see that $g_\pm$ is the germ of a para-Hermitian metric or a pseudo-Hermitian metric on $\mathbb{R}^m$ and that the associated curvatures at the origin $A_\pm := \Xi_\pm(\Theta_\pm)$ belong to $\mathfrak{R}_-^{\mathcal{U}_\pm}$. The (possibly) non-zero components of the curvature tensors are determined by the following equations up to the usual $\mathbb{Z}_2$ symmetries:

$$A_\pm(\partial_{x_1}, \partial_{x_2}, \partial_{x_2}, \partial_{x_1}) = \varrho_1, \quad A_\pm(\partial_{y_1}, \partial_{x_2}, \partial_{x_2}, \partial_{y_1}) = \mp\varrho_1,$$
$$A_\pm(\partial_{x_1}, \partial_{y_2}, \partial_{y_2}, \partial_{x_1}) = \pm\varrho_1, \quad A_\pm(\partial_{y_1}, \partial_{y_2}, \partial_{y_2}, \partial_{y_1}) = -\varrho_1,$$
$$A_\pm(\partial_{y_1}, \partial_{x_1}, \partial_{y_2}, \partial_{y_1}) = \mp\varrho_2, \quad A_\pm(\partial_{x_1}, \partial_{y_1}, \partial_{x_2}, \partial_{x_1}) = \varrho_2.$$

We may therefore compute:

$$\rho(A_\pm)(\partial_{x_2}, \partial_{x_2})$$
$$= \varepsilon_{\pm,11}\{A_\pm(\partial_{x_1}, \partial_{x_2}, \partial_{x_2}, \partial_{x_1}) \mp A_\pm(\partial_{y_1}, \partial_{x_2}, \partial_{x_2}, \partial_{y_1})\}$$

$$= 2\varepsilon_{\pm,11}\varrho_1,$$

$$\rho_{J_\pm}(A)(\partial_{x_2}, \partial_{y_1}) = \varepsilon_{\pm,11}A(\partial_{x_2}, \partial_{x_1}, J_\pm\partial_{x_1}, J_\pm\partial_{y_1}) = \mp\varepsilon_{\pm,11}\rho_2.$$

We take $\rho_1 = \rho_2 = 1$. Since $A_\pm \in \mathfrak{R}_-^{\mathcal{U}_\pm}$,

$$0 \neq \rho(A_\pm) \in S_\pm^{2,\mathcal{U}_\pm} \quad \text{and} \quad 0 \neq \rho_{J_\pm}(A_\pm) \in \Lambda_\pm^{2,\mathcal{U}_\pm}$$

by Lemma 7.1.2. Assertion (3) and Assertion (4) follow as $S_\pm^{2,\mathcal{U}_\pm}$ and $\Lambda_\pm^{2,\mathcal{U}_\pm}$ are inequivalent irreducible modules for the group $\mathcal{U}_\pm^\star$.

Let $m \geq 6$. We clear the previous notation and consider

$$\begin{aligned} g_\pm := \varepsilon_\pm - ((x^1)^2 \pm (y^1)^2) \\ \times (dx^2 \otimes dx^3 + dx^3 \otimes dx^2 \mp dy^2 \otimes dy^3 \mp dy^3 \otimes dy^2). \end{aligned}$$

Since the perturbation $\Theta_\pm$ defining $g_\pm$ belongs to $S_\mp^{2,\mathcal{U}_\pm} \otimes S_\pm^{2,\mathcal{U}_\pm}$, $g_\pm$ is the germ of a para-Hermitian metric (+) or a pseudo-Hermitian metric (−) on $\mathbb{R}^m$. The associated curvatures at the origin $A_\pm := \Xi_\pm(\Theta_\pm)$ belong to $\mathfrak{R}_-^{\mathcal{U}_\pm}$. The (possibly) non-zero components of $A_\pm$ are determined by the following equations up to the usual $\mathbb{Z}_2$ symmetries:

$$\begin{aligned} &A_\pm(\partial_{x_1}, \partial_{x_2}, \partial_{x_3}, \partial_{x_1}) = 1, \quad A_\pm(\partial_{x_1}, \partial_{y_2}, \partial_{y_3}, \partial_{x_1}) = \mp 1, \\ &A_\pm(\partial_{y_1}, \partial_{x_2}, \partial_{x_3}, \partial_{y_1}) = \pm 1, \quad A_\pm(\partial_{y_1}, \partial_{y_2}, \partial_{y_3}, \partial_{y_1}) = -1. \end{aligned}$$

We have $\rho(A_\pm) = \rho_{J_\pm}(A) = 0$ so $0 \neq A_\pm \in W_{\pm,10}^{\mathfrak{R}}$. □

The usual $\mathbb{Z}_2$ symmetries show that if $A \in W_{\pm,7}^{\mathfrak{R}}$, then

$$A(J_\pm x, y, z, w) = A(x, J_\pm y, z, w) = A(x, y, J_\pm z, w) = A(x, y, z, J_\pm w).$$

It is now immediate that $W_{\pm,7}^{\mathfrak{R}} \subset \mathfrak{R}_+^{\mathcal{U}_\pm}$.

Lemma 7.1.4 *$W_{\pm,7}^{\mathfrak{R}} \cap \mathfrak{G}_\pm = \{0\}$, $W_{\pm,7}^{\mathfrak{R}} \subset \ker(\rho \oplus \rho_{J_\pm})$, and $W_{\pm,7}^{\mathfrak{R}} \neq \{0\}$.*

Proof. Suppose $A_\pm \in W_{\pm,7}^{\mathfrak{R}} \cap \mathfrak{G}_\pm$. We show $A = 0$ by computing:

$$\begin{aligned} 0 &= A(x, y, z, w) + A(J_\pm x, J_\pm y, J_\pm z, J_\pm w) \\ &\quad \pm A(J_\pm x, J_\pm y, z, w) \pm A(x, y, J_\pm z, J_\pm w) \pm A(J_\pm x, y, J_\pm z, w) \\ &\quad \pm A(x, J_\pm y, z, J_\pm w) \pm A(J_\pm x, y, z, J_\pm w) \pm A(x, J_\pm y, J_\pm z, w) \\ &= A(x, y, z, w) + A(J_\pm^4 x, y, z, w) \pm 6A(J_\pm^2 x, y, z, w) \\ &= 8A(x, y, z, w). \end{aligned}$$

Let $A \in W_{\pm,7}^{\mathfrak{R}}$. We show $\rho(A) = \rho_{J_\pm}(A) = 0$ by checking:

$$\begin{aligned}\rho(A)(x,y) &= \varepsilon^{ij}A(e_i,x,y,e_j) = (J_\pm^*\varepsilon)^{ij}A(J_\pm e_i,x,y,J_\pm e_j)\\ &= \mp\varepsilon^{ij}A(J_\pm e_i,x,y,J_\pm e_j) = \mp\varepsilon^{ij}A(J_\pm^2 e_i,x,y,e_j)\\ &= -\varepsilon^{ij}A(e_i,x,y,e_j) = -\rho(A)(x,y),\\ \rho_{J_\pm}(A)(x,y) &= \varepsilon^{ij}A(e_i,x,J_\pm y,J_\pm e_j) = \varepsilon^{ij}A(J_\pm J_\pm e_i,x,y,e_j)\\ &= \pm\varepsilon^{ij}A(e_i,x,y,e_j) = \pm\rho(A)(x,y) = 0.\end{aligned}$$

Let $A_\pm$ be determined up to the usual $\mathbb{Z}_2$ symmetries by:

$$\begin{aligned}&A_\pm(J_\pm e_1,e_2,e_2,e_1) = A(e_1,J_\pm e_2,e_2,e_1) = 1,\\ &A_\pm(J_\pm e_1,J_\pm e_2,J_\pm e_2,e_1) = A(J_\pm e_1,J_\pm e_2,e_2,J_\pm e_1) = \pm 1.\end{aligned}$$

It is clear by inspection that $A_\pm \in W_{\pm,7}^{\mathfrak{R}}$. □

Let ρ_0 be the part of ρ having trace zero. Similarly, let $\rho_{J_\pm,0}$ be the part of the tensor $\rho_{J_\pm}$ having zero trace.

Lemma 7.1.5 *Let $m \geq 6$. Then:*

(1) $\rho_0 \oplus \rho_{J_\pm,0} : \mathfrak{G}_\pm \cap \operatorname{Range}(\Xi_\pm) \cap \mathfrak{R}_+^{\mathcal{U}_\pm} \to S_{0,\mp}^{2,\mathcal{U}_\pm} \oplus S_{0,\mp}^{2,\mathcal{U}_\pm} \to 0$.

(2) $\tau \oplus \tau_{J_\pm} : \mathfrak{G}_\pm \cap \operatorname{Range}(\Xi_\pm) \to \mathbb{R} \oplus \mathbb{R} \to 0$.

Proof. We clear the previous notation and set:

$$\begin{aligned}g_\pm = \varepsilon_\pm &- \varrho_1((x^1)^2 \mp (y^1)^2)\\ &\qquad \times(dx^1\otimes dx^2 + dx^2\otimes dx^1 \mp dy^1\otimes dy^2 \mp dy^2\otimes dy^1)\\ &-\varrho_2((x^1)^2 \mp (y^1)^2)\\ &\qquad \times(dx^2\otimes dx^3 + dx^3\otimes dx^2 \mp dy^2\otimes dy^3 \mp dy^3\otimes dy^2).\end{aligned}$$

We use Equation (7.1.b). Since the perturbation $\Theta_\pm$ defining $g_\pm$ belongs to $S_\mp^{2,\mathcal{U}_\pm} \otimes S_\mp^{2,\mathcal{U}_\pm}$, $g_\pm$ is the germ of a para-Hermitian metric (+) or of a pseudo-Hermitian metric (−) with curvature $A_\pm := \Xi_\pm(\Theta_\pm) \in \mathfrak{G}_\pm \cap \mathfrak{R}_+^{\mathcal{U}_\pm}$. Consequently by Lemma 7.1.2, $\rho(A_\pm)$ and $\rho_{J_\pm}(A_\pm)$ are symmetric and in particular lie in $S_\mp^{2,\mathcal{U}_\pm}$. The (possibly) non-zero curvature components are determined by the following equations up to the usual $\mathbb{Z}_2$ symmetries:

$$\begin{aligned}&A_\pm(\partial_{x_1},\partial_{y_1},\partial_{y_2},\partial_{x_1}) = A_\pm(\partial_{y_1},\partial_{x_1},\partial_{x_2},\partial_{y_1}) = \mp\varrho_1,\\ &A_\pm(\partial_{x_1},\partial_{x_2},\partial_{x_3},\partial_{x_1}) = A_\pm(\partial_{y_1},\partial_{y_2},\partial_{y_3},\partial_{y_1}) = \varrho_2,\\ &A_\pm(\partial_{y_1},\partial_{x_2},\partial_{x_3},\partial_{y_1}) = A_\pm(\partial_{x_1},\partial_{y_2},\partial_{y_3},\partial_{x_1}) = \mp\varrho_2.\end{aligned}$$

It is clear by inspection that $\tau(A_\pm) = \tau_{J_\pm}(A_\pm) = 0$. We compute:

$$\begin{aligned}
&\rho(A_\pm)(\partial_{x_1},\partial_{x_2}) = \mp\varepsilon_{\pm,11}A_\pm(\partial_{y_1},\partial_{x_1},\partial_{x_2},\partial_{y_1}) = \varepsilon_{\pm,11}\varrho_1,\\
&\rho(A_\pm)(\partial_{x_2},\partial_{x_3}) = \varepsilon_{\pm,11}\{A(\partial_{x_1},\partial_{x_2},\partial_{x_3},\partial_{x_1}) \mp A(\partial_{y_1},\partial_{x_2},\partial_{x_3},\partial_{y_1})\}\\
&\quad = 2\varepsilon_{\pm,11}\varrho_2,\\
&\rho_{J_\pm}(A_\pm)(\partial_{x_1},\partial_{x_2}) = \mp\varepsilon_{\pm,11}A(\partial_{y_1},\partial_{x_1},J_\pm\partial_{x_2},J_\pm\partial_{y_1})\\
&\quad = -\varepsilon_{\pm,11}A(\partial_{y_1},\partial_{x_1},\partial_{y_2},\partial_{x_1}) = \mp\varepsilon_{\pm,11}\varrho_1,\\
&\rho_{J_\pm}(A_\pm)(\partial_{x_2},\partial_{x_3}) = 0.
\end{aligned}$$

Since ϱ_1 and ϱ_2 are arbitrary parameters, and since $S^{2,\mathcal{U}_\pm}_{0,\pm}$ is an irreducible module for the group $\mathcal{U}_\pm$, Assertion (1) follows.

We clear the previous notation and consider:

$$\begin{aligned}
g_\pm = \varepsilon_\pm - &\{\varrho_1((x^2)^2 \mp (y^2)^2) + \varrho_2((x^1)^2 \mp (y^1)^2)\}\\
&\times(dx^1 \otimes dx^1 \mp dy^1 \otimes dy^1).
\end{aligned}$$

As before, $g_\pm$ is the germ of a para-Hermitian metric (+) or a pseudo-Hermitian metric (−) on $\mathbb{R}^m$ defined by a perturbation $\Theta_\pm$ belonging to $S^{2,\mathcal{U}_\pm}_{\mp} \otimes S^{2,\mathcal{U}_\pm}_{\mp}$. Thus the curvatures at the origin $A_\pm := \Xi_\pm(\Theta_\pm)$ belong to $\mathfrak{G}_\pm \cap \mathfrak{R}^{\mathcal{U}_\pm}_{+}$. The (possibly) non-zero curvature components are determined by the following equations and the usual $\mathbb{Z}_2$ symmetries:

$$\begin{aligned}
&A_\pm(\partial_{x_1},\partial_{x_2},\partial_{x_2},\partial_{x_1}) = \varrho_1, &&A(\partial_{x_1},\partial_{y_2},\partial_{y_2},\partial_{x_1}) = \mp\varrho_1,\\
&A_\pm(\partial_{y_1},\partial_{x_2},\partial_{x_2},\partial_{y_1}) = \mp\varrho_1, &&A_\pm(\partial_{y_1},\partial_{y_2},\partial_{y_2},\partial_{y_1}) = \varrho_1,\\
&A_\pm(\partial_{x_1},\partial_{y_1},\partial_{y_1},\partial_{x_1}) = \mp 2\varrho_2.
\end{aligned}$$

We complete the proof by verifying that

$$\begin{aligned}
\tau(A_\pm) &= \mp 2A(\partial_{x_1},\partial_{y_1},\partial_{y_1},\partial_{x_1})\\
&\qquad +4\varepsilon_{\pm,11}\varepsilon_{\pm,22}\{A(\partial_{x_1},\partial_{x_2},\partial_{x_2},\partial_{x_1}) \mp A(\partial_{x_1},\partial_{y_2},\partial_{y_2},\partial_{x_1})\}\\
&= 4\varrho_2 + 8\varepsilon_{\pm,11}\varepsilon_{\pm,22}\varrho_1,\\
\tau_{J_\pm}(A) &= \mp A_\pm(\partial_{x_1},\partial_{y_1},J_\pm\partial_{y_1},J_\pm\partial_{x_1}) \mp A_\pm(\partial_{y_1},\partial_{x_1},J_\pm\partial_{x_1},\partial_{y_1})\\
&= -A_\pm(\partial_{x_1},\partial_{y_1},\partial_{x_1},\partial_{y_1}) - A_\pm(\partial_{y_1},\partial_{x_1},\partial_{y_1},\partial_{x_1}) = \mp 4\varrho_2. \quad \square
\end{aligned}$$

Let $\pi^{\mathfrak{R}}_{\pm,6}$ be orthogonal projection on $W^{\mathfrak{R}}_{\pm,6}$.

Lemma 7.1.6 *If $m \geq 8$, then $\pi^{\mathfrak{R}}_{\pm,6}\{\text{Range}(\Xi_\pm)\} \neq \{0\}$.*

Proof. We take

$$\begin{aligned}
g_\pm = \varepsilon_\pm - &2\{x^1x^2 \mp y^1y^2\}\\
&\times(dx^3 \otimes dx^4 + dx^4 \otimes dx^3 \mp dy^3 \otimes dy^4 \mp dy^4 \otimes dy^3).
\end{aligned}$$

We apply Equation (7.1.b). Since the perturbation $\Theta_\pm$ defining $g_\pm$ belongs to $S^{2,\mathcal{U}_\pm}_\mp \otimes S^{2,\mathcal{U}}_\mp$, $g_\pm$ is the germ of a para-Hermitian metric (+) or a pseudo-Hermitian metric (−) whose associated curvatures at the origin satisfy $A_\pm := \Xi_\pm(\Theta_\pm) \in \mathfrak{G}_\pm \cap \mathfrak{R}^{\mathcal{U}_\pm}_+$. The non-zero components of these tensors are determined by the following equations and the usual $\mathbb{Z}_2$ symmetries:

$$\begin{aligned}
A_\pm(\partial_{x_1},\partial_{x_3},\partial_{x_4},\partial_{x_2}) &= A_\pm(\partial_{x_1},\partial_{x_4},\partial_{x_3},\partial_{x_2}) = 1,\\
A_\pm(\partial_{y_1},\partial_{x_3},\partial_{x_4},\partial_{y_2}) &= A_\pm(\partial_{y_1},\partial_{x_4},\partial_{x_3},\partial_{y_2}) = \mp 1,\\
A_\pm(\partial_{x_1},\partial_{y_3},\partial_{y_4},\partial_{x_2}) &= A_\pm(\partial_{x_1},\partial_{y_4},\partial_{y_3},\partial_{x_2}) = \mp 1,\\
A_\pm(\partial_{y_1},\partial_{y_3},\partial_{y_4},\partial_{y_2}) &= A_\pm(\partial_{y_1},\partial_{y_4},\partial_{y_3},\partial_{y_2}) = 1.
\end{aligned}$$

It is immediate that $\rho(A_\pm) = \rho_{J_\pm}(A_\pm) = 0$. Furthermore, since

$$A_\pm(\partial_{y_1},\partial_{y_3},\partial_{x_4},\partial_{x_2}) = 0 \quad \text{and} \quad A_\pm(\partial_{x_1},\partial_{x_3},\partial_{x_4},\partial_{x_2}) \neq 0,$$

we have that $A_\pm \notin \mathfrak{K}^{\mathfrak{R}}_\pm$. □

We now examine the role played by $W^{\mathfrak{R}}_{\pm,3} = \mathfrak{K}^{\mathfrak{R}}_\pm \cap \ker(\rho \oplus \rho_{J_\pm})$.

Lemma 7.1.7

(1) If $A_\pm \in \mathfrak{K}^{\mathfrak{R}}_\pm$ then $A_\pm \in \mathfrak{G}_\pm$ and $\rho(A_\pm) = \mp\rho_{J_\pm}(A_\pm)$.

(2) $W^{\mathfrak{R}}_{\pm,3} \cap \operatorname{Range}(\Xi_\pm) \neq \{0\}$.

(3) $\rho : \mathfrak{K}^{\mathfrak{R}}_\pm \cap \operatorname{Range}(\Xi_\pm) \to S^{2,\mathcal{U}_\pm}_\mp \to 0$.

Proof. Let $A_\pm \in \mathfrak{K}^{\mathfrak{R}}_\pm$. We have

$$A_\pm(J_\pm x, y, *, *) = \mp A_\pm(J_\pm J_\pm x, J_\pm y, *, *) = -A_\pm(x, J_\pm y, *, *).$$

We verify $A_\pm \in \mathfrak{G}_\pm$ by computing:

$$\begin{aligned}
& A_\pm(x,y,z,w) + A_\pm(J_\pm x, J_\pm y, J_\pm z, J_\pm w)\\
& \pm A_\pm(J_\pm x, J_\pm y, z, w) \pm A_\pm(x, y, J_\pm z, J_\pm w) \pm A_\pm(J_\pm x, y, J_\pm z, w)\\
& \pm A_\pm(x, J_\pm y, z, J_\pm w) \pm A_\pm(J_\pm x, y, z, J_\pm w) \pm A_\pm(x, J_\pm y, J_\pm z, w)\\
= \;& A_\pm(x,y,z,w) \mp A_\pm(x,y,J_\pm z, J_\pm w)\\
& - A_\pm(x,y,z,w) \pm A_\pm(x,y,J_\pm z,J_\pm w) \mp A_\pm(x, J_\pm y, J_\pm z, w)\\
& \pm A_\pm(x, J_\pm y, z, J_\pm w) \mp A_\pm(x, J_\pm y, z, J_\pm w) \pm A_\pm(x, J_\pm y, J_\pm z, w)\\
= \;& 0.
\end{aligned}$$

We complete the proof of Assertion (1) by computing:

$$\begin{aligned}
\rho_{J_\pm}(A_\pm)(x,y) &= \varepsilon^{ij} A_\pm(e_i, x, J_\pm y, J_\pm e_j) = \varepsilon^{ij} A_\pm(J_\pm e_j, J_\pm y, x, e_i)\\
&= \mp\varepsilon^{ij} A_\pm(e_j, y, x, e_i) = \mp\rho(x,y).
\end{aligned}$$

We now establish Assertion (2). Let $\delta = -\varepsilon_{11}\varepsilon_{22}$. We clear the previous notation and define:

$$\begin{aligned} g_\pm = \varepsilon_\pm - \{(x^1)^2 \mp (y^1)^2 + \delta(x^2)^2 \mp \delta(y^2)^2\} \\ \times(dx^1 \otimes dx^2 + dx^2 \otimes dx^1 \mp dy^1 \otimes dy^2 \mp dy^2 \otimes dy^1). \end{aligned}$$

Again, we apply Equation (7.1.b). As the perturbation $\Theta_\pm$ defining $g_\pm$ belongs to $S_\mp^{2,\mathcal{U}_\pm} \otimes S_\mp^{2,\mathcal{U}_\pm}$, $g_\pm$ is the germ of a para-Hermitian metric $(+)$ or of a pseudo-Hermitian metric $(-)$ on $\mathbb{R}^m$. Consequently, the associated curvatures $A_\pm := \Xi_\pm(\Theta_\pm)$ of $g_\pm$ at the origin belong to $\mathfrak{G}_\pm \cap \mathfrak{R}_+^{\mathcal{U}_\pm}$. The non-zero components of the curvature tensors are defined by the following relationships and the usual $\mathbb{Z}_2$ symmetries:

$$\begin{aligned} A_\pm(\partial_{x_1}, \partial_{y_1}, \partial_{y_2}, \partial_{x_1}) = A_\pm(\partial_{y_1}, \partial_{x_1}, \partial_{x_2}, \partial_{y_1}) = \mp 1, \\ A_\pm(\partial_{x_2}, \partial_{y_1}, \partial_{y_2}, \partial_{x_2}) = A_\pm(\partial_{y_2}, \partial_{x_1}, \partial_{x_2}, \partial_{y_2}) = \mp\delta. \end{aligned}$$

We verify by inspection that $A_\pm \in \mathfrak{K}_\pm^{\mathfrak{R}}$ and that $\rho(A_\pm) = 0$. Consequently $\rho_{J_\pm}(A_\pm) = 0$ by Assertion (1). This shows that $0 \neq A_\pm \in W_{\pm,3}^{\mathfrak{R}}$ and establishes Assertion (2).

To prove Assertion (3), we clear the previous notation and set:

$$\begin{aligned} g_\pm := \varepsilon_\pm - \varrho_1((x^1)^2 \mp (y^1)^2) \times (dx^1 \otimes dx^1 \mp dy^1 \otimes dy^1) \\ -\varrho_2((x^2)^2 \mp (y^2)^2) \times (dx^2 \otimes dx^2 \mp dy^2 \otimes dy^2). \end{aligned}$$

Because the perturbation $\Theta_\pm$ defining $g_\pm$ belongs to $S_\mp^{2,\mathcal{U}_\pm} \otimes S_\mp^{2,\mathcal{U}_\pm}$, applying Equation (7.1.b) yields that $g_\pm$ is the germ of a para-Hermitian metric $(+)$ or of a pseudo-Hermitian metric $(-)$ on $\mathbb{R}^m$. The associated curvatures at the origin $A_\pm := \Xi_\pm(\Theta_\pm)$ belong to $\mathfrak{G}_\pm \cap \mathfrak{R}_+^{\mathcal{U}_\pm}$. The (possibly) non-zero components of the curvatures are determined by the following equations and the usual $\mathbb{Z}_2$ symmetries:

$$A_\pm(\partial_{x_1}, \partial_{y_1}, \partial_{y_1}, \partial_{x_1}) = \mp 2\varrho_1, \quad A(\partial_{x_2}, \partial_{y_2}, \partial_{y_2}, \partial_{x_2}) = \mp 2\varrho_2.$$

Consequently

$$\begin{aligned} \rho(A_\pm)(\partial_{x_1}, \partial_{x_1}) = \mp\rho(A_\pm)(\partial_{y_1}, \partial_{y_1}) = 2\varrho_1\varepsilon_{\pm,11}, \\ \rho(A_\pm)(\partial_{x_2}, \partial_{x_2}) = \mp\rho(A_\pm)(\partial_{y_2}, \partial_{y_2}) = 2\varrho_2\varepsilon_{\pm,22}. \end{aligned}$$

We verify by inspection that $A_\pm \in \mathfrak{K}_\pm^{\mathfrak{R}}$ and that $\tau(A_\pm) = 4\varrho_1 + 4\varrho_2$. We take $\varrho_1 = \varrho_2 \neq 0$ to see that:

$$\tau : \mathfrak{K}_\pm^{\mathfrak{R}} \cap \operatorname{Range}(\Xi_\pm) \to \mathbb{R} \to 0.$$

Let ρ_0 be the part of the Ricci tensor which has zero trace. We may set $\varrho_1 = -\varrho_2 \neq 0$. We then have $\tau(A_\pm) = 0$ but $\rho(A_\pm) \neq 0$. Thus

$$\rho_0 : \mathfrak{K}_\pm^{\mathfrak{R}} \cap \text{Range}(\Xi_\pm) \to S_{0,\mp}^{2,\mathcal{U}_\pm} \to 0. \qquad \square$$

Proof of Theorem 1.8.1. Suppose $m \geq 8$. By Lemma 7.1.2,

$$(\rho \oplus \rho_{J_\pm})(\mathfrak{R}) \subset 2 \cdot \mathbb{R} \oplus 2 \cdot S_{0,\mp}^{2,\mathcal{U}_\pm} \oplus S_\pm^{2,\mathcal{U}_\pm} \oplus \Lambda_\pm^{2,\mathcal{U}_\pm}. \tag{7.1.d}$$

By Theorem 5.4.1, the modules $\{\mathbb{R}, S_{0,\mp}^{2,\mathcal{U}_\pm}, S_\pm^{2,\mathcal{U}_\pm}, \Lambda_\pm^{2,\mathcal{U}_\pm}\}$ are inequivalent and irreducible with the structure groups $\mathcal{U}_-$ and $\mathcal{U}_\pm^\star$. Thus by Lemma 7.1.3 and Lemma 7.1.5, we have surjective maps:

$$\rho \oplus \rho_{J_\pm} : \text{Range}(\Xi_\pm) \to 2 \cdot \mathbb{R} \oplus 2 \cdot S_{0,\mp}^{2,\mathcal{U}_\pm} \oplus S_\pm^{2,\mathcal{U}_\pm} \oplus \Lambda_\pm^{2,\mathcal{U}_\pm} \to 0. \tag{7.1.e}$$

By Lemma 2.1.5, we have module isomorphisms:

$$\mathfrak{R} \approx \ker(\rho \oplus \rho_{J_\pm}) \oplus \text{Range}(\rho \oplus \rho_{J_\pm}).$$

We apply Lemma 7.1.3, Lemma 7.1.4, Lemma 7.1.6, and Lemma 7.1.7. Since $m \geq 8$, the modules

$$\{W_{\pm,3}^{\mathfrak{R}},\ W_{\pm,6}^{\mathfrak{R}},\ W_{\pm,7}^{\mathfrak{R}},\ W_{\pm,10}^{\mathfrak{R}}\}$$

are non-trivial. By definition, $W_{\pm,3}^{\mathfrak{R}} \cap W_{\pm,6}^{\mathfrak{R}} = \{0\}$. As $W_{\pm,3}^{\mathfrak{R}} \oplus W_{\pm,6}^{\mathfrak{R}} \subset \mathfrak{G}_\pm$,

$$\{W_{\pm,3}^{\mathfrak{R}} \oplus W_{\pm,6}^{\mathfrak{R}}\} \cap W_{\pm,7}^{\mathfrak{R}} = \{0\}.$$

Since $W_{\pm,3}^{\mathfrak{R}} \oplus W_{\pm,6}^{\mathfrak{R}} \oplus W_{\pm,7}^{\mathfrak{R}} \subset \mathfrak{R}_+^{\mathcal{U}_\pm}$ and since $W_{\pm,10}^{\mathfrak{R}} \subset \mathfrak{R}_-^{\mathcal{U}_\pm}$,

$$\{W_{\pm,3}^{\mathfrak{R}} \oplus W_{\pm,6}^{\mathfrak{R}} \oplus W_{\pm,7}^{\mathfrak{R}}\} \cap W_{\pm,10}^{\mathfrak{R}} = \{0\}.$$

Consequently we have a direct sum of modules. These belong to $\ker(\rho \oplus \rho_{J_\pm})$ by definition and by Lemma 7.1.7. We summarize:

$$W_{\pm,3}^{\mathfrak{R}} \oplus W_{\pm,6}^{\mathfrak{R}} \oplus W_{\pm,7}^{\mathfrak{R}} \oplus W_{\pm,10}^{\mathfrak{R}} \subset \ker(\rho \oplus \rho_{J_\pm}\}.$$

We set

$$\begin{array}{ll} W_{\pm,1}^{\mathfrak{R}} = W_{\pm,4}^{\mathfrak{R}} = \mathbb{R}, & W_{\pm,2}^{\mathfrak{R}} = W_{\pm,5}^{\mathfrak{R}} = S_{0,\mp}^{2,\mathcal{U}_\pm}, \\ W_{\pm,8}^{\mathfrak{R}} = S_\pm^{2,\mathcal{U}_\pm}, & W_{\pm,9}^{\mathfrak{R}} = \Lambda_\pm^{2,\mathcal{U}_\pm}. \end{array} \tag{7.1.f}$$

We will identify these subspaces of $\mathfrak{R}$ explicitly in Theorem 7.2.1. However, all we need for the moment is their isomorphism class. Let

$$W^{\mathcal{U}_\pm} := \bigoplus_{i=1}^{10} W_{\pm,i}^{\mathfrak{R}};$$

these modules are isomorphic to submodules of $\mathfrak{R}$. These modules are all non-trivial; two appear with multiplicity two and the remaining six appear with multiplicity one. Thus we may apply Lemma 2.2.2 to estimate

$$\dim\{\mathcal{I}_2^{\mathcal{U}_-}(\mathfrak{R})\} \geq 3+3+6=12, \quad \dim\{\mathcal{I}_2^{\mathcal{U}_\pm^\star}(\mathfrak{R})\} \geq 3+3+6=12.$$

The reverse inequality is provided by Lemma 7.1.1. Thus equality holds and by Lemma 2.2.2, we have that $\mathfrak{R} = W^{\mathcal{U}_\pm}$; this establishes Remark 7.1.1. We also have that the modules $W_{\pm,i}^{\mathfrak{R}}$ are irreducible and, except for the isomorphisms

$$W_{\pm,1}^{\mathfrak{R}} \approx W_{\pm,4}^{\mathfrak{R}} \quad \text{and} \quad W_{\pm,2}^{\mathfrak{R}} \approx W_{\pm,5}^{\mathfrak{R}},$$

the modules $W_{\pm,i}^{\mathfrak{R}}$ are inequivalent modules for the groups $\mathcal{U}_-$ and $\mathcal{U}_\pm^\star$. This establishes Theorem 1.8.1 if $m \geq 8$.

We now deal with the exceptional dimensions. Let $m=4$ or $m=6$. We can define a vector space U of dimension $\tilde{m}=8$ so that $U = V \oplus V_1$ is an orthogonal direct sum decomposition preserved by $J_\pm$. We use the metric to identify $U = U^*$, $V = V^*$, and $V_1 = V_1^*$. The inclusion of V^* in U^* induces natural inclusions

$$W_{\pm,i}^{\mathfrak{R}}(V, \langle\cdot,\cdot\rangle, J_\pm) \to W_{\pm,i}^{\mathfrak{R}}(U, \langle\cdot,\cdot\rangle, J_\pm).$$

Lemma 2.3.1 shows that the natural dual maps

$$\mathcal{I}_2^{\mathcal{U}_-(U,\langle\cdot,\cdot\rangle,J_\pm)}(W_{\pm,i}^{\mathfrak{R}}(U,\langle\cdot,\cdot\rangle,J_\pm)) \to \mathcal{I}_2^{\mathcal{U}_-(V,\langle\cdot,\cdot\rangle,J_\pm)}(W_{\pm,i}^{\mathfrak{R}}(V,\langle\cdot,\cdot\rangle,J_\pm)),$$
$$\mathcal{I}_2^{\mathcal{U}_\pm^\star(U,\langle\cdot,\cdot\rangle,J_\pm)}(W_{\pm,i}^{\mathfrak{R}}(U,\langle\cdot,\cdot\rangle,J_\pm)) \to \mathcal{I}_2^{\mathcal{U}_\pm^\star(V,\langle\cdot,\cdot\rangle,J_\pm)}(W_{\pm,i}^{\mathfrak{R}}(V,\langle\cdot,\cdot\rangle,J_\pm))$$

are surjective. Consequently

$$\dim\{\mathcal{I}_2^{\mathcal{U}_-(V,\langle\cdot,\cdot\rangle,J_\pm)}(W_{\pm,i}^{\mathfrak{R}}(V,\langle\cdot,\cdot\rangle,J_\pm))\} \leq 1,$$
$$\dim\{\mathcal{I}_2^{\mathcal{U}_\pm^\star(V,\langle\cdot,\cdot\rangle,J_\pm)}(W_{\pm,i}^{\mathfrak{R}}(V,\langle\cdot,\cdot\rangle,J_\pm))\} \leq 1.$$

It now follows that the modules $W_{\pm,i}^{\mathfrak{R}}$ are either trivial or irreducible in dimensions $m=4$ and $m=6$ as well. Furthermore, analytic continuation shows that

$$\dim\{\mathcal{I}_2^{\mathcal{U}_\pm^\star(V,\langle\cdot,\cdot\rangle,J_\pm)}(W_{\pm,i}^{\mathfrak{R}}(V,\langle\cdot,\cdot\rangle,J_\pm))\}$$

is independent of the signature and equal to

$$\dim\{\mathcal{I}_2^{\mathcal{U}_-(V,\langle\cdot,\cdot\rangle,J_-)}(W_{-,i}^{\mathfrak{R}}(V,\langle\cdot,\cdot\rangle,J_-))\}.$$

We apply [Tricerri and Vanhecke (1981)]. In the positive definite setting, the only additional relation that must be imposed is that $W^{\mathfrak{R}}_{\pm,6} = 0$ and in dimension $m = 4$, we must set $W^{\mathfrak{R}}_{\pm,5} = W^{\mathfrak{R}}_{\pm,6} = W^{\mathfrak{R}}_{\pm,10} = 0$. Thus these are the only additional relations in either the para-Hermitian or the pseudo-Hermitian settings as well. □

Proof of Theorem 1.10.3. By Lemma 7.1.7, $\rho = \mp\rho_{J_\pm}$ on $\mathfrak{K}^{\mathfrak{R}}_{\pm}$. Lemma 7.1.7 also shows that $\rho(\mathfrak{K}^{\mathfrak{R}}_{\pm}) = S^{2,\mathcal{U}_\pm}_{\mp}$. Because

$$W^{\mathfrak{R}}_{\pm,3} = \mathfrak{K}^{\mathfrak{R}}_{\pm} \cap \ker(\rho \oplus \rho_{J_\pm}),$$

we may apply Lemma 2.1.5 to construct an isomorphism

$$\mathfrak{K}^{\mathfrak{R}}_{\pm} \approx S^{2,\mathcal{U}_\pm}_{\mp} \oplus W^{\mathfrak{R}}_{\pm,3}.$$

We expand $S^{2,\mathcal{U}_\pm}_{\mp} \approx \mathbb{R} \oplus S^{2,\mathcal{U}_\pm}_{\mp,0}$ to complete the proof. □

7.2 The Submodules of $\mathfrak{R}$ Arising from the Ricci Tensors

We now study the submodules of $\mathfrak{R}$ of Equation (7.1.f) in more detail motivated by [Tricerri and Vanhecke (1981)]. We begin with:

Definition 7.2.1 Let $S \in V^* \otimes V^*$. Set:

(1) $\pi_1(x,y,z,w) := \langle x,z\rangle\langle y,w\rangle - \langle y,z\rangle\langle x,w\rangle$.

(2) $\pi_{\pm,2}(x,y,z,w) := 2\langle J_\pm x,y\rangle\langle J_\pm z,w\rangle + \langle J_\pm x,z\rangle\langle J_\pm y,w\rangle - \langle J_\pm y,z\rangle\langle J_\pm x,w\rangle$.

(3) $\Phi(S)(x,y,z,w) := \langle x,z\rangle S(y,w) + \langle y,w\rangle S(x,z) - \langle x,w\rangle S(y,z) - \langle y,z\rangle S(x,w)$.

(4) $\Psi_\pm(S)(x,y,z,w) := 2\langle x,J_\pm y\rangle S(z,J_\pm w) + 2\langle z,J_\pm w\rangle S(x,J_\pm y) + \langle x,J_\pm z\rangle S(y,J_\pm w) + \langle y,J_\pm w\rangle S(x,J_\pm z) - \langle x,J_\pm w\rangle S(y,J_\pm z) - \langle y,J_\pm z\rangle S(x,J_\pm w)$.

Lemma 7.2.1 *Adopt the notation established above.*

(1) If $S \in S^2$, then $\Phi(S) \in \mathfrak{R}$.

(2) $\pi_1 \in \mathfrak{R}$ and $\pi_{\pm,2} \in \mathfrak{R}$.

(3) If $S(x,J_\pm y) + S(y,J_\pm x) = 0$, then $\Psi_\pm(S) \in \mathfrak{R}$.

(4) If $S \in S^{2,\mathcal{U}_\pm}_{\mp} \oplus \Lambda^{2,\mathcal{U}_\pm}_{\pm}$, then $\Psi_\pm(S) \in \mathfrak{R}$.

Proof. Let $S \in S^2$. We adopt the notation of Definition 4.1.1 to express $\Phi(S) = -\sigma_2(S) - \sigma_3(S)$ and thus Lemma 4.1.3 implies $\Phi(S) \in \mathfrak{A}$. Since clearly $\Phi(S)(x, y, z, w) = \Phi(S)(z, w, x, y)$, $\Phi(S) \in \mathfrak{R}$. This establishes Assertion (1). Let $S(x, y) = \langle x, y\rangle$. Since $\pi_1(x, y, z, w) = \frac{1}{2}\Phi(S)(x, y, z, w)$, $\pi_1 \in \mathfrak{R}$. Let $\psi \in \Lambda^2$. We adopt the notation of Definition 1.6.1 and set:

$$A_\psi(x, y, z, w) := \psi(x, w)\psi(y, z) - \psi(x, z)\psi(y, w) - 2\psi(x, y)\psi(z, w).$$

We showed in Section 6.1 that $A_\psi \in \mathfrak{R}$. Let $\Omega_\pm(\cdot, \cdot) = \langle \cdot, J_\pm \cdot\rangle$ be the (para)-Kähler form. We complete the proof of Assertion (2) by noting that $\pi_{\pm,2} = -A_{\Omega_\pm} \in \mathfrak{R}$.

Let $\psi_i \in \Lambda^2$. We polarize A_ψ to define $\Psi_1(\psi_1, \psi_2) := \partial_\varepsilon\{A_{\psi_1+\varepsilon_1\psi_2}\} \in \mathfrak{R}$:

$$\begin{aligned}
&\Psi_1(\psi_1, \psi_2)(x, y, z, w)\\
&= \psi_1(x, w)\psi_2(y, z) - \psi_1(x, z)\psi_2(y, w) - 2\psi_1(x, y)\psi_2(z, w)\\
&\quad +\psi_2(x, w)\psi_1(y, z) - \psi_2(x, z)\psi_1(y, w) - 2\psi_2(x, y)\psi_1(z, w).
\end{aligned}$$

Suppose $S(x, J_\pm y) + S(y, J_\pm x) = 0$. We set $\psi_1(x, y) := S(x, J_\pm y)$. Then $\psi_1 \in \Lambda^2$ and

$$\begin{aligned}
&\Psi_1(\psi_1, \Omega_\pm)\\
&= S(x, J_\pm w)\langle y, J_\pm z\rangle - S(x, J_\pm z)\langle y, J_\pm w\rangle - 2S(x, J_\pm y)\langle z, J_\pm w\rangle\\
&\quad +\langle x, J_\pm w\rangle S(y, J_\pm z) - \langle x, J_\pm z\rangle S(y, J_\pm w) - 2\langle x, J_\pm y\rangle S(z, J_\pm w)\\
&= -\Psi_\pm(S).
\end{aligned}$$

Assertion (3) follows. Assertion (4) follows from Assertion (3) and from the following computation:

$$\begin{aligned}
S \in S^{2,\mathcal{U}_\pm}_{\mp} &\Rightarrow S(y, J_\pm x) = S(J_\pm x, y)\\
&\qquad = \mp S(J_\pm J_\pm x, J_\pm y) = -S(x, J_\pm y),\\
S \in \Lambda^{2,\mathcal{U}_\pm}_{\pm} &\Rightarrow S(y, J_\pm x) = -S(J_\pm x, y)\\
&\qquad = \mp S(J_\pm J_\pm x, J_\pm y) = -S(x, J_\pm y).
\end{aligned}$$ □

We continue our study with the following technical result:

Lemma 7.2.2

(1) If $(a, b) \neq (0, 0)$, *then* $a\pi_1 + b\pi_{\pm,2} \neq 0$.

(2) Let $m \geq 6$.

  (a) If $(a, b) \neq (0, 0)$, *then* $(a\Phi + b\Psi_\pm)(S^{2,\mathcal{U}_\pm}_{0,\mp}) \neq \{0\}$.

  (b) $\Phi(S^{2,\mathcal{U}_\pm}_{0,\mp}) \not\subset \mathfrak{K}^{\mathfrak{R}}_\pm$.

(3) $0 \neq \pi_1 \mp \pi_{\pm,2} \in \mathfrak{K}_\pm^{\mathfrak{R}}$ *and* $0 \neq (\Phi \mp \Psi_\pm)(S_{0,\mp}^{2,\mathcal{U}_\pm}) \subset \mathfrak{K}_\pm^{\mathfrak{R}}$.

(4) $0 \neq \pi_1 \notin \mathfrak{K}_\pm^{\mathfrak{R}}$.

(5) $\Phi(S_\pm^{2,\mathcal{U}_\pm}) \neq \{0\}$ *and* $\Psi_\pm(\Lambda_\pm^{2,\mathcal{U}_\pm}) \neq \{0\}$.

Proof. Let $\{e_1, ..., e_{\bar{m}}, f_1, ... f_{\bar{m}}\}$ be an orthonormal basis for V with $J_\pm e_i = f_i$ and $J_\pm f_i = \pm e_i$. Let $\{e^i, f^i\}$ be the corresponding orthonormal basis for V^*. Let $\varepsilon_{\pm,ii} := \langle e_i, e_i\rangle$ for $i = 1, 2$. We establish Assertion (1) by computing:

$$\pi_1(e_1, e_2, e_2, e_1) = -\varepsilon_{\pm,11}\varepsilon_{\pm,22}, \qquad \pi_1(e_1, e_2, f_2, f_1) = 0,$$
$$\pi_{\pm,2}(e_1, e_2, e_2, e_1) = 0, \quad \pi_{\pm,2}(e_1, e_2, f_2, f_1) = -\varepsilon_{\pm,11}\varepsilon_{\pm,22}.$$

Let $m \geq 6$. Let $\delta_\pm := -\varepsilon_{\pm,11}\varepsilon_{\pm,22}$. Define $\phi_\mp \in S^2$ by setting:

$$\phi_\mp := e^1 \otimes e^1 \mp f^1 \otimes f^1 + \delta_\pm e^2 \otimes e^2 \mp \delta_\pm f^2 \otimes f^2.$$

Clearly $J_\pm^* \phi_\mp = \mp\phi_\mp$. We show $\phi_\mp \in S_{0,\mp}^{2,\mathcal{U}_\pm}$ by computing:

$$\begin{aligned}\tau(\phi_\mp) &= \langle e_1, e_1\rangle \mp \langle f_1, f_1\rangle + \delta_\pm\langle e_2, e_2\rangle \mp \delta_\pm\langle f_2, f_2\rangle \\ &= 2\varepsilon_{\pm,11} + 2\delta_\pm\varepsilon_{\pm,22} = 0.\end{aligned}$$

We examine:

$$\begin{aligned}&\Phi(\phi_\mp)(e_1, f_1, f_1, e_1) = \pm 2\varepsilon_{\pm,11}, && \Psi_\pm(\phi_\mp)(e_1, f_1, f_1, e_1) = -6\varepsilon_{\pm,11},\\ &\Phi(\phi_\mp)(e_1, e_3, f_3, f_1) = 0, && \Psi_\pm(\phi_\mp)(e_1, e_3, f_3, f_1) = -\varepsilon_{\pm,33},\\ &\Phi(\phi_\mp)(e_1, e_3, e_3, e_1) = \pm\varepsilon_{\pm,33}, && \Psi_\pm(\phi_\mp)(e_1, e_3, e_3, e_1) = 0.\end{aligned}$$

Since $\Phi(\phi_\mp)(f_1, f_3, e_3, e_1) \neq \Phi(\phi_\mp)(e_1, e_3, e_3, e_1)$, $\Phi(\phi_\mp)$ is not (para)-Kähler; this establishes Assertion (2b). Assertion (2a) follows by inspection.

Let $S_\pm \in S_\mp^{2,\mathcal{U}_\pm}$. Since $\pi_{\pm,2} = \frac{1}{2}\Psi_\pm\varepsilon_\pm$, Assertion (3) follows from Assertion (1) by checking:

$$\begin{aligned}\{(\Phi \mp \Psi_\pm)(S_\mp)\}(x, y, z, w) &= \langle x, z\rangle S_\mp(y, w) + \langle y, w\rangle S_\mp(x, z)\\ &\quad -\langle x, w\rangle S_\mp(y, z) - \langle y, z\rangle S_\mp(x, w)\\ &\quad \mp 2\langle x, J_\pm y\rangle S_\mp(z, J_\pm w) \mp 2\langle z, J_\pm w\rangle S_\mp(x, J_\pm y)\\ &\quad \mp\langle x, J_\pm z\rangle S_\mp(y, J_\pm w) \mp \langle y, J_\pm w\rangle S_\mp(x, J_\pm z)\\ &\quad \pm\langle x, J_\pm w\rangle S_\mp(y, J_\pm z) \pm \langle y, J_\pm z\rangle S_\mp(x, J_\pm w),\end{aligned}$$

$$\begin{aligned}&\{(\Phi \mp \Psi_\pm)(S_\mp)\}(J_\pm x, J_\pm y, z, w)\\ &\quad = \langle J_\pm x, z\rangle S_\mp(J_\pm y, w) + \langle J_\pm y, w\rangle S_\mp(J_\pm x, z)\\ &\qquad -\langle J_\pm x, w\rangle S_\mp(J_\pm y, z) - \langle J_\pm y, z\rangle S_\mp(J_\pm x, w)\end{aligned}$$

$$\begin{aligned}
&\mp 2\langle J_\pm x, J_\pm J_\pm y\rangle S_\mp(z, J_\pm w) \mp 2\langle z, J_\pm w\rangle S_\mp(J_\pm x, J_\pm J_\pm y)\\
&\mp\langle J_\pm x, J_\pm z\rangle S_\mp(J_\pm y, J_\pm w) \mp \langle J_\pm y, J_\pm w\rangle S_\mp(J_\pm x, J_\pm z)\\
&\pm\langle J_\pm x, J_\pm w\rangle S_\mp(J_\pm y, J_\pm z) \pm \langle J_\pm y, J_\pm z\rangle S_\mp(J_\pm x, J_\pm w),
\end{aligned}$$

$$\{(\Phi \mp \Psi_\pm)(S_\mp)\}(J_\pm x, J_\pm y, z, w) = \mp\{(\Phi \mp \Psi_\pm)(S_\mp)\}(x, y, z, w).$$

We argue as follows to establish Assertion (4):

$$\pi_1(e_1, e_2, e_2, e_1) = -1 \quad \text{and} \quad \pi_1(f_1, f_2, e_1, e_2) = 0.$$

We clear the previous notation and set:

$$\begin{aligned}
\phi_\pm &:= e^1 \otimes e^2 + e^2 \otimes e^1 \pm f^1 \otimes f^2 \pm f^2 \otimes f^1 \in S_\pm^{2,\mathcal{U}_\pm},\\
\psi_\pm &:= e^1 \otimes e^2 - e^2 \otimes e^1 \pm f^1 \otimes f^2 \mp f^2 \otimes f^1 \in \Lambda_\pm^{2,\mathcal{U}_\pm}.
\end{aligned}$$

We establish Assertion (5) and complete the proof by checking:

$$\begin{aligned}
&\Phi(\phi_\pm)(e_1, f_1, f_2, e_1) = -\langle e_1, e_1\rangle \phi_\pm(f_1, f_2) = \mp\varepsilon_{\pm,11},\\
&\Psi_\pm(\psi_\pm)(e_1, f_1, e_2, f_1)\\
&\quad = 2\langle e_1, J_\pm f_1\rangle \psi_\pm(e_2, J_\pm f_1) - \langle e_1, J_\pm f_1\rangle \phi_\pm(f_1, J_\pm e_2)\\
&\quad = -3\varepsilon_{\pm,11}.
\end{aligned}$$

□

We can now identify the submodules of $\mathfrak{R}$ that are described by the Ricci tensors in either the pseudo-Hermitian or in the para-Hermitian settings:

Theorem 7.2.1

(1) $W_{\pm,1}^{\mathfrak{R}} = (\pi_1 \mp \pi_{\pm,2}) \cdot \mathbb{R}$.

(2) $W_{\pm,1}^{\mathfrak{R}} \oplus W_{\pm,4}^{\mathfrak{R}} = \operatorname{Span}\{\pi_1, \pi_{\pm,2}\}$.

(3) $W_{\pm,2}^{\mathfrak{R}} = (\Phi \mp \Psi_\pm)(S_{0,\mp}^{2,\mathcal{U}_\pm})$.

(4) If $m \geq 6$, *then* $W_{\pm,2}^{\mathfrak{R}} \oplus W_{\pm,5}^{\mathfrak{R}} = \operatorname{Span}\{\Phi(S), \Psi_\pm(S)\}_{S \in S_{0,\mp}^{2,\mathcal{U}_\pm}}$.

(5) $W_{\pm,8}^{\mathfrak{R}} = \Phi(S_\pm^{2,\mathcal{U}_\pm})$ *and* $W_{\pm,9}^{\mathfrak{R}} = \Psi_\pm(\Lambda_\pm^{2,\mathcal{U}_\pm})$.

Proof. We apply Theorem 1.8.1 and Lemma 7.2.2. By Lemma 7.2.2, $(\pi_1 \mp \pi_{\pm,2}) \cdot \mathbb{R}$ is a linear submodule of $\mathfrak{K}_\pm^{\mathfrak{R}}$ and hence may be identified with $W_{\pm,1}^{\mathfrak{R}}$. This proves Assertion (1). Assertion (1) of Lemma 7.2.2 shows π_1 and $\pi_{\pm,2}$ are linearly independent. Consequently $\operatorname{Span}\{\pi_1, \pi_{\pm,2}\}$ is a trivial module of dimension $r = 2$. Assertion (2) follows since the trivial module appears with multiplicity two in $\mathfrak{R}$ and since $W_{\pm,1}^{\mathfrak{R}} \oplus W_{\pm,4}^{\mathfrak{R}}$ represents that module.

We prove Assertion (3) by noting that $S^{2,\mathcal{U}_\pm}_{0,\mp}$ is an irreducible module for the group $\mathcal{U}^\star_\pm$, that $S^{2,\mathcal{U}_\pm}_{0,\mp}$ appears with multiplicity one in $\mathfrak{K}^{\mathfrak{R}}_\pm$, and that

$$0 \neq (\Phi \mp \Psi_\pm) S^{\mathcal{U}_\pm}_{0,+} \subset W^{\mathfrak{R}}_{\pm,2}.$$

Let $m \geq 6$. Assertion (4) follows from Assertion (2) of Lemma 7.2.2 since $S^{2,\mathcal{U}_\pm}_{0,\mp}$ appears with multiplicity two in $\mathfrak{R}$ as a module for the group $\mathcal{U}^\star_\pm$ and since

$$\Phi(S^{2,\mathcal{U}_\pm}_{0,\pm}) \not\subset \mathfrak{K}^{\mathfrak{R}}_\pm.$$

Since $S^{2,\mathcal{U}_\pm}_\pm$ and $\Lambda^{2,\mathcal{U}_\pm}_\pm$ are irreducible modules for the group $\mathcal{U}^\star_\pm$ appearing with multiplicity one in $\mathfrak{R}$, since $\Phi(S^{2,\mathcal{U}_\pm}_\pm) \neq \{0\}$, and since $\Psi_\pm(\Lambda^{2,\mathcal{U}_\pm}_\pm) \neq \{0\}$,

$$W^{\mathfrak{R}}_{\pm,8} = \Phi(S^{2,\mathcal{U}_\pm}_\pm) \text{ and } W^{\mathfrak{R}}_{\pm,9} = \Psi_\pm(\Lambda^{2,\mathcal{U}_\pm}_\pm). \qquad \square$$

Remark 7.2.1 [Tricerri and Vanhecke (1981)] showed in the positive definite setting that

$$W^{\mathfrak{R}}_{-,4} = (3\pi_1 - \pi_{-,2}) \cdot \mathbb{R} \quad \text{and} \quad W^{\mathfrak{R}}_{-,5} = (3\Phi - \Psi_-)(S^{2,\mathcal{U}_-}_{0,+}).$$

We use analytic continuation to see this relation extends to the indefinite setting as well and that in the para-Hermitian setting we have

$$W^{\mathfrak{R}}_{+,4} = (3\pi_1 + \pi_{+,2}) \cdot \mathbb{R} \quad \text{and} \quad W^{\mathfrak{R}}_{+,5} = (3\Phi + \Psi_+)(S^{2,\mathcal{U}_+}_{0,-}).$$

We omit the proof of these relationships as they will play no role in our development.

We continue our study:

Lemma 7.2.3

(1) $\tau(\pi_1 \mp \pi_{\pm,2}) = -m(m+2)$.
(2) If $\phi_\mp \in S^{2,\mathcal{U}_\pm}_{0,\mp}$, *then* $\rho(\Phi \mp \Psi_\pm)\phi_\mp = -(m+4)\phi_\mp$.

Proof. We use the relation $\varepsilon^{il}\langle e_i, x\rangle e_l = x$ to prove Assertion (1):

$$\begin{aligned}
\tau(\pi_1 \mp \pi_{\pm,2}) &= \varepsilon^{il}\varepsilon^{jk}\{\langle e_i, e_k\rangle\langle e_j, e_l\rangle - \langle e_j, e_k\rangle\langle e_i, e_l\rangle\} \\
&\quad \mp \varepsilon^{il}\varepsilon^{jk}\{2\langle e_i, J_\pm e_j\rangle\langle e_k, J_\pm e_l\rangle + \langle e_i, J_\pm e_k\rangle\langle e_j, J_\pm e_l\rangle \\
&\quad -\langle e_j, J_\pm e_k\rangle\langle e_i, J_\pm e_l\rangle\} \\
&= \varepsilon^{jk}\{\langle e_j, e_k\rangle - m\langle e_j, e_k\rangle \mp 2\langle e_k, J_\pm J_\pm e_j\rangle \mp \langle J_\pm J_\pm e_j, e_k\rangle - 0\} \\
&= m - m^2 - 2m - m = -m^2 - 2m.
\end{aligned}$$

We establish Assertion (2) by computing:

$$\begin{aligned}\rho(\Phi(\phi_\mp))(y,z) &= \varepsilon^{il}\{\langle e_i,z\rangle\phi_\mp(y,e_l)+\langle y,e_l\rangle\phi_\mp(e_i,z)\}\\&\quad+\varepsilon^{il}\{-\langle e_i,e_l\rangle\phi_\mp(y,z)-\langle y,z\rangle\phi_\mp(e_i,e_l)\}\\&=\{1+1-m-0\}\phi_\mp(y,z),\\ \rho(\Psi_\pm(\phi_\mp))(y,z) &= \varepsilon^{il}\{2\langle e_i,J_\pm y\rangle\phi_\mp(z,J_\pm e_l)+2\langle z,J_\pm e_l\rangle\phi_\mp(e_i,J_\pm y)\}\\&\quad+\varepsilon^{il}\{\langle e_i,J_\pm z\rangle\phi_\mp(y,J_\pm e_l)+\langle y,J_\pm e_l\rangle\phi_\mp(e_i,J_\pm z)\}\\&\quad+\varepsilon^{il}\{-\langle e_i,J_\pm e_l\rangle\phi_\mp(y,J_\pm z)-\langle y,J_\pm z\rangle\phi_\mp(e_i,J_\pm e_l)\}\\&=(\pm2\pm2\pm1\pm1+0+0)\phi_\mp(y,z).\end{aligned}$$ □

We will need the following result when we study Kähler geometry in Section 7.5. Further information concerning the orthogonal projections on the modules $W^{\mathfrak{R}}_{\pm,i}$ is available in [Tricerri and Vanhecke (1981)]. Let ρ_0 be the part of the Ricci tensor which has zero trace;

$$\rho_0(A)=\rho(A)-\tfrac{1}{m}\tau(A)\langle\cdot,\cdot\rangle.$$

Lemma 7.2.4 *Let $A_\pm\in\mathfrak{K}^{\mathfrak{R}}_\pm$. Let $\pi^{\mathfrak{R}}_{\pm,i}$ be orthogonal projection on $W^{\mathfrak{R}}_{\pm,i}$.*

(1) $\pi^{\mathfrak{R}}_{\pm,1}(A_\pm)=-(m(m+2))^{-1}\tau(A_\pm)(\pi_1\mp\pi_{\pm,2})$.

(2) $\pi^{\mathfrak{R}}_{\pm,2}(A_\pm)=-(m+4)^{-1}(\Phi\mp\Psi_\pm)(\rho_0(A_\pm))$.

Proof. We expand $A_\pm=A_{\pm,1}+A_{\pm,2}+A_{\pm,3}$ where $A_{\pm,i}\in W^{\mathfrak{R}}_{\pm,i}$. The modules $\{W^{\mathfrak{R}}_{\pm,1},W^{\mathfrak{R}}_{\pm,2},W^{\mathfrak{R}}_{\pm,3}\}$ are inequivalent and irreducible modules for the group $\mathcal{U}^\star_\pm$. It now follows that

$$\tau(A_\pm)=\tau(A_{\pm,1})\quad\text{and}\quad\rho_0(A_\pm)=\rho_0(A_{\pm,2}).$$

By Theorem 7.2.1, we may express

$$A_{\pm,1}=c_\pm(A_\pm)\cdot(\pi_1\mp\pi_{\pm,2}).$$

We use Lemma 7.2.3 to derive the following relations:

$$\tau(A_\pm)=\tau(A_{\pm,1})=c_\pm(A_\pm)\cdot\tau(\pi_1\mp\pi_{\pm,2})=-c_\pm(A_\pm)\cdot m(m+2).$$

We solve these relations to determine $c_\pm(A_\pm)$ and establish Assertion (1). Similarly, by Theorem 7.2.1, there are tensors $\phi_\mp(A_\pm)$ in $S^{2,\mathcal{U}_\pm}_{0,\mp}$ so

$$A_{\pm,2}=(\Phi\mp\Psi_\pm)(\phi_\mp(A_\pm)).$$

We have $\rho_0(A_\pm)=\rho_0(A_{\pm,2})\in S^{2,\mathcal{U}_\pm}_{0,\mp}$. We use Lemma 7.2.3 to derive the following relations and determine $\phi_\mp(A_\pm)$. This will complete the proof:

$$\rho_0(A_\pm)=\rho_0(A_{\pm,2})=\rho_0(\Phi\mp\Psi_\pm)(\phi_\mp)=-(m+4)\phi_\mp(A_\pm).$$ □

7.3 Para-Hermitian and Pseudo-Hermitian Geometry

Recall that $\mathfrak{G}_\pm$ is the subspace of curvature tensors satisfying the Gray identity in the para-Hermitian setting (+) or in the pseudo-Hermitian setting (−). By Theorem 1.9.1, the curvature tensor R of a para-Hermitian manifold (+) or of a pseudo-Hermitian manifold (−) belongs to $\mathfrak{G}_\pm$. In Section 7.3, we shall establish Theorem 1.9.3 showing:

$$\mathfrak{G}_\pm = (W^{\mathfrak{R}}_{\pm,7})^\perp.$$

Combined with Theorem 1.8.1, this gives the decomposition of $\mathfrak{G}_-$ as a module for the group $\mathcal{U}_-$ and of $\mathfrak{G}_\pm$ as a module for the group $\mathcal{U}^\star_\pm$. We will then use this curvature decomposition to establish Theorem 1.9.2; Theorem 1.9.2 shows any element of $\mathfrak{G}_\pm$ is geometrically realizable by a para-Hermitian manifold (+) of constant scalar curvature τ or a pseudo-Hermitian manifold (−) of constant scalar curvature τ which also has $\star$-scalar curvature $\tau_{J_\pm}$.

Proof of Theorem 1.9.3. We suppose that $m \geq 8$; the exceptional dimensions $m = 4$ and $m = 6$ are handled similarly. We use Lemma 7.1.3 and Lemma 7.1.5 to see that the modules detected by ρ and by $\rho_{J_\pm}$ are geometrically realizable by a para-Hermitian manifold (+) or by a pseudo-Hermitian manifold (−), or, equivalently, that:

$$W^{\mathfrak{R}}_{\pm,1} \oplus W^{\mathfrak{R}}_{\pm,2} \oplus W^{\mathfrak{R}}_{\pm,4} \oplus W^{\mathfrak{R}}_{\pm,5} \oplus W^{\mathfrak{R}}_{\pm,8} \oplus W^{\mathfrak{R}}_{\pm,9} \subset \text{Range}(\Xi_\pm).$$

We have also shown in Lemma 7.1.3, Lemma 7.1.6, and Lemma 7.1.7 that

$$\pi^{\mathfrak{R}}_{\pm,3}\Xi_- \neq 0, \qquad \pi^{\mathfrak{R}}_{\pm,6}\Xi_- \neq 0, \qquad \pi^{\mathfrak{R}}_{\pm,10}\Xi_- \neq 0.$$

Since these modules are irreducible modules occurring with multiplicity one in $\mathfrak{R}$, Lemma 2.1.7 shows that

$$W^{\mathfrak{R}}_{\pm,3} \oplus W^{\mathfrak{R}}_{\pm,6} \oplus W^{\mathfrak{R}}_{\pm,10} \subset \text{Range}(\Xi_\pm).$$

Consequently, one has the inclusions:

$$\bigoplus_{i\neq 7} W^{\mathfrak{R}}_{\pm,i} \subset \text{Range}(\Xi_\pm) \subset \mathfrak{G}_\pm.$$

On the other hand, $W^{\mathfrak{R}}_{\pm,7} \cap \mathfrak{G}_\pm = \{0\}$. Since $W^{\mathfrak{R}}_{\pm,7}$ is irreducible and occurs with multiplicity one in $\mathfrak{R}$, Lemma 2.1.7 permits us to conclude:

$$\mathfrak{G}_\pm \subset W_{\pm,7}^\perp \subset \bigoplus_{i\neq 7} W_{\pm,i}^{\mathfrak{R}} \subset \operatorname{Range}(\Xi_\pm) \subset \mathfrak{G}_\pm .$$

Thus we have equality in all the containments of the above equation:

$$\operatorname{Image}(\Xi_\pm) = \mathfrak{G}_\pm = (W_{\pm,7}^{\mathfrak{R}})^\perp = \bigoplus_{i\neq 7} W_{\pm,i}^{\mathfrak{R}}. \tag{7.3.a}$$

This establishes Theorem 1.9.3 □

Proof of Theorem 1.9.2. Let $A_\pm$ belong to $\mathfrak{G}_\pm$. We follow the discussion in [Brozos-Vázquez et al. (2009a)]; see also [Brozos-Vázquez et al. (2010)]. We take the integrable complex and para-complex structures given in Equation (7.1.c). By Equation (7.3.a) we have $\mathfrak{G}_\pm = \operatorname{Image}(\Xi_\pm)$. Thus we can find $\Theta_\pm \in S^2_\mp \otimes S^2$ so that the associated metrics

$$g_{\pm,ij} = \varepsilon_{\pm,ij} + \Theta_{\pm,ijkl} u^k u^l$$

yield curvatures at the origin by $A_\pm$. We perturb these examples to adjust the scalar curvature and use the Cauchy–Kovalevskaya Theorem of Section 6.3. Let $v := (u^1, \ldots, u^{m-1})$ so $u = (v, y^{\bar m})$. Let

$$\begin{aligned} \xi_\pm(v,0) &= 0, & \eta_\pm(v,0) &= 0,\\ \partial_{y_{\bar m}}\xi_\pm(v,0) &= 0, & \partial_{y_{\bar m}}\eta_\pm(v,0) &= 0. \end{aligned}$$

We consider the germ of a para-Hermitian metric (+) or of a pseudo-Hermitian metric (−):

$$\begin{aligned} h_\pm := g_\pm &+ \xi_\pm(dx^1 \otimes dx^1 \mp dy^1 \otimes dy^1)\\ &+\eta_\pm(dx^{\bar m} \otimes dx^{\bar m} \mp dy^{\bar m} \otimes dy^{\bar m}). \end{aligned}$$

Let $\eta_{\pm,\bar m\bar m} := \partial^2_{y_{\bar m}}\eta_\pm$ and $\xi_{\pm,\bar m\bar m} := \partial^2_{y_{\bar m}}\xi_\pm$. Then the non-zero curvatures of interest are determined, up to the usual $\mathbb{Z}_2$ symmetries, by the following relations (where we omit terms not of interest in the Cauchy–Kovalevskaya Theorem):

$$\begin{aligned} &R_\pm(\partial_{x_1}, \partial_{y_{\bar m}}, \partial_{y_{\bar m}}, \partial_{x_1}) = -\xi_{\pm,\bar m\bar m} + \ldots,\\ &R_\pm(\partial_{y_1}, \partial_{y_{\bar m}}, \partial_{y_{\bar m}}, \partial_{y_1}) = \pm\xi_{\pm,\bar m\bar m} + \ldots,\\ &R_\pm(\partial_{x_{\bar m}}, \partial_{y_{\bar m}}, \partial_{y_{\bar m}}, \partial_{x_{\bar m}}) = -\eta_{\pm,\bar m\bar m} + \ldots,\\ &\tau(R_\pm) = \pm 4\xi_{\pm,\bar m\bar m} \pm 2\eta_{\pm,\bar m\bar m} + \ldots,\\ &\tau_{J_\pm} = -2\eta_{\pm,\bar m\bar m} + \ldots. \end{aligned}$$

Consequently the vector valued version of the Cauchy–Kovalevskaya Theorem implies we can solve the equations

$$\tau^{h_\pm} - \tau^{g_\pm}(0) = 0, \qquad \tau_{J_\pm}^{h_\pm} - \tau_{J_\pm}^{g_\pm}(0) = 0.$$

The only possible non-zero second order jets of the metric at the origin are $\{\eta_{\pm,mm}, \xi_{\pm,mm}\}$; these are seen to be zero by the defining equation. Thus

$$\eta_\pm = O(|(x,y)|^3) \quad \text{and} \quad \xi_\pm = O(|(x,y)|^3).$$

Consequently, the curvatures at the origin are unchanged. □

7.4 Almost Para-Hermitian and Almost Pseudo-Hermitian Geometry

In Section 7.4, we will establish Theorem 1.8.3. We wish to show that any almost para-Hermitian curvature model (+) and that any almost pseudo-Hermitian curvature model (−) is geometrically realizable by an almost para-Hermitian manifold (+) or by an almost pseudo-Hermitian manifold (−) with τ constant and with $\tau_{J_\pm}$ constant. We will show that any almost pseudo-Hermitian curvature model is geometrically realizable by an almost pseudo-Hermitian manifold and that any almost para-Hermitian curvature model is geometrically realizable by an almost para-Hermitian manifold. The Cauchy–Kovalevskaya Theorem will then be used exactly as was done in Section 7.3 to adjust the scalar curvatures to be constant. Thus to prove Theorem 1.8.3, it suffices to show:

Lemma 7.4.1 *Let $\mathfrak{C} = (V, \langle\cdot,\cdot\rangle, J_\pm, A)$ be a para-Hermitian curvature model (+) or be a pseudo-Hermitian curvature model (−). There exists a real analytic manifold $\mathcal{C} = (M, g, J_\pm)$ and a point P of M such that $\mathfrak{C}$ is isomorphic to $(T_PM, g_P, J_{\pm,P}, R_P)$.*

Proof. Let $(V, \langle\cdot,\cdot\rangle, A)$ be a curvature model. We use Lemma 6.1.1 to choose an analytic pseudo-Riemannian metric g so that $g(0) = \langle\cdot,\cdot\rangle$ and $R(0) = A$. Let $J_\pm$ be a para-Hermitian complex structure (+) or be a pseudo-Hermitian complex structure (−) on $(V, \langle\cdot,\cdot\rangle)$. The difficulty now is to extend $J_\pm$ to be a suitable structure on the tangent bundle TM. First extend $J_\pm$ and $\langle\cdot,\cdot\rangle$ to a neighborhood of 0 by requiring the extension to be constant with respect to the coordinate frame. Apply Lemma 3.3.5 to express $g = \psi^*\langle\cdot,\cdot\rangle$ for the germ of some real analytic map ψ. Set

$J_{\pm,1} := \psi J_\pm \psi^{-1} = \psi^* J_\pm$. We show that $(M, g, J_{\pm,1})$ provides the required structure by computing:

$$J^2_{\pm,1} = (\psi^* J_\pm)^2 = \psi^*(J^2_\pm) = \pm \mathrm{Id},$$
$$J^*_{\pm,1} g = (\psi^* J_\pm)^* \{\psi^* \langle \cdot, \cdot \rangle\} = \psi^* \{J^*_\pm \langle \cdot, \cdot \rangle\} = \mp \psi^* \langle \cdot, \cdot \rangle = \mp g. \quad \square$$

7.5 Kähler Geometry in the Riemannian Setting III

In Section 7.5, we report on joint work with E. Merino [Brozos-Vázquez, Gilkey, and Merino (2010)] to prove Theorem 1.10.2. We will first show that any (para)-Kähler curvature model can be geometrically realized by a (para)-Kähler manifold. We will then use the Cauchy–Kovalevskaya theorem to adjust the (para)-Kähler metric to have constant scalar curvature.

We recall some notation established previously. Let $\{e_1, \dots, e_m\}$ be the standard basis for $\mathbb{R}^m$ defining dual coordinates

$$(u^1, \dots, u^m) = (x^1, \dots, x^{\bar{m}}, y^1, \dots, y^{\bar{m}})$$

on $\mathbb{R}^m$. Let $J_\pm$ be the canonical (para)-complex structure on $\mathbb{R}^m$ described in Equation (7.1.a) for the coordinates $\{u^1, ..., u^m\}$ and in Equation (7.1.c) for the coordinates $\{x^1, y^1, ..., x^{\bar{m}}, y^{\bar{m}}\}$. Consider $\Theta_\pm \in S^{2,\mathcal{U}_\pm}_\mp \otimes S^{2,\mathcal{U}_\pm}_\mp$. We use the construction of Definition 7.1.1 to define the germ of a metric and associated (para)-Kähler form:

$$\begin{aligned} g_\pm(x, y) &:= \varepsilon_\pm(x, y) - \Theta_\pm(x, y, e_i, e_j) u^i u^j, \\ \Omega_\pm(x, y) &= \varepsilon_\pm(x, J_\pm y) - \Theta_\pm(x, J_\pm y, e_i, e_j) u^i u^j. \end{aligned} \tag{7.5.a}$$

Let ψ be a form of degree two. We may express

$$\begin{aligned} d\psi &= \partial_{u_i} \psi_{jk} du^i \wedge du^j \wedge du^k \\ &= \sum_{i<j<k} \{\partial_{u_i} \psi_{jk} + \partial_{u_j} \psi_{ki} + \partial_{u_k} \psi_{ij}\} du^i \wedge du^j \wedge du^k, \end{aligned}$$

and

$$\begin{aligned} &d\psi(\partial_{u_i}, \partial_{u_j}, \partial_{u_k}) \\ &\quad = \partial_{u_i} \psi(\partial_{u_j}, \partial_{u_k}) + \partial_{u_j} \psi(\partial_{u_k}, \partial_{u_i}) + \partial_{u_k} \psi(\partial_{u_i}, \partial_{u_j}). \end{aligned}$$

Let x, y, z be coordinate vector fields. We then have:

$$\begin{aligned} &d\Omega_\pm(x, y, z)(P) \\ &= -2\{\Theta_\pm(x, J_\pm y, z, P) + \Theta_\pm(y, J_\pm z, x, P) + \Theta_\pm(z, J_\pm x, y, P)\}. \end{aligned}$$

We define:

$$\begin{aligned}\mathcal{K}_\pm^{\mathfrak{R}} := \{\Theta_\pm \in S_\mp^{2,\mathcal{U}_\pm} \otimes S_\mp^{2,\mathcal{U}_\pm} : \Theta_\pm(x, J_\pm y, z, w) + \Theta_\pm(y, J_\pm z, x, w)\\ +\Theta_\pm(z, J_\pm x, y, w) = 0\}.\end{aligned}$$

It is then clear that $(\mathbb{R}^m, g_\pm, J_\pm)$ is a (para)-Kähler manifold if and only if the defining tensor $\Theta_\pm$ in Equation (7.5.a) belongs to $\mathcal{K}_\pm^{\mathfrak{R}}$. We have the following preliminary geometric realization result showing every (para)-Kähler curvature model is geometrically realizable by a (para)-Kähler manifold:

Lemma 7.5.1 $\Xi_\pm : \mathcal{K}_\pm^{\mathfrak{R}} \to \mathfrak{K}_\pm^{\mathfrak{R}} \to 0$.

Proof. $\mathcal{K}_\pm^{\mathfrak{R}}$ and $\mathfrak{K}_\pm^{\mathfrak{R}}$ are modules for the group $\mathcal{U}_\pm^\star$. Furthermore, by Lemma 7.1.3, $\Xi_\pm$ is a module morphism. The discussion given above shows that $\Xi_\pm \mathcal{K}_\pm^{\mathfrak{R}} \subset \mathfrak{K}_\pm^{\mathfrak{R}}$. We complete the proof by showing $\Xi_\pm \mathcal{K}_\pm^{\mathfrak{R}} = \mathfrak{K}_\pm^{\mathfrak{R}}$. We take

$$g_\pm := \varepsilon_\pm - ((x^1)^2 \mp (y^1)^2)(dx^1 \otimes dx^1 \mp dy^1 \otimes dy^1).$$

The defining perturbation $\Theta_\pm \in S_\mp^{2,\mathcal{U}_\pm} \otimes S_\mp^{2,\mathcal{U}_\pm}$ is admissible. Being a perturbation of dimension $m = 2$, the metric is necessarily a Kähler metric; the defining relations can also be checked directly. Let $A_\pm = \Xi_\pm(\Theta_\pm)$ be the curvature at the origin. We decompose $A_\pm = A_{\pm,1} + A_{\pm,2} + A_{\pm,3}$ where $A_{\pm,i} := \pi_{\pm,i}^{\mathfrak{R}} A \in W_{\pm,i}^{\mathfrak{R}}$. We use Lemma 7.2.4 to show $A_{\pm,1} \neq 0$:

$$\begin{aligned}&A_\pm(e_1, f_1, f_1, e_1) = \mp 2, && \tau(A_\pm) = 4,\\ &\pi_1^{\varepsilon_\pm}(e_2, f_2, f_2, e_2) = \pm 1, && \pi_{\pm,2}(e_2, f_2, f_2, e_2) = -3,\\ &A_{\pm,1}(e_2, f_2, f_2, e_2) = -\tfrac{1}{m(m+2)}\tau(A_\pm)(\pi_1 \mp \pi_{\pm,2})(e_2, f_2, f_2, e_2)\\ &\quad = \mp\tfrac{16}{m(m+2)}.\end{aligned}$$

Note that $\rho(A_\pm) = 0$ on $\operatorname{Span}\{e_2, f_2\}$. We now study $A_{\pm,2}$:

$$\begin{aligned}&\Phi(\rho(A_\pm))(e_2, f_2, f_2, e_2) = 0, && \Psi_\pm(\rho(A_\pm))(e_2, f_2, f_2, e_2) = 0,\\ &\Phi(\langle\cdot,\cdot\rangle)(e_2, f_2, f_2, e_2) = \pm 2, && \Psi_\pm(\langle\cdot,\cdot\rangle)(e_2, f_2, f_2, e_2) = -6,\\ &A_{\pm,2}(e_2, f_2, f_2, e_2) = -\tfrac{1}{m+4}(\Phi \mp \Psi_\pm)(\rho_0(A_\pm))(e_2, f_2, f_2, e_2)\\ &\quad = -\tfrac{1}{m+4}(\Phi \mp \Psi_\pm)(\rho(A_\pm) - \tfrac{4}{m}\langle\cdot,\cdot\rangle)(e_2, f_2, f_2, e_2)\\ &\quad = \pm\tfrac{32}{m(m+4)}.\end{aligned}$$

Consequently, $A_{\pm,2} \neq 0$. Finally, we have

$$\begin{aligned}A_{\pm,3}(e_2, f_2, f_2, e_2) = A_\pm(e_2, f_2, f_2, e_2) - A_{\pm,1}(e_2, f_2, f_2, e_2)\\ -A_{\pm,2}(e_2, f_2, f_2, e_2)\end{aligned}$$

$$= 0 \pm \tfrac{16}{m(m+2)} \mp \tfrac{32}{m(m+4)} = \tfrac{\pm 16(m+4) \mp 32(m+2)}{m(m+2)(m+4)}$$
$$= \mp \tfrac{16}{(m+2)(m+4)} \neq 0 \quad \text{so} \quad A_{\pm,3} \neq 0.$$

Thus $A_\pm$ has non-zero components in $W^{\mathfrak{R}}_{\pm,i}$ for $i = 1,2,3$. The desired result now follows as the modules are inequivalent and irreducible. □

We complete the proof of Theorem 1.10.2 by showing the realization in question can be chosen to have constant scalar curvature. We apply an argument similar to that used in Section 7.3; the major difference is that now we study a quasi-linear partial differential equation of order four rather than a quasi-linear partial differential equation of order two. We apply Theorem 6.3.2. Let $1 \leq i \leq \bar{m}$. In the holomorphic setting, we adopt the notation of Section 3.6 and define:

$$\begin{aligned} z^i &:= x^i + \sqrt{-1}y^i, & \bar{z}^i &:= x^i - \sqrt{-1}y^i, \\ dz^i &:= dx^i + \sqrt{-1}dy^i, & d\bar{z}^i &:= dx^i - \sqrt{-1}dy^i, \\ \partial_{z_i} &:= \tfrac{1}{2}\{\partial_{x_i} - \sqrt{-1}\partial_{y_i}\}, & \partial_{\bar{z}_i} &:= \tfrac{1}{2}\{\partial_{x_i} + \sqrt{-1}\partial_{y_i}\}. \end{aligned}$$

If Φ is a real analytic function (which is called the *Kähler potential*), form

$$\kappa_{-,\Phi} := \left\{\partial_{z_i}\partial_{\bar{z}_j}\Phi\right\}(dz^i \otimes d\bar{z}^j + d\bar{z}^j \otimes dz^i).$$

If we disentangle the notation, we have

$$\begin{aligned} &\kappa_{-,\Phi}(\partial_{z_i}, \partial_{z_j}) = \kappa_{-,\Phi}(\partial_{\bar{z}_i}, \partial_{\bar{z}_j}) = 0, \quad \kappa_{-,\Phi}(\partial_{z_i}, \partial_{\bar{z}_j}) = \partial_{z_i}\partial_{\bar{z}_j}\phi, \\ &J_-\partial_{z_i} = \sqrt{-1}\partial_{z_i}, \quad J_-\partial_{\bar{z}_i} = -\sqrt{-1}\partial_{\bar{z}_i}, \\ &\kappa_{-,\Phi}(J_-\partial_{z_i}, J_-\partial_{\bar{z}_j}) = (\sqrt{-1})(-\sqrt{-1})\kappa_{-,\Phi}(\partial_{z_i}, \partial_{\bar{z}_j}) = \kappa_{-,\Phi}(\partial_{z_i}, \partial_{\bar{z}_j}), \\ &\partial_{x_i} = \partial_{z_i} + \partial_{\bar{z}_i}, \qquad \partial_{y_i} = \sqrt{-1}(\partial_{z_i} - \partial_{\bar{z}_i}), \\ &\kappa_{-,\Phi}(\partial_{x_i}, \partial_{x_j}) = \kappa_{-,\Phi}(\partial_{y_i}, \partial_{y_j}) = (\partial_{z_i}\partial_{\bar{z}_j} + \partial_{z_j}\partial_{\bar{z}_i})\Phi \\ &\qquad = 2\operatorname{Re}\{\partial_{z_i}\partial_{\bar{z}_j}\Phi\} = \tfrac{1}{2}\{\partial_{x_i}\partial_{x_j} + \partial_{y_i}\partial_{y_j}\}\Phi \in \mathbb{R}, \\ &\kappa_{-,\Phi}(\partial_{x_i}, \partial_{y_j}) = -\sqrt{-1}\{\partial_{z_i}\partial_{\bar{z}_j} - \partial_{z_j}\partial_{\bar{z}_i}\}\Phi = 2\operatorname{Im}\{\partial_{z_i}\partial_{\bar{z}_j}\Phi\} \\ &\qquad = \tfrac{1}{2}\{\partial_{x_i}\partial_{y_j} - \partial_{y_i}\partial_{x_j}\}\Phi \in \mathbb{R}. \end{aligned}$$

This shows that $\kappa_{-,\Phi}$ is a real symmetric bilinear form on the tangent space and that $J_-^* g_{-,\Phi} = g_{-,\Phi}$. The associated Kähler form is then given by:

$$\Omega_{-,\Phi}(x, y) := \kappa_{-,\Phi}(x, J_- y).$$

Similarly, in the para-holomorphic setting, we define:

$$\begin{aligned} w^i &:= x^i + \iota y^i, & \bar{w}^i &:= x^i - \iota y^i, \\ dw^i &:= dx^i + \iota dy^i, & d\bar{w}^i &:= dx^i - \iota dy^i, \\ \partial_{w_i} &:= \tfrac{1}{2}\{\partial_{x_i} + \iota\partial_{y_i}\}, & \partial_{\bar{w}_i} &:= \tfrac{1}{2}\{\partial_{x_i} - \iota\partial_{y_i}\}. \end{aligned}$$

We then form:

$$\kappa_{+,\Phi} := \left\{\partial_{w_i}\partial_{\bar{w}_j}\Phi\right\}(dw^i \otimes d\bar{w}^j + d\bar{w}^j \otimes dw^i).$$

Again, we have:

$$\begin{aligned}
&\kappa_{+,\Phi}(\partial_{w_i}, \partial_{w_j}) = \kappa_{+,\Phi}(\partial_{\bar{w}_i}, \partial_{\bar{w}_j}) = 0,\ \kappa_{+,\Phi}(\partial_{w_i}, \partial_{\bar{w}_j}) = \partial_{w_i}\partial_{\bar{w}_j}\Phi,\\
&J_+\partial_{w_i} = \iota\partial_{w_i}, \quad J_+\partial_{\bar{w}_i} = -\iota\partial_{\bar{w}_i},\\
&\kappa_{+,\Phi}(J_+\partial_{w_i}, J_+\partial_{\bar{w}_j}) = \iota(-\iota)\kappa_{+,\Phi}(\partial_{w_i}, \partial_{\bar{w}_j}) = -\kappa_{+,\Phi}(\partial_{w_i}, \partial_{\bar{w}_j}),\\
&\partial_{x_i} = \partial_{w_i} + \partial_{\bar{w}_i}, \quad \partial_{y_i} = \iota(\partial_{w_i} - \partial_{\bar{w}_i}),\\
&\kappa_{+,\Phi}(\partial_{x_i}, \partial_{x_j}) = -\kappa_{+,\Phi}(\partial_{y_i}, \partial_{y_j}) = (\partial_{w_i}\partial_{\bar{w}_j} + \partial_{w_j}\partial_{\bar{w}_i})\Phi\\
&\qquad = 2\,\mathrm{Re}\{\partial_{w_i}\partial_{\bar{w}_j}\Phi\} = \tfrac{1}{2}\{\partial_{x_i}\partial_{x_j} - \partial_{y_i}\partial_{y_j}\}\Phi \in \mathbb{R},\\
&\kappa_{+,\Phi}(\partial_{x_i}, \partial_{y_j}) = -\iota\{\partial_{w_i}\partial_{\bar{w}_j} - \partial_{w_j}\partial_{\bar{w}_i}\}\Phi = -2\,\mathrm{Im}\{\partial_{w_i}\partial_{\bar{w}_j}\Phi\}\\
&\qquad = \tfrac{1}{2}\{\partial_{x_i}\partial_{y_j} - \partial_{y_i}\partial_{x_j}\}\Phi \in \mathbb{R}.
\end{aligned}$$

This shows that $\kappa_{+,\Phi}$ is a real symmetric bilinear form on the tangent space and that $J_+^*\kappa_{+,\Phi} = -\kappa_{+,\Phi}$. The associated para-Kähler form is then given by:

$$\Omega_{+,\Phi}(x, y) := \kappa_{+,\Phi}(x, J_+y).$$

Lemma 7.5.2 $d\Omega_{\pm,\Phi} = 0$.

Proof. Our previous computations permit us to expand:

$$\begin{aligned}
&\Omega_{-,\Phi} = -\sqrt{-1}\partial_{z_i}\partial_{\bar{z}_j}\Phi \cdot dz^i \wedge d\bar{z}^j,\\
&d\Omega_{-,\Phi} = -\sqrt{-1}(\partial_{z_k}\partial_{z_i}\partial_{\bar{z}_j}\Phi \cdot dz^k + \partial_{\bar{z}_k}\partial_{z_i}\partial_{\bar{z}_j}\Phi \cdot d\bar{z}^k) \wedge dz^i \wedge d\bar{z}^j = 0,\\
&\Omega_{+,\Phi} = -\iota\partial_{w_i}\partial_{\bar{w}_j}\Phi \cdot dw^i \wedge d\bar{w}^j,\\
&d\Omega_{+,\Phi} = -\iota(\partial_{w_k}\partial_{w_i}\partial_{\bar{w}_j}\Phi \cdot dw^k + \partial_{\bar{w}_k}\partial_{w_i}\partial_{\bar{w}_j}\Phi \cdot d\bar{w}^k) \wedge dw^i \wedge d\bar{w}^j = 0. \quad \square
\end{aligned}$$

Proof of Theorem 1.10.2. If $(V, \langle\cdot,\cdot\rangle, J_\pm)$ is a para-Hermitian vector space $(+)$ or a pseudo-Hermitian vector space $(-)$, let $A_\pm \in \mathfrak{K}_\pm^{\mathfrak{R}}$. We use Lemma 7.5.1 to choose a (para)-Kähler metric $g_\pm$ so that $R(0) = A_\pm$ and $g_\pm(0) = \langle\cdot,\cdot\rangle$. Let $h_{\pm,\Phi} := g_\pm + \kappa_{\pm,\Phi}$. We impose the Cauchy initial data given by Equation (6.3.b) so $h_{\pm,\Phi}(0) = \varepsilon_\pm$ and thus $h_{\pm,\Phi}$ is non-degenerate on some neighborhood of the origin. The scalar curvature $\tau_{\pm,\Phi}$ of $h_{\pm,\Phi}$ is given by a quasi-linear equation of order four of the form given by Equation (6.3.a). We suppress terms not involving maximal derivatives in $\partial_{y_{\bar{m}}}$ to write:

$$\begin{aligned}
&h_{\pm,\Phi}(\partial_{x_{\bar m}},\partial_{x_{\bar m}}) = \mp h_{\pm,\Phi}(\partial_{y_{\bar m}},\partial_{y_{\bar m}}) = \mp\tfrac{1}{2}\partial^2_{y_{\bar m}}\Phi+\dots,\\
&R_{\pm,\Phi}(\partial_{x_{\bar m}},\partial_{y_{\bar m}},\partial_{y_{\bar m}},\partial_{x_{\bar m}}) = \pm\tfrac{1}{4}\partial^4_{y_{\bar m}}\Phi+\dots,\\
&\tau_\Phi = -\tfrac{1}{2}\partial^4_{y_{\bar m}}\Phi+\dots.
\end{aligned}$$

Thus the non-degeneracy condition of Theorem 6.3.2 is satisfied and we can solve the equation $\tau_{\pm,\Phi}-c_\pm=0$ with vanishing Cauchy initial data. All the derivatives of order four of Φ vanish except possibly for $\partial^4_{y_{\bar m}}\Phi$ – the equation $\tau_{\pm,\Phi}-c_\pm=0$ implies this vanishes as well. Thus $\Phi=O(|u|^5)$ so Φ makes no contribution to the curvature tensor at the origin. This shows the (para)-Kähler manifold in question can be chosen to have constant scalar curvature. This completes the proof of Theorem 1.10.2. □

7.6 Complex Weyl Geometry

In Section 7.6, we will establish Theorem 1.11.1 and Theorem 1.11.4. Let $(M,g,\nabla,J_\pm)$ be a complex Weyl structure; here ∇ is an affine connection on a pseudo-Riemannian manifold (M,g) with $\nabla g=-2\phi\otimes g$ and $J_\pm$ is an almost (para)-complex structure on M with $J_\pm^* g=\mp g$. Let $(V,\langle\cdot,\cdot\rangle,J_\pm)$ be a para-Hermitian vector space $(+)$ or a pseudo-Hermitian vector space $(-)$. Let $\mathfrak{W}$ be the space of all Weyl tensors. We begin by summarizing the decomposition results we have established so far. We adopt the notation of Equation (6.5.d) and of Definition 7.2.1 so that if ψ is a tensor of degree two, then:

$$\begin{aligned}
\Psi_\pm(\psi)(x,y,z,w) &= 2\langle x,J_\pm y\rangle\psi(z,J_\pm w)+2\langle z,J_\pm w\rangle\psi(x,J_\pm y)\\
&\quad+\langle x,J_\pm z\rangle\psi(y,J_\pm w)+\langle y,J_\pm w\rangle\psi(x,J_\pm z)\\
&\quad-\langle x,J_\pm w\rangle\psi(y,J_\pm z)-\langle y,J_\pm z\rangle\psi(x,J_\pm w),\\
\{\sigma^{\mathfrak{W}}\psi\}_{ijkl} &= 2\psi_{ij}\varepsilon_{kl}+\psi_{ik}\varepsilon_{jl}-\psi_{jk}\varepsilon_{il}-\psi_{il}\varepsilon_{jk}+\psi_{jl}\varepsilon_{ik}.
\end{aligned}$$

We have $W^{\mathfrak{R}}_{\pm,9}=\Psi_\pm(\Lambda^{2,\mathcal{U}_\pm}_\pm)$. Set

$$W^{\mathfrak{W}}_{\pm,11}:=\sigma^{\mathfrak{W}}(\Omega_\pm\mathbb{R}),\quad W^{\mathfrak{W}}_{\pm,12}:=\sigma^{\mathfrak{W}}(\Lambda^{2,\mathcal{U}_\pm}_{0,\mp}),\quad W^{\mathfrak{W}}_{\pm,13}:=\sigma^{\mathfrak{W}}(\Lambda^{2,\mathcal{U}_\pm}_\pm).$$

Theorem 1.11.3, Theorem 6.5.3, and Theorem 7.2.1, yield:

Theorem 7.6.1 *Let $(V,\langle\cdot,\cdot\rangle,J_\pm)$ be a para-Hermitian vector space $(+)$ of dimension $m\geq 6$ or be a pseudo-Hermitian vector space $(-)$ of dimension $m\geq 6$. We have an orthogonal direct sum decomposition of $\mathfrak{W}$ as a module for the group $\mathcal{U}^\star_\pm$ in the form:*

$$\mathfrak{W}=W^{\mathfrak{R}}_{\pm,1}\oplus\dots\oplus W^{\mathfrak{R}}_{\pm,10}\oplus W^{\mathfrak{W}}_{\pm,11}\oplus W^{\mathfrak{W}}_{\pm,12}\oplus W^{\mathfrak{W}}_{\pm,13}.$$

We have $W^{\mathfrak{R}}_{\pm,1} \approx W^{\mathfrak{R}}_{\pm,4}$, $W^{\mathfrak{R}}_{\pm,2} \approx W^{\mathfrak{R}}_{\pm,5}$, and $W^{\mathfrak{R}}_{\pm,9} \approx W^{\mathfrak{W}}_{\pm,13}$; otherwise these are inequivalent modules for the group $\mathcal{U}^{\star}_{\pm}$.

The following is a useful technical result.

Lemma 7.6.1 *If $(V, \langle\cdot,\cdot\rangle, J_\pm)$ is a para-Hermitian vector space (+) or a pseudo-Hermitian vector space (−) of dimension $m \geq 6$, then the map $\psi \to \psi \wedge \Omega_\pm$ is an injective map from $\Lambda^2 \to \Lambda^4$.*

Proof. Let $\{e_1, ..., e_{\bar{m}}, f_1, ..., f_{\bar{m}}\}$ be an orthonormal basis for V so that $J_\pm e_i = f_i$ and $J_\pm f_i = \pm e_i$. Let $\varepsilon_i := \langle e_i, e_i\rangle = \pm 1$ and let $\omega \in \Lambda^2$. We expand:

$$\begin{aligned}\Omega_\pm &= \pm \sum_{1\leq i\leq \bar{m}} \varepsilon^i (e_i \otimes f_i - f_i \otimes e_i),\\ \omega &= \sum_{1\leq i<j\leq \bar{m}} \{a_{ij}(e_i\otimes e_j - e_j \otimes e_i) + b_{ij}(f_i \otimes f_j - f_j \otimes f_i)\}\\ &+ \sum_{1\leq i\leq \bar{m}, 1\leq j\leq \bar{m}} c_{ij}(e_i \otimes f_j - f_j \otimes e_i).\end{aligned}$$

Let $\{i, j, k\}$ be distinct indices between 1 and $\bar{m}$. We temporarily drop the Einstein convention and do not sum over repeated indices. In $\omega \wedge \Omega_\pm$, the coefficient of $e_i \otimes e_j \otimes e_k \otimes f_k$ arises only from a_{ij}, the coefficient of $f_i \otimes f_j \otimes e_k \otimes f_k$ arises only from b_{ij}, and the coefficient of $e_i \otimes f_j \otimes e_k \otimes f_k$ arises only from c_{ij}. Thus $\omega \wedge \Omega_\pm = 0$ implies these coefficients vanish and only the coefficients c_{ii} are at issue. The coefficient of $e_i \otimes f_i \otimes e_j \otimes f_j$ is $c_{ii} + c_{jj}$ and thus $c_{ii} + c_{jj} = 0$. Similarly $c_{ii} + c_{kk} = 0$ and $c_{jj} + c_{kk} = 0$. This implies $c_{ii} = 0$ as well. □

Proof of Theorem 1.11.1. We shall follow the treatment of [Pedersen, Poon, and Swann (1993)] using some of the results contained in [Vaisman (1982)] and [Vaisman (1983)]. Let $(M, g, J_\pm, \nabla)$ be a Kähler–Weyl manifold (−). We apply Lemma 3.4.1 to see that $\nabla(J_\pm) = 0$ implies $J_\pm$ is an integrable (para)-complex structure. Let

$$\Omega_{\pm,g}(x, y) := g(x, J_\pm y)$$

be the associated *(para)-Kähler form.* We use the identity $\nabla g = -2\phi \otimes g$

and the identity $\nabla J_\pm = J_\pm \nabla$ to compute:

$$\begin{aligned}(\nabla_z \Omega_{\pm,g})(x,y) &= zg(x, J_\pm y) - g(\nabla_z x, J_\pm y) - g(x, J_\pm \nabla_z y)\\ &= zg(x, J_\pm y) - g(\nabla_z x, J_\pm y) - g(x, \nabla_z J_\pm y)\\ &= (\nabla_z g)(x, J_\pm y) = -2\phi(z)\Omega_{\pm,g}(x,y).\end{aligned}$$

Let $\{e_i\}$ be a local frame for TM and let $\{e^i\}$ be the dual frame for the cotangent bundle T^*M. As the connection ∇ is affine, Lemma 3.3.4 yields:

$$d\Omega_{\pm,g} = e^i \wedge \nabla_{e_i} \Omega_{\pm,g}.$$

Consequently

$$\begin{aligned}d\Omega_{\pm,g} &= -2\phi(e_i)e^i \wedge \Omega_{\pm,g} = -2\phi \wedge \Omega_{\pm,g},\\ 0 &= d^2\Omega_{\pm,g} = -2d\phi \wedge \Omega_{\pm,g}.\end{aligned}$$

By Lemma 7.6.1, $d\phi \wedge \Omega_{\pm,g} = 0$ implies that $d\phi = 0$. □

This argument fails if $m = 4$; we can only conclude from this that $d\phi \perp \Omega_{\pm,g}$ and in fact there exist Kähler–Weyl manifolds where the Weyl structure is non-trivial if $n = 4$ (see [Calderbank and Pedersen (2000)] and [Gilkey and Nikčević (2011a)]).

Proof of Theorem 1.11.4. Let $(V, h, J_\pm)$ be a para-Hermitian vector space (+) of dimension $m \geq 6$ or be a pseudo-Hermitian vector space (−) of dimension $m \geq 6$. Recall that we defined

$$\mathfrak{K}_\pm^{\mathfrak{W}} = \{A \in \mathfrak{W} : A(x,y,z,w) = \mp A(x,y,J_\pm z, J_\pm w)\}$$

to be the space of all Weyl tensors satisfying the (para)-Kähler identity. Let:

$$\mathfrak{K}_{\pm,1}^{\mathfrak{W}} := \left\{\oplus_{4\leq i\leq 13} W_{\pm,i}^{\mathfrak{R}/\mathfrak{W}}\right\} \cap \mathfrak{K}_\pm^{\mathfrak{W}}.$$

By Theorem 1.10.3, $\mathfrak{R} \cap \mathfrak{K}_\pm^{\mathfrak{W}} = W_{\pm,1}^{\mathfrak{R}} \oplus W_{\pm,2}^{\mathfrak{R}} \oplus W_{\pm,3}^{\mathfrak{R}}$. Consequently

$$\mathfrak{K}_\pm^{\mathfrak{W}} = W_{\pm,1}^{\mathfrak{R}} \oplus W_{\pm,2}^{\mathfrak{R}} \oplus W_{\pm,3}^{\mathfrak{R}} \oplus \mathfrak{K}_{\pm,1}^{\mathfrak{W}}.$$

Let $i \in \{4,5,6,7,8,10\}$. Then $W_{\pm,i}^{\mathfrak{R}}$ appears with multiplicity one in $\mathfrak{K}_{\pm,1}^{\mathfrak{W}}$. Thus by Lemma 2.1.7,

$$\text{either}\quad W_{\pm,i}^{\mathfrak{R}} \subset \mathfrak{K}_{\pm,1}^{\mathfrak{W}} \quad\text{or}\quad W_{\pm,i}^{\mathfrak{R}} \perp \mathfrak{K}_{\pm,1}^{\mathfrak{W}}.$$

By Theorem 1.10.3, $W_{\pm,i}^{\mathfrak{R}} \cap \mathfrak{K}_\pm^{\mathfrak{R}} = \{0\}$. Consequently

$$\mathfrak{K}_{\pm,1}^{\mathfrak{W}} = \{W_{\pm,9}^{\mathfrak{R}} \oplus W_{\pm,11}^{\mathfrak{W}} \oplus W_{\pm,12}^{\mathfrak{W}} \oplus W_{\pm,13}^{\mathfrak{W}}\} \cap \mathfrak{K}_\pm^{\mathfrak{W}}.$$

We apply Theorem 7.6.1. Choose an orthonormal basis $\{e^i\}$ for V so

$$J_\pm e_{2i-1} = e_{2i} \quad \text{and} \quad J_\pm e_{2i} = \pm e_{2i-1} \quad \text{for} \quad 1 \le i \le \bar{m}.$$

We study $W_{\pm,11}^{\mathfrak{W}} = \sigma^{\mathfrak{W}}(\Omega_\pm \cdot \mathbb{R})$:

$$\begin{aligned}&\sigma^{\mathfrak{W}}(\Omega_\pm)(e_1, e_4, e_3, e_1) = -\langle e_4, J_\pm e_3\rangle\langle e_1, e_1\rangle = -\varepsilon_{11}\varepsilon_{44},\\ &\mp\sigma^{\mathfrak{W}}(\Omega_\pm)(e_1, e_4, J_\pm e_3, J_\pm e_1) = \pm\langle e_1, J_\pm J_\pm e_1\rangle\langle e_4, J_\pm e_3\rangle = \varepsilon_{11}\varepsilon_{44}.\end{aligned}$$

This shows that $\sigma^{\mathfrak{W}}(\mathbb{R} \cdot \Omega_\pm) \not\subset \mathfrak{K}_{\pm,1}^{\mathfrak{W}}$ and consequently:

$$\mathfrak{K}_{\pm,1}^{\mathfrak{W}} = \{W_{\pm,9}^{\mathfrak{R}} \oplus W_{\pm,12}^{\mathfrak{W}} \oplus W_{\pm,13}^{\mathfrak{W}}\} \cap \mathfrak{K}_{\pm}^{\mathfrak{W}}.$$

We study the module $W_{\pm,12}^{\mathfrak{W}}$. Let

$$\psi_{\pm,0} := e^1 \otimes e^2 - e^2 \otimes e^1 + \delta_\pm\{e^3 \otimes e^4 - e^4 \otimes e^3\}$$

where $\delta_\pm$ is chosen to ensure that $\psi_{\pm,0} \perp \Omega_\pm$. We have $J_\pm^*\psi_{\pm,0} = \mp\psi_{\pm,0}$ and thus $\psi_{\pm,0} \in \Lambda_{0,\pm}^{2,\mathcal{U}_\pm}$. We show that $W_{\pm,12} \not\subset \mathfrak{K}_{\pm,1}^{\mathfrak{W}}$ by computing:

$$\begin{aligned}&\sigma^{\mathfrak{W}}(\psi_{\pm,0})(e_5, e_1, e_2, e_5) = -\psi_{\pm,0}(e_1, e_2)\langle e_5, e_5\rangle = -\varepsilon_{55},\\ &\mp\sigma^{\mathfrak{W}}(\psi_{\pm,0})(e_5, e_1, J_\pm e_2, J_\pm e_5) = -\psi_{\pm,0}(e_5, J_\pm e_5)\langle e_1, J_\pm e_2\rangle = 0.\end{aligned}$$

We also study the module $W_{\pm,9}^{\mathfrak{R}} \oplus W_{\pm,13}^{\mathfrak{W}}$. Let

$$\psi_\pm := e^1 \otimes e^3 - e^3 \otimes e^1 \pm e^2 \otimes e^4 \mp e^4 \otimes e^2.$$

Then $J_\pm^*\psi_\pm = \pm\psi_\pm$ so $\psi_\pm \in \Lambda_\pm^{2,\mathcal{U}_\pm}$. We compute:

(1) $\sigma^{\mathfrak{W}}(\psi_\pm)(e_5, e_1, e_3, e_5) = -\psi_\pm(e_1, e_3)\langle e_5, e_5\rangle = -\varepsilon_{55}$.
(2) $\sigma^{\mathfrak{W}}(\psi_\pm)(e_5, e_1, e_4, e_6) = 0$.
(3) $\Psi_\pm(\psi_\pm)(e_5, e_1, e_3, e_5) = 0$.
(4) $\Psi_\pm(\psi_\pm)(e_5, e_1, e_4, e_6) = -\langle e_5, J_\pm e_6\rangle\psi_\pm(e_1, J_\pm e_4) = -\varepsilon_{55}$.
(5) $\sigma^{\mathfrak{W}}(\psi_\pm)(e_5, e_6, e_1, e_4) = 0$.
(6) $\sigma^{\mathfrak{W}}(\psi_\pm)(e_5, e_6, J_\pm e_1, J_\pm e_4) = 0$.
(7) $\Psi_\pm(\psi_\pm)(e_5, e_6, e_1, e_4) = 2\langle e_5, J_\pm e_6\rangle\psi_\pm(e_1, J_\pm e_4) = 2\varepsilon_{55}$.
(8) $\Psi_\pm(\psi_\pm)(e_5, e_6, J_\pm e_1, J_\pm e_4) = 2\langle e_5, J_\pm e_6\rangle\psi_\pm(J_\pm e_1, J_\pm J_\pm e_4) = \pm 2\varepsilon_{55}$.

For $(a, b) \neq (0, 0)$, let

$$\xi(a, b) := \text{Range}\{a\sigma^{\mathfrak{W}} + b\Psi\} \subset W_{\pm,9}^{\mathfrak{R}} \oplus W_{\pm,13}^{\mathfrak{W}}.$$

We suppose $\xi(a, b) \cap \mathfrak{K}_{\pm,1}^{\mathfrak{W}} \neq \{0\}$. This implies that $\xi(a, b) \subset \mathfrak{K}_{\pm,1}^{\mathfrak{W}}$. Assertions (1)–(4) then yield $a = \mp b$ while Assertions (5)–(8) yield $b = 0$. We apply Lemma 2.2.2 and Lemma 2.5.4 to see that every non-trivial proper submodule for the group $\mathcal{U}_\pm^\star$ of $W_{\pm,9}^{\mathfrak{R}} \oplus W_{\pm,13}^{\mathfrak{W}}$ is isomorphic to $\xi(a,b)$ for some $(a,b) \neq 0$. Thus

$$\{W_{\pm,9}^{\mathfrak{R}} \oplus W_{\pm,13}^{\mathfrak{W}}\} \cap \mathfrak{K}_{\pm,1}^{\mathfrak{W}} = \{0\}.$$

This shows that $\mathfrak{K}_{\pm,1}^{\mathfrak{W}} = \{0\}$ and thus $\mathfrak{K}_\pm^{\mathfrak{W}} \subset \mathfrak{R}$ as desired. □

7.7 The Covariant Derivative of the Kähler Form II

In Section 7.7, we establish the results of Section 1.12. We refer to [Gadea and Masque (1991)], [Gray (1965)], [Gray (1969a)], [Gray and Hervella (1980)], and [Koto (1960)] for related work. We recall for the convenience of the reader the notation established previously in Definition 1.12.1. Let $(V, \langle\cdot,\cdot\rangle, J_\pm)$ be a para-Hermitian vector space (+) or a pseudo-Hermitian vector space (−).

(1) If $H \in \otimes^3 V^*$, define $(\tau_1 H)(x) := \varepsilon^{ij} H(x, e_i; e_j)$.

(2) If $\kappa \in \mathrm{GL}$ and if $\phi \in V^*$, define $\sigma_\kappa(\phi) \in \otimes^3 V^*$ by setting:
$\sigma_\kappa(\phi)(x,y;z) = \phi(\kappa x)\langle y,z\rangle - \phi(\kappa y)\langle x,z\rangle + \phi(x)\langle \kappa y,z\rangle - \phi(y)\langle \kappa x,z\rangle$.

(3) $\mathfrak{H}_\pm = \Lambda_\pm^{2,\mathcal{U}_\pm} \otimes V^*$.

(4) $U_{\pm,3} = \{H_\pm \in \mathfrak{H}_\pm : H_\pm(x,y;z) = \mp H_\pm(x, J_\pm y; J_\pm z)\}$.

(5) $W_{\pm,1}^{\mathfrak{H}} = \{H \in \mathfrak{H}_\pm : H(x,y;z) + H(x,z;y) = 0\}$.

(6) $W_{\pm,2}^{\mathfrak{H}} = \{H \in \mathfrak{H}_\pm : H(x,y;z) + H(y,z;x) + H(z,x;y) = 0\}$.

(7) $W_{\pm,3}^{\mathfrak{H}} = U_{\pm,3} \cap \ker(\tau_1)$.

(8) $W_{\pm,4}^{\mathfrak{H}} = \mathrm{Range}(\sigma_{J_\pm})$.

We begin our study in the geometric setting. Let $\nabla\Omega_\pm(x,y;z)$ be the components of the covariant derivative of the (para)-Kähler form:

$$\nabla\Omega_\pm(x,y;z) = (\nabla_z\Omega_\pm)(x,y) := z\Omega_\pm(x,y) - \Omega_\pm(\nabla_z x, y) - \Omega_\pm(x, \nabla_z y).$$

Lemma 7.7.1 *Let $g(x,y;z) := zg(x,y)$.*

(1) $\nabla\Omega_\pm(x,y;z) = g(x, (\nabla_z J_\pm - J_\pm\nabla_z)y) = g(x, \nabla_z J_\pm y) + g(J_\pm x, \nabla_z y)$.

(2) If $J_\pm$ is an integrable (para)-complex structure and if $\{x,y,z\}$ are constant vector fields in a system of (para)-holomorphic coordinates, then:

$$\nabla\Omega_\pm(x,y;z) = \tfrac{1}{2}\{g(x,z;J_\pm y) - g(J_\pm y,z;x) + g(J_\pm x,z;y) - g(y,z;J_\pm x)\}.$$

(3) Let $g = \varepsilon$ be a flat metric. Then $\nabla\Omega_\pm(x,y;z) = \langle x, z(J_\pm)y\rangle$.

Proof. Since $\nabla g = 0$ and since $J_\pm$ is skew-adjoint with respect to g, we may prove Assertion (1) by computing:

$$\begin{aligned}\nabla\Omega_\pm(x, y; z) &= zg(x, J_\pm y) - g(\nabla_z x, J_\pm y) - g(x, J_\pm \nabla_z y)\\ &= zg(x, J_\pm y) - g(\nabla_z x, J_\pm y) - g(x, \nabla_z J_\pm y) + g(x, \nabla_z J_\pm y)\\ &\quad -g(x, J_\pm \nabla_z y)\\ &= (\nabla g)(x, J_\pm y; z) + g(x, (\nabla_z J_\pm - J_\pm \nabla z)y)\\ &= 0 + g(x, (\nabla_z J_\pm - J_\pm \nabla_z)y) = g(x, \nabla_z J_\pm y) + g(J_\pm x, \nabla_z y).\end{aligned}$$

Let $J_\pm$ be an integrable (para)-complex structure. We choose local (para)-holomorphic coordinates so that the matrix of $J_\pm$ is constant relative to this coordinate system. If $\{x, y, z\}$ are coordinate vector fields, then Equation (3.3.a) implies:

$$g(\nabla_x y, z) = \tfrac{1}{2}\{g(y, z; x) + g(x, z; y) - g(x, y; z)\}.$$

We use Assertion (1) to establish Assertion (2) by computing:

$$\begin{aligned}\nabla\Omega_\pm(x, y; z) = \tfrac{1}{2}\{&g(x, J_\pm y; z) + g(x, z; J_\pm y) - g(J_\pm y, z; x)\}\\ &+\tfrac{1}{2}\{g(J_\pm x, y; z) + g(J_\pm x, z; y) - g(y, z; J_\pm x)\}\end{aligned}$$

and noting that

$$g(x, J_\pm y; z) + g(J_\pm x, y; z) = z\{g(x, J_\pm y) + g(J_\pm x, y)\} = 0.$$

Finally, if ε is a flat metric, then ∇ is ordinary partial differentiation so:

$$\nabla_z J_\pm - J_\pm \nabla_z = z(J_\pm). \qquad \square$$

Let $\mathcal{H}_\pm \subset \otimes^3 V^*$ be the set of all tensors $\nabla\Omega_\pm(0)$ arising as the covariant derivative of the (para)-Kähler form of the germ of some suitable structure on V at 0.

Lemma 7.7.2 $\mathcal{H}_\pm \subset \mathfrak{H}_\pm$.

Proof. Since $\Omega_\pm \in C^\infty(\Lambda^2)$, $\nabla\Omega_\pm \in C^\infty(\Lambda^2 \otimes V^*)$. We complete the proof by studying the action of $J_\pm^*$:

$$\begin{aligned}&\nabla\Omega_\pm(J_\pm x, J_\pm y; z)\\ &= zg(J_\pm x, J_\pm J_\pm y) - g(\nabla_z J_\pm x, J_\pm J_\pm y) - g(J_\pm x, J_\pm \nabla_z J_\pm y)\\ &= \mp zg(x, J_\pm y) \mp g(\nabla_z J_\pm x, y) \pm g(x, \nabla_z J_\pm y)\\ &= \mp zg(x, J_\pm y) \mp zg(J_\pm x, y) \pm g(J_\pm x, \nabla_z y) \pm zg(x, J_\pm y)\\ &\quad \mp g(\nabla_z x, J_\pm y)\end{aligned}$$

$$= \pm z g(x, J_\pm y) \mp g(x, J_\pm \nabla_z y) \mp g(\nabla_z x, J_\pm y)$$
$$= \pm \nabla \Omega_\pm(x, y; z). \qquad \square$$

We now examine the space of quadratic invariants. We introduce three orthonormal bases $\{e_{i_1}\}$, $\{e_{i_2}\}$, $\{e_{i_3}\}$ for V. Let $(\mu_1, \mu_2, \mu_3, \mu_4, \mu_5, \mu_6)$ be a reordering of the indices $(1, 1, 2, 2, 3, 3)$. If $H \in \mathfrak{H}_\pm$, we define

$$\mathcal{I}\{(\mu_1, \mu_2; \mu_3)(\mu_4, \mu_5; \mu_6)\}(H)$$
$$:= \sum \varepsilon_I H(e_{i_{\mu_1}}, e_{i_{\mu_2}}; e_{i_{\mu_3}}) H(e_{i_{\mu_4}}, e_{i_{\mu_5}}; e_{i_{\mu_6}}).$$

We can also use $J_\pm$ to decorate some of the indices. Define the following invariants:

$$\psi^{\mathfrak{H}}_{\pm,1} := \mathcal{I}\{(1,2;3)(1,2;3)\}, \quad \psi^{\mathfrak{H}}_{\pm,2} := \mathcal{I}\{(1,2;3)(1,3;2)\},$$
$$\psi^{\mathfrak{H}}_{\pm,3} := \mathcal{I}\{(1,2;1)(3,2;3)\}, \quad \psi^{\mathfrak{H}}_{\pm,4} := \mathcal{I}\{(1, J_\pm 2; J_\pm 3)(1,2;3)\}.$$

Lemma 7.7.3 *Let $(V, \langle\cdot,\cdot\rangle, J_\pm)$ be a para-Hermitian vector space $(+)$ or be a pseudo-Hermitian vector space $(-)$.*

(1) The invariants $\{\psi^{\mathfrak{H}}_{\pm,1}, \psi^{\mathfrak{H}}_{\pm,2}, \psi^{\mathfrak{H}}_{\pm,3}, \psi^{\mathfrak{H}}_{\pm,4}\}$ span $\mathcal{I}_2^{\mathcal{U}^\star_\pm}(\mathfrak{H}_\pm)$ and $\mathcal{I}_2^{\mathcal{U}^\star_-}(\mathfrak{H}_-)$.

(2) $\dim\{\mathcal{I}_2^{\mathcal{U}^\star_\pm}(\mathfrak{H}_\pm)\} \le 4$ and $\dim\{\mathcal{I}_2^{\mathcal{U}^\star_-}(\mathfrak{H}_-)\} \le 4$.

Remark 7.7.1 Let $m \ge 6$. Observation 7.7.1 will show equality holds in Lemma 7.7.3. Consequently, the invariants of Assertion (1) are a basis for the associated spaces of invariants in this setting.

Proof. We construct invariants as follows and clear all previous notation. We will stratify the invariants by the number of times $J_\pm$ appears; this gives rise to four basic cases. Within a given case, we consider the two subcases where the indices decouple and where all indices appear with multiplicity one in each monomial. Each subcase can be divided into various possibilities. We use the fact that

$$H_\pm(x, y; z) = \pm H_\pm(J_\pm x, J_\pm y; z) = -H_\pm(y, x; z).$$

Thus, for example, we may assume we do not have $(J_\pm *, J_\pm *; *)$.

(1) $J_\pm$ does not appear. This gives rise to three invariants:

  (a) Each index appears in each variable:

    i. $\psi^{\mathfrak{H}}_{\pm,1} := \mathcal{I}\{(1,2;3)(1,2;3)\}$.
    ii. $\psi^{\mathfrak{H}}_{\pm,2} := \mathcal{I}\{(1,2;3)(1,3;2)\}$.

(b) Only one index appears in both variables:

i. $\psi^{\mathfrak{H}}_{\pm,3} := \mathcal{I}\{(1,2;1)(3,2;3)\}$.

(2) $J_\pm$ appears once. There are several possibilities all of which yield 0.

(a) Each index appears in each variable:

i. $\mathcal{I}\{(J_\pm 1,2;3)(1,2;3)\} = -\mathcal{I}\{(1,2;3)(J_\pm 1,2;3)\}$ so this vanishes.

ii. $\mathcal{I}\{(1,2;J_\pm 3)(1,2;3)\} = -\mathcal{I}\{(1,2;3)(1,2,J_\pm 3)\}$ so this vanishes.

iii. $\mathcal{I}\{(J_\pm 1,2;3)(1,3;2)\} = -\mathcal{I}\{(1,2;3)(J_\pm 1,3;2)\}$
$= -\mathcal{I}\{(1,3;2)(J_\pm 1,2;3)\}$ so this vanishes.

iv. $\mathcal{I}\{(1,2;J_\pm 3)(1,3;2)\} = -\mathcal{I}\{(1,2;3)(1,J_\pm 3;2)\}$
$= \mp\mathcal{I}\{(1,2;3)(J_\pm 1,J_\pm J_\pm 3;2)\} = -\mathcal{I}\{(1,2;3)(J_\pm 1,3;2)\} = 0$
by (2)(a)(iii).

(b) Only one index appears in both variables.

i. $\mathcal{I}\{(1,J_\pm 2;1)(3,2;3)\} = -\mathcal{I}\{(1,2;1)(3,J_\pm 2;3)\} = 0$.

ii. $\mathcal{I}\{(J_\pm 1,2;1)(3,2;3)\} = \pm\mathcal{I}\{(J_\pm J_\pm 1,J_\pm 2;1)(3,2;3)\}$
$= \mathcal{I}\{(1,J_\pm 2;1)(3,2;3)\} = 0$ by (2)(b)(i).

(3) $J_\pm$ appears twice. We may assume the variable is not $(J_\pm *, J_\pm *; *)$.

(a) Each index appears in each variable:

i. $\psi^{\mathfrak{H}}_{\pm,4} = \mathcal{I}\{(1,J_\pm 2;J_\pm 3)(1,2;3)\}$.

ii. $\mathcal{I}\{(1,J_\pm 2;3)(1,J_\pm 3;2)\} = \mathcal{I}\{(J_\pm 1,J_\pm J_\pm 2;3)(J_\pm 1,J_\pm J_\pm 3;2)\}$
$= \mathcal{I}\{(J_\pm 1,2;3)(J_\pm 1,3;2)\} = \mp\mathcal{I}\{(1,2;3)(1,3;2)\} = \mp\psi^{\mathfrak{H}}_{\pm,2}$.

(b) Only one index appears in both variables:

i. $\mathcal{I}\{(J_\pm 1,2;1)(J_\pm 3,2;3)\} = \mathcal{I}\{(1,J_\pm 2;1)(3,J_\pm 2;3)\} = \mp\psi^{\mathfrak{H}}_{\pm,3}$.

(4) $J_\pm$ touches each variable. There are no relevant new invariants.

We have enumerated all the possibilities. Four invariants have resulted from our analysis. □

Let $d\Omega_\pm \in C^\infty(\Lambda^3(M))$ be the exterior derivative of the (para)-Kähler form and let $\delta\Omega_\pm \in T^*M$ be the co-derivative of $\Omega_\pm$.

Lemma 7.7.4 *Let $(V,\langle\cdot,\cdot\rangle,J_\pm)$ be a para-Hermitian vector space (+) or a pseudo-Hermitian vector space (−).*

(1) $\nabla\Omega_\pm(V,e^{2f}\langle\cdot,\cdot\rangle,J_\pm) = -e^{2f}\sigma^{\mathfrak{H}}_{J_\pm}(df)$.

(2) If $J_\pm$ is an integrable (para)-complex structure, then $\nabla\Omega_\pm(M,g,J_\pm)$ belongs to $U^{\mathfrak{H}}_{\pm,3}$.

(3) $W^{\mathfrak{H}}_{\pm,4} \subset U^{\mathfrak{H}}_{\pm,3}$.

(4) $\delta\Omega_\pm = \tau_1\nabla\Omega_\pm$ *and* $d\Omega_\pm = e^i \wedge \nabla\Omega_\pm(\cdot,\cdot;e_i)$.

Proof. The constant structure $J_\pm$ on TV is an integrable (para)-complex structure. Thus we may use Lemma 7.7.1 to prove Assertion (1) by computing:

$$\begin{aligned}\nabla\Omega_\pm(x,y;z) &= e^{2f}\{df(J_\pm y)\langle x,z\rangle - df(x)\langle J_\pm y,z\rangle \\ &\quad +df(y)\langle J_\pm x,z\rangle - df(J_\pm x)\langle y,z\rangle\} \\ &= -e^{2f}\sigma^{\mathfrak{H}}_{J_\pm}(df).\end{aligned}$$

Let $(M,g,J_\pm)$ be a para-Hermitian manifold $(+)$ or a pseudo-Hermitian manifold $(-)$. We may then normalize the coordinates so $J_\pm$ is constant on the coordinate frame. We use Lemma 7.7.1 to verify Assertion (2) by computing:

$$\begin{aligned}&\nabla\Omega_\pm(x,J_\pm y;J_\pm z) \\ &\quad= \tfrac{1}{2}\{g(x,J_\pm z;J_\pm J_\pm y) - g(J_\pm J_\pm y,J_\pm z;x) + g(J_\pm x,J_\pm z;J_\pm y) \\ &\quad\quad -g(J_\pm y,J_\pm z;J_\pm x)\} \\ &\quad= \tfrac{1}{2}\{\pm g(x,J_\pm z;y) \mp g(y,J_\pm z;x) \mp g(x,z;J_\pm y) \pm g(y,z;J_\pm x)\} \\ &\quad= \tfrac{1}{2}\{\mp g(J_\pm x,z;y) \pm g(J_\pm y,z;x) \mp g(x,z;J_\pm y) \pm g(y,z;J_\pm x)\} \\ &\quad= \mp\nabla\Omega_\pm(x,y;z).\end{aligned}$$

By Assertion (1), we may express

$$W^{\mathfrak{H}}_{\pm,4} = \operatorname{Range}_{f\in C^\infty(V)}\{\nabla\Omega_\pm(V,e^{2f}\langle\cdot,\cdot\rangle,J_\pm)\}.$$

By Assertion (2), $\nabla\Omega_\pm(V,e^{2f}\langle\cdot,\cdot\rangle,J_\pm) \in U^{\mathfrak{H}}_{\pm,3}$. This proves Assertion (3); Assertion (4) follows from Lemma 3.3.4. □

We now turn to purely algebraic considerations.

Definition 7.7.1

(1) $(\pi^{\mathfrak{H}}_{\pm,1}H_\pm)(x,y;z) := \frac{1}{6}\{H_\pm(x,y;z) + H_\pm(y,z;x) + H_\pm(z,x;y)$
$\pm H_\pm(x,J_\pm y;J_\pm z) \pm H_\pm(y,J_\pm z;J_\pm x) \pm H_\pm(z,J_\pm x;J_\pm y)\}$.

(2) $(\pi^{\mathfrak{H}}_{\pm,2}H_\pm)(x,y;z) := \frac{1}{6}\{2H_\pm(x,y;z) - H_\pm(y,z;x) - H_\pm(z,x;y)$
$\pm 2H_\pm(x,J_\pm y;J_\pm z) \mp H_\pm(y,J_\pm z;J_\pm x) \mp H_\pm(z,J_\pm x;J_\pm y)\}$.

(3) $(\pi^{\mathfrak{H}}_{\pm,3}H_\pm)(x,y;z) := \frac{1}{2}\{H_\pm(x,y;z) \mp H_\pm(x,J_\pm y;J_\pm z)\}$.

(4) $\pi^{\mathfrak{H}}_{\pm,4} := \pm\frac{1}{m-2}\sigma_{J_\pm}(J_\pm)^*\tau_1$.

Lemma 7.7.5

(1) $\pi^{\mathfrak{H}}_{\pm,1}$ *is a projection from* $\mathfrak{H}_\pm$ *onto* $W^{\mathfrak{H}}_{\pm,1}$.
(2) $\pi^{\mathfrak{H}}_{\pm,2}$ *is a projection from* $\mathfrak{H}_\pm$ *onto* $W^{\mathfrak{H}}_{\pm,2}$.
(3) $\pi^{\mathfrak{H}}_{\pm,3}$ *is a projection from* $\mathfrak{H}_\pm$ *onto* $U^{\mathfrak{H}}_{\pm,3}$.
(4) $\pi^{\mathfrak{H}}_{\pm,4}$ *is a projection from* $\mathfrak{H}_\pm$ *onto* $W^{\mathfrak{H}}_{\pm,4}$.

Proof. Let $H_\pm \in \mathfrak{H}_\pm$. We have

$$H_\pm(x, J_\pm y; z) = \pm H_\pm(J_\pm x, J_\pm J_\pm y; z) = H_\pm(J_\pm x, y; z). \tag{7.7.a}$$

Thus although $J_\pm$ is skew-adjoint with respect to $\langle\cdot,\cdot\rangle$, $J_\pm$ is self-adjoint with respect to $H_\pm$. Set:

$$\begin{aligned}(\kappa^1_\pm H_\pm)(x,y;z) &:= \pm H_\pm(x, J_\pm y; J_\pm z),\\ (\kappa^2_\pm H_\pm)(x,y;z) &:= H_\pm(y,z;x) + H_\pm(z,x;y)\\ &\qquad \pm H_\pm(y, J_\pm z; J_\pm x) \pm H_\pm(z, J_\pm x; J_\pm y).\end{aligned}$$

We use Equation (7.7.a) to see that $\kappa^1_\pm H_\pm$ and $\kappa^2_\pm H_\pm$ are anti-symmetric in x and y. We show that $\kappa^1_\pm\mathfrak{H}_\pm \subset \mathfrak{H}_\pm$ and that $\kappa^2_\pm\mathfrak{H}_\pm \subset \mathfrak{H}_\pm$ by checking:

$$\begin{aligned}(\kappa^1_\pm H_\pm)(J_\pm x, J_\pm y; z) &= \pm H_\pm(J_\pm x, J_\pm J_\pm y; J_\pm z) = H_\pm(x, J_\pm y; J_\pm z)\\ &= \pm(\kappa^1_\pm H_\pm)(x,y;z),\end{aligned}$$

and

$$\begin{aligned}(\kappa^2_\pm H_\pm)(J_\pm x, J_\pm y; z) &= H_\pm(J_\pm y, z; J_\pm x) + H_\pm(z, J_\pm x; J_\pm y)\\ &\quad \pm H_\pm(J_\pm y, J_\pm z; J_\pm J_\pm x) \pm H_\pm(z, J_\pm J_\pm x; J_\pm J_\pm y)\\ &= H_\pm(y, J_\pm z; J_\pm x) + H_\pm(z, J_\pm x; J_\pm y) \pm H_\pm(y,z;x) \pm H_\pm(z,x;y)\\ &= \pm(\kappa^2_\pm H_\pm)(x,y;z).\end{aligned}$$

We see $\pi^{\mathfrak{H}}_{\pm,1}\mathfrak{H}_\pm \subset \mathfrak{H}_\pm$, $\pi^{\mathfrak{H}}_{\pm,2}\mathfrak{H}_\pm \subset \mathfrak{H}_\pm$, and $\pi^{\mathfrak{H}}_{\pm,3}\mathfrak{H}_\pm \subset \mathfrak{H}_\pm$ by expressing:

$$\begin{aligned}&\pi^{\mathfrak{H}}_{\pm,1} = \tfrac{1}{6}\{\mathrm{Id} + \kappa^1_\pm + \kappa^2_\pm\}, \quad \pi^{\mathfrak{H}}_{\pm,2} = \tfrac{1}{6}\{2\,\mathrm{Id} + 2\kappa^1_\pm - \kappa^2_\pm\},\\ &\pi^{\mathfrak{H}}_{\pm,3} = \tfrac{1}{2}\{\mathrm{Id} - \kappa^1_\pm\}.\end{aligned}$$

Summing then yields:

$$\pi^{\mathfrak{H}}_{\pm,1} + \pi^{\mathfrak{H}}_{\pm,2} + \pi^{\mathfrak{H}}_{\pm,3} = \mathrm{Id}\,.$$

Let $H_\pm \in \mathfrak{H}_\pm$. We compute:

$$\begin{aligned}
(\pi^{\mathfrak{H}}_{\pm,1}H_\pm)(x,z;y) &:= \tfrac{1}{6}\{H_\pm(x,z;y)+H_\pm(z,y;x)+H_\pm(y,x;z)\\
&\quad \pm H_\pm(x,J_\pm z;J_\pm y)\pm H_\pm(z,J_\pm y;J_\pm x)\pm H_\pm(y,J_\pm x;J_\pm z)\}\\
&= -\pi^{\mathfrak{H}}_{\pm,1}H_\pm(x,y;z),
\end{aligned}$$

$$\begin{aligned}
&(\pi^{\mathfrak{H}}_{\pm,2}H_\pm)(x,y;z)+(\pi^{\mathfrak{H}}_{\pm,2}H_\pm)(y,z;x)+(\pi^{\mathfrak{H}}_{\pm,2}H_\pm)(z,x;y)\\
&\quad = \tfrac{1}{6}\{2H_\pm(x,y;z)-H_\pm(y,z;x)-H_\pm(z,x;y)\\
&\qquad \pm 2H_\pm(x,J_\pm y;J_\pm z)\mp H_\pm(y,J_\pm z;J_\pm x)\mp H_\pm(z,J_\pm x;J_\pm y)\\
&\qquad +2H_\pm(y,z;x)-H_\pm(z,x;y)-H_\pm(x,y;z)\\
&\qquad \pm 2H_\pm(y,J_\pm z;J_\pm x)\mp H_\pm(z,J_\pm x;J_\pm y)\mp H_\pm(x,J_\pm y;J_\pm z)\\
&\qquad +2H_\pm(z,x;y)-H_\pm(x,y;z)-H_\pm(y,z;x)\\
&\qquad \pm 2H_\pm(z,J_\pm x;J_\pm y)\mp H_\pm(x,J_\pm y;J_\pm z)\mp H_\pm(y,J_\pm z;J_\pm x)\}\\
&\quad = 0,
\end{aligned}$$

$$\begin{aligned}
&(\pi^{\mathfrak{H}}_{\pm,3}H_\pm)(x,J_\pm y;J_\pm z)\\
&\quad = \tfrac{1}{2}\{H_\pm(x,J_\pm y;J_\pm z)\mp H_\pm(x,J_\pm J_\pm y;J_\pm J_\pm z)\}\\
&\quad = \tfrac{1}{2}\{H_\pm(x,J_\pm y;J_\pm z)\mp H_\pm(x,y;z)\} = \mp(\pi^{\mathfrak{H}}_{\pm,3}H_\pm)(x,y;z).
\end{aligned}$$

This shows that:

$$\pi^{\mathfrak{H}}_{\pm,1}H_\pm \in W^{\mathfrak{H}}_{\pm,1},\quad \pi^{\mathfrak{H}}_{\pm,2}H_\pm \in W^{\mathfrak{H}}_{\pm,2},\quad \pi^{\mathfrak{H}}_{\pm,3}H_\pm \in U^{\mathfrak{H}}_{\pm,3}.$$

Let $H_{\pm,i}\in W^{\mathfrak{H}}_{\pm,i}$ for $i=1,2,3$. We compute:

$$\begin{aligned}
(\pi^{\mathfrak{H}}_{\pm,1}H_{\pm,1})(x,y;z) &:= \tfrac{1}{6}\{H_{\pm,1}(x,y;z)+H_{\pm,1}(y,z;x)+H_{\pm,1}(z,x;y)\\
&\quad \pm H_{\pm,1}(x,J_\pm y;J_\pm z)\pm H_{\pm,1}(y,J_\pm z;J_\pm x)\pm H_{\pm,1}(z,J_\pm x;J_\pm y)\}\\
&= \tfrac{1}{6}\{H_{\pm,1}(x,y;z)-H_{\pm,1}(y,x;z)-H_{\pm,1}(x,z;y)\\
&\quad \mp H_{\pm,1}(J_\pm z,J_\pm y;x)\mp H_{\pm,1}(J_\pm x,J_\pm z;y)\mp H_{\pm,1}(J_\pm y,J_\pm x;z)\}\\
&= \tfrac{1}{6}\{H_{\pm,1}(x,y;z)+H_{\pm,1}(x,y;z)+H_{\pm,1}(x,y;z)\\
&\quad -H_{\pm,1}(z,y;x)-H_{\pm,1}(x,z;y)-H_{\pm,1}(y,x;z)\}\\
&= H_{\pm,1}(x,y;z),
\end{aligned}$$

$$\begin{aligned}
(\pi^{\mathfrak{H}}_{\pm,2}H_{\pm,2})(x,y;z) &= \tfrac{1}{6}\{2H_{\pm,2}(x,y;z)-H_{\pm,2}(y,z;x)-H_{\pm,2}(z,x;y)\\
&\quad \pm 2H_{\pm,2}(x,J_\pm y;J_\pm z)\mp H_{\pm,2}(y,J_\pm z;J_\pm x)\mp H_{\pm,2}(z,J_\pm x;J_\pm y)\}\\
&= \tfrac{1}{6}\{3H_{\pm,2}(x,y;z)\pm H_{\pm,2}(J_\pm x,y;J_\pm z)\pm H_{\pm,2}(x,J_\pm y;J_\pm z)\\
&\quad \mp H_{\pm,2}(y,J_\pm z;J_\pm x)\mp H_{\pm,2}(J_\pm z,x;J_\pm y)\}\\
&= \tfrac{1}{6}\{3H_{\pm,2}(x,y;z)\mp H_{\pm,2}(J_\pm z,J_\pm x;y)\mp H_{\pm,2}(y,J_\pm z;J_\pm x)\\
&\quad \mp H_{\pm,2}(J_\pm y,J_\pm z;x)\mp H_{\pm,2}(J_\pm z,x;J_\pm y)
\end{aligned}$$

$$\begin{aligned}
&\mp H_{\pm,2}(y, J_\pm z; J_\pm x) \mp H_{\pm,2}(J_\pm z, x; J_\pm y)\} \\
&= \tfrac{1}{6}\{3H_{\pm,2}(x, y; z) - H_{\pm,2}(z, x; y) - H_{\pm,2}(y, z; x) \\
&\quad \mp 2H_{\pm,2}(J_\pm y, z; J_\pm x) \mp 2H_{\pm,2}(z, J_\pm x; J_\pm y)\} \\
&= \tfrac{1}{6}\{4H_{\pm,2}(x, y; z) \pm 2H_{\pm,2}(J_\pm x, J_\pm y; z)\} = H_{\pm,2}(x, y; z),
\end{aligned}$$

$$\begin{aligned}
(\pi^{\mathfrak{H}}_{\pm,3} H_{\pm,3})(x, y; z) &= \tfrac{1}{2}\{H_{\pm,3}(x, y; z) \mp H_{\pm,3}(x, J_\pm y; J_\pm z\} \\
&= \tfrac{1}{2}\{H_{\pm,3}(x, y; z) + H_{\pm,3}(x, y; z)\} = H_{\pm,3}(x, y; z).
\end{aligned}$$

This shows that

$$\pi^{\mathfrak{H}}_{\pm,1} H_{\pm,1} = H_{\pm,1}, \quad \pi^{\mathfrak{H}}_{\pm,2} H_{\pm,2} = H_{\pm,2}, \quad \pi^{\mathfrak{H}}_{\pm,3} H_{\pm,3} = H_{\pm,3}.$$

The first three assertions now follow.

We now turn to the final assertion. We compute:

$$\begin{aligned}
\tau_1(\sigma_{J_\pm}(\phi))(x) &= \varepsilon^{ij}\{\phi(J_\pm x)\langle e_i, e_j\rangle - \phi(J_\pm e_i)\langle x, e_j\rangle\} \\
&\quad + \varepsilon^{ij}\{\phi(x)\langle(J_\pm e_i, e_j\rangle - \phi(e_i)\langle J_\pm x, e_j\rangle\} \\
&= m\phi(J_\pm x) - \phi(J_\pm x) + 0 - \phi(J_\pm x) = (m-2)((J_\pm)^*\phi)(x).
\end{aligned}$$

It is immediate that $\pi^{\mathfrak{H}}_{\pm,4}$ takes values in $W^{\mathfrak{H}}_{\pm,4}$. We complete the proof by checking $\pi^{\mathfrak{H}}_{\pm,4}$ is the identity on Range$\{\sigma_{J_\pm}\}$:

$$\begin{aligned}
\pi^{\mathfrak{H}}_{\pm,4}\sigma_{J_\pm}\phi &= \pm\tfrac{1}{m-2}(\sigma_{J_\pm}(J_\pm)^*\tau_1)(\sigma_{J_\pm}\phi) \\
&= \pm\sigma_{J_\pm}(J_\pm)^*(J_\pm)^*\phi = \sigma_{J_\pm}\phi.
\end{aligned}$$

□

We continue the analysis of these modules by establishing:

Lemma 7.7.6

(1) $W^{\mathfrak{H}}_{\pm,1} + W^{\mathfrak{H}}_{\pm,2} \subset \ker(\pi^{\mathfrak{H}}_{\pm,3})$.

(2) $W^{\mathfrak{H}}_{\pm,1} \cap W^{\mathfrak{H}}_{\pm,2} = \{0\}$.

(3) $W^{\mathfrak{H}}_{\pm,1} \oplus W^{\mathfrak{H}}_{\pm,2} \oplus W^{\mathfrak{H}}_{\pm,3} \oplus W^{\mathfrak{H}}_{\pm,4}$ *is a submodule of* $\mathfrak{H}_\pm$ *with structure group* $\mathcal{U}^\star_\pm$.

Proof. Suppose first that $H_{\pm,1} \in W^{\mathfrak{H}}_{\pm,1}$. Then

$$\begin{aligned}
(\pi^{\mathfrak{H}}_{\pm,3} H_{\pm,1})(x, y; z) &= \tfrac{1}{2}\{H_{\pm,1}(x, y; z) \mp H_{\pm,1}(x, J_\pm y; J_\pm z)\} \\
&= \tfrac{1}{2}\{H_{\pm,1}(x, y; z) \mp H_{\pm,1}(J_\pm x, y; J_\pm z)\} \\
&= \tfrac{1}{2}\{H_{\pm,1}(x, y; z) \pm H_{\pm,1}(J_\pm x, J_\pm z; y)\} \\
&= \tfrac{1}{2}\{H_{\pm,1}(x, y; z) + H_{\pm,1}(x, z; y)\} = 0.
\end{aligned}$$

Next suppose that $H_{\pm,2} \in W^{\mathfrak{H}}_{\pm,2}$. We have

$$\begin{aligned}
(\pi^{\mathfrak{H}}_{\pm,3}H_{\pm,2})(x,y;z) &= \tfrac{1}{2}\{H_{\pm,2}(x,y;z) \mp H_{\pm,2}(x,J_\pm y;J_\pm z)\}\\
&= \tfrac{1}{2}\{H_{\pm,2}(x,y;z) \pm H_{\pm,2}(J_\pm y,J_\pm z;x) \pm H_{\pm,2}(J_\pm z,x;J_\pm y)\}\\
&= \tfrac{1}{2}\{H_{\pm,2}(x,y;z) + H_{\pm,2}(y,z;x) \pm H_{\pm,2}(z,J_\pm x;J_\pm y)\}\\
&= \tfrac{1}{2}\{H_{\pm,2}(x,y;z) + H_{\pm,2}(y,z;x) \mp H_{\pm,2}(J_\pm x,J_\pm y;z)\\
&\quad \mp H_{\pm,2}(J_\pm y,z;J_\pm x)\}\\
&= \tfrac{1}{2}\{H_{\pm,2}(y,z;x) \mp H_{\pm,2}(J_\pm y,z;J_\pm x)\}\\
&= -\tfrac{1}{2}\{H_{\pm,2}(z,y;x) \mp H_{\pm,2}(z,J_\pm y;J_\pm x)\}\\
&= -(\pi^{\mathfrak{H}}_{\pm,3}H_{\pm,2})(z,y;x).
\end{aligned}$$

Consequently $H_{\pm,1} := \pi^{\mathfrak{H}}_{\pm,3}H_{\pm,2} \in W^{\mathfrak{H}}_{\pm,1}$. Thus:

$$\pi^{\mathfrak{H}}_{\pm,3}H_{\pm,2} = \pi^{\mathfrak{H}}_{\pm,3}\pi^{\mathfrak{H}}_{\pm,3}H_{\pm,2} = \pi^{\mathfrak{H}}_{\pm,3}H_{\pm,1} = 0.$$

Let $H_\pm \in W^{\mathfrak{H}}_{\pm,1} \cap W^{\mathfrak{H}}_{\pm,2}$. We establish Assertion (2) by checking:

$$\begin{aligned}
0 &= H_\pm(x,y;z) + H_\pm(y,z;x) + H_\pm(z,x;y)\\
&= H_\pm(x,y;z) - H_\pm(y,x;z) - H_\pm(x,z;y)\\
&= 3H_\pm(x,y,z).
\end{aligned}$$

If $\pi_\pm : \mathfrak{H}_\pm \to \mathfrak{H}_\pm$ satisfies $\pi_\pm^2 = \pi_\pm$, then Lemma 2.1.5 shows

$$\mathfrak{H}_\pm = \ker(\pi_\pm) \oplus \operatorname{Range}(\pi_\pm).$$

By Lemma 7.7.5, we can apply this observation to $\pi^{\mathfrak{H}}_{\pm,3}$ and to $\pi^{\mathfrak{H}}_{\pm,4}$. By Assertion (1) and by Assertion (2),

$$W^{\mathfrak{H}}_{\pm,1} \cap W^{\mathfrak{H}}_{\pm,2} = \{0\} \quad \text{so} \quad W^{\mathfrak{H}}_{\pm,1} \oplus W^{\mathfrak{H}}_{\pm,2} \subset \ker(\pi^{\mathfrak{H}}_{\pm,3}).$$

By Lemma 7.7.5, we have $U^{\mathfrak{H}}_{\pm,3} = \operatorname{Range}(\pi^{\mathfrak{H}}_{\pm,3})$. Consequently

$$W^{\mathfrak{H}}_{\pm,1} \oplus W^{\mathfrak{H}}_{\pm,2} \oplus U^{\mathfrak{H}}_{\pm,3}$$

is a submodule of $\mathfrak{H}_\pm$. By Lemma 7.7.5,

$$W^{\mathfrak{H}}_{\pm,4} = \operatorname{Range}(\pi^{\mathfrak{H}}_{\pm,4}) \subset U^{\mathfrak{H}}_{\pm,3}.$$

Since $W^{\mathfrak{H}}_{\pm,4} = \pi^{\mathfrak{H}}_{\pm,4}U^{\mathfrak{H}}_{\pm,3}$, $W^{\mathfrak{H}}_{\pm,3} \oplus W^{\mathfrak{H}}_{\pm,4}$ is a submodule of $U^{\mathfrak{H}}_{\pm,3}$ with structure group $\mathcal{U}^\star_\pm$. □

Remark 7.7.2 We can apply Lemma 7.7.4, Lemma 7.7.5, and Lemma 7.7.6 to draw some geometric consequences:

(1) Since $(d\Omega_\pm)(x,y,z) = \{\nabla\Omega_\pm(x,y;z) + \nabla\Omega_\pm(y,z;x) + \nabla\Omega_\pm(z,x;y)\}$, we have $d\Omega_\pm = 0$ if and only if $\nabla\Omega_\pm \in W^{\mathfrak{H}}_{\pm,2}$.

(2) If $J_\pm$ is an integrable (para)-complex structure and if $d\Omega_\pm = 0$, then $\nabla\Omega \in U^{\mathfrak{H}}_{\pm,3} \cap W^{\mathfrak{H}}_{\pm,2} = \{0\}$ and hence $\nabla\Omega_\pm = 0$. This gives another proof of one implication of Theorem 1.10.1 (1).
(3) Since $\delta\Omega_\pm = \tau_1\nabla\Omega_\pm$, $\Omega_\pm$ is co-closed if and only if $\tau_1\nabla\Omega_\pm = 0$.

There is a useful ansatz for constructing examples. Fix $(V, \langle\cdot,\cdot\rangle, J_\pm)$. Let θ be the germ of a smooth map from $(V, 0)$ to $(\mathrm{GL}, \mathrm{Id})$, and let Θ be the germ of a smooth map from $(V, 0)$ to $(\mathcal{O}, \mathrm{Id})$. Set:

$$g^\vartheta(x, y) := \langle\vartheta x, \vartheta y\rangle \quad \text{and} \quad J_\pm^{\vartheta,\Theta} := \vartheta^{-1}\Theta^{-1}J_\pm\Theta\vartheta.$$

Because ϑ is invertible, g^ϑ is a pseudo-Riemannian metric. We show that $(M, g^\vartheta, J_\pm^{\vartheta,\Theta})$ is an almost para-Hermitian manifold $(+)$ or is an almost pseudo-Hermitian manifold $(-)$ by checking:

$$\begin{aligned}
J_\pm^{\vartheta,\Theta}J_\pm^{\vartheta,\Theta} &= \vartheta^{-1}\Theta^{-1}J_\pm\Theta\vartheta\vartheta^{-1}\Theta^{-1}J_\pm\Theta\vartheta = \vartheta^{-1}\Theta^{-1}J_\pm J_\pm\Theta\vartheta \\
&= \pm\vartheta^{-1}\Theta^{-1}\Theta\vartheta = \pm\,\mathrm{Id}, \\
g(J_\pm^{\vartheta,\Theta}x, J_\pm^{\vartheta,\Theta}y) &= \langle\vartheta\vartheta^{-1}\Theta^{-1}J_\pm\Theta\vartheta x, \vartheta\vartheta^{-1}\Theta^{-1}J_\pm\Theta\vartheta y\rangle \\
&= \langle\Theta^{-1}J_\pm\Theta\vartheta x, \Theta^{-1}J_\pm\Theta\vartheta y\rangle = \langle J_\pm\Theta\vartheta x, J_\pm\Theta\vartheta y\rangle \\
&= \mp\langle\Theta\vartheta x, \Theta\vartheta y\rangle = \mp\langle\vartheta x, \vartheta y\rangle = \mp g(x, y).
\end{aligned}$$

Lemma 7.7.7 *If $(V, g, J_\pm)$ is the germ of an almost para-Hermitian manifold $(+)$ or is the germ of an almost pseudo-Hermitian manifold $(-)$ at $0 \in V$ with $g(0) = \langle\cdot,\cdot\rangle$ and $J_\pm(0) = J_\pm$, then*

$$g^\vartheta = g \quad \textit{and} \quad J_\pm = J_\pm^{\vartheta,\Theta} \quad \textit{for some} \quad \vartheta, \Theta.$$

Proof. We apply Lemma 3.3.5 to choose ϑ so $g = g^\vartheta$. Set $\check{J}_\pm^\vartheta = \vartheta J_\pm\vartheta^{-1}$. Then $\check{J}_\pm^\vartheta\check{J}_\pm^\vartheta = \pm\,\mathrm{Id}$ and

$$\langle\check{J}_\pm^\vartheta x, \check{J}_\pm^\vartheta y\rangle = g(J_\pm\vartheta^{-1}x, J_\pm\vartheta^{-1}y) = \mp g(\vartheta^{-1}x, \vartheta^{-1}y) = \mp\langle x, y\rangle.$$

Consequently $(V, \langle\cdot,\cdot\rangle, \check{J}_\pm^\vartheta)$ is the germ of an almost para-Hermitian manifold $(+)$ or an almost pseudo-Hermitian manifold $(-)$. We will express $\check{J}_\pm^\vartheta = \Theta^{-1}J_\pm\Theta$ for some Θ. It will then follow that $J_\pm = \vartheta^{-1}\Theta^{-1}J_\pm\Theta\vartheta$. Thus we may assume without loss of generality that $g = \langle\cdot,\cdot\rangle$ in completing the proof.

We choose the germ of a smooth orthonormal frame for $(TV, \langle\cdot,\cdot\rangle)$ of the form $\{\xi_1, \ldots, \xi_m\}$ so that

$$J_\pm\xi_1 = \xi_2,\ J_\pm\xi_2 = \pm\xi_1,\ \ldots,\ J_\pm\xi_{m-1} = \xi_m,\ J_\pm\xi_m = \pm\xi_{m-1}.$$

We may express $\xi_i = \Theta^{-1}\xi_i(0)$ where Θ is the germ of a smooth map to the orthogonal group with $\Theta(0) = \text{Id}$. It is immediate from the construction that $\Theta^{-1}J_\pm = \check{J}^\vartheta_\pm\Theta^{-1}$ and so $\check{J}^\vartheta_\pm = \Theta^{-1}J_\pm\Theta$. □

Remark 7.7.3 The maps ϑ and Θ are not unique; if ϑ and Θ take values in $\mathcal{U}_\pm$, then $g^\vartheta = \langle\cdot,\cdot\rangle$ and $J^{\vartheta,\Theta}_\pm = J_\pm$.

Fix the linear structure $(V, \langle\cdot,\cdot\rangle, J_\pm)$ henceforth. Let

$$\Xi_\pm(\vartheta,\Theta) := \nabla\Omega^{\vartheta,\Theta}_\pm(0) \in \mathcal{H}_\pm.$$

We note that $\Xi_\pm(\vartheta,\Theta)$ is a linear function of first order jets of ϑ and Θ at 0 with coefficients depending analytically on $(\langle\cdot,\cdot\rangle, J_\pm)$. The Lie algebra $\mathfrak{gl}$ of the general linear group is the matrix algebra $M_m = M_m(V)$. The Lie algebra $\mathfrak{so}$ of the orthogonal group is subspace of skew-adjoint (with respect to $\langle\cdot,\cdot\rangle$) matrices. Let χ be the representation of $\mathcal{U}^\star_\pm$ into $\mathbb{Z}_2$ described previously in Definition 1.2.1. We then have a module morphism:

$$\Xi_\pm : \{\mathfrak{gl} \oplus \mathfrak{so}\} \otimes V^* \otimes \chi \to \text{Range}(\Xi_\pm) \subset \mathfrak{H}_\pm.$$

It follows from Lemma 7.7.2 and Lemma 7.7.7 that:

$$\text{Range}(\Xi_\pm) = \mathcal{H}_\pm.$$

Thus in particular $\mathcal{H}_\pm$ is a module for the group $\mathcal{U}^\star_\pm$.

Lemma 7.7.8

(1) If $m \geq 6$, then $\pi^{\mathfrak{H}}_{\pm,1}\Xi_\pm\{\mathfrak{gl}\oplus 0\} \neq \{0\}$ and $\pi^{\mathfrak{H}}_{\pm,3}\Xi_\pm\{\mathfrak{gl}\oplus 0\} \neq \{0\}$.
(2) $\pi^{\mathfrak{H}}_{\pm,2}\Xi_\pm\{\mathfrak{gl}\oplus 0\} \neq \{0\}$ and $\pi^{\mathfrak{H}}_{\pm,4}\Xi_\pm\{\mathfrak{gl}\oplus 0\} \neq \{0\}$.

Proof. We take the metric to be flat and set $g = \langle\cdot,\cdot\rangle$; thus we take $\vartheta = \text{Id}$ and only vary the almost complex structure. Let $f \in C^\infty(V)$ be a smooth function and let $\phi(z) = df(z)(0)$. We take the canonical orthonormal coordinate frame and suppose $\{\partial_{x_1}, \partial_{x_2}\}$ are spacelike vectors. This is always the case in the para-Hermitian setting, of course. Minor modifications of the argument replacing cos and sin by cosh and sinh can be made in the pseudo-Hermitian setting if needed. Define:

$$\begin{aligned} \Theta : \partial_{x_1} &\to \cos f\partial_{x_1} + \sin f\partial_{x_2}, & \Theta_z(0)\partial_{x_1} &= \phi(z)\partial_{x_2}, \\ \Theta : \partial_{x_2} &\to -\sin f\partial_{x_1} + \cos f\partial_{x_2}, & \Theta_z(0)\partial_{x_2} &= -\phi(z)\partial_{x_1}. \end{aligned}$$

We have that $z(J_\Theta) = -\Theta_z(0)J_\pm + J_\pm\Theta_z(0)$. Thus

$$\begin{aligned}
z(J_\Theta)\partial_{x_1} &= J_\pm\Theta_z(0)\partial_{x_1} = \phi(z)J_\pm\partial_{x_2} = \phi(z)\partial_{y_2},\\
z(J_\Theta)\partial_{x_2} &= J_\pm\Theta_z(0)\partial_{x_2} = -\phi(z)J_\pm\partial_{x_1} = -\phi(z)\partial_{y_1},\\
z(J_\Theta)\partial_{y_1} &= -\Theta_z(0)J_\pm\partial_{y_1} = \mp\Theta_z(0)\partial_{x_1} = \mp\phi(z)\partial_{x_2},\\
z(J_\Theta)\partial_{y_2} &= -\Theta_z(0)J_\pm\partial_{y_2} = \mp\Theta_z(0)\partial_{x_2} = \pm\phi(z)\partial_{x_1}.
\end{aligned}$$

Suppose $m \geq 6$. Let $f = x_3$. The non-zero components of $\nabla\Theta$ are determined up to the $\mathbb{Z}_2$ symmetry in the first two indices by:

$$\begin{aligned}
&\nabla\Omega_\pm(\partial_{y_2},\partial_{x_1};\partial_{x_3}) = \mp 1, \quad \nabla\Omega_\pm(\partial_{x_1},\partial_{y_2};\partial_{x_3}) = \pm 1,\\
&\nabla\Omega_\pm(\partial_{y_1},\partial_{x_2};\partial_{x_3}) = \pm 1, \quad \nabla\Omega_\pm(\partial_{x_2},\partial_{y_1};\partial_{x_3}) = \mp 1.
\end{aligned}$$

Clearly $\tau_1\nabla\Omega_\pm = 0$; thus $\pi^{\mathfrak{H}}_{\pm,3}\nabla\Omega_\pm \in W^{\mathfrak{H}}_{\pm,3}$. We prove Assertion (1) by computing:

$$\pi^{\mathfrak{H}}_{\pm,1}\nabla\Omega_\pm(\partial_{y_2},\partial_{x_1};\partial_{x_3}) = \mp\tfrac{1}{6}, \quad \pi^{\mathfrak{H}}_{\pm,3}\nabla\Omega_\pm(\partial_{y_2},\partial_{x_1};\partial_{x_3}) = \mp\tfrac{1}{2}.$$

If $m = 4$, we let $f = x_2$. The non-zero components of $\nabla\Theta$ are then given, up to the $\mathbb{Z}_2$ symmetry in the first two indices, by:

$$\begin{aligned}
&\nabla\Omega_\pm(\partial_{y_2},\partial_{x_1};\partial_{x_2}) = \mp 1, \qquad \nabla\Omega_\pm(\partial_{x_1},\partial_{y_2};\partial_{x_2}) = \pm 1,\\
&\nabla\Omega_\pm(\partial_{y_1},\partial_{x_2};\partial_{x_2}) = \pm 1, \qquad \nabla\Omega_\pm(\partial_{x_2},\partial_{y_1};\partial_{x_2}) = \mp 1.
\end{aligned}$$

Thus $\tau_1\nabla\Omega_\pm(\partial_{y_1}) = \pm 1$ so the component of $\nabla\Omega_\pm$ in $W^{\mathfrak{H}}_{\pm,4}$ is non-zero. We complete the proof by computing:

$$\begin{aligned}
&(\pi^{\mathfrak{H}}_{\pm,2}\nabla\Omega)(\partial_{y_2},\partial_{x_1};\partial_{x_2})\\
&\quad = \tfrac{1}{6}\{2H(\partial_{y_2},\partial_{x_1};\partial_{x_2}) \mp H_\pm(\partial_{x_1},J_\pm\partial_{x_2};J_\pm\partial_{y_2})\}\\
&\quad = \tfrac{1}{6}\{2H(\partial_{y_2},\partial_{x_1};\partial_{x_2}) - H_\pm(\partial_{x_1},\partial_{y_2};\partial_{x_2})\} = \mp\tfrac{1}{2}. \qquad \square
\end{aligned}$$

Proof of Theorem 1.12.3. We suppose $m \geq 6$; the exceptional dimension $m = 4$ can be handled using the same techniques as those used to handle the exceptional dimensions $m = 4$ and $m = 6$ in the proof of Theorem 1.8.1. We make the following:

Observation 7.7.1

(1) By Lemmas 7.7.6 and 7.7.8, $\oplus_{1\leq i\leq 4}W^{\mathfrak{H}}_{\pm,i}$ is a submodule of $\mathfrak{H}_\pm$ with four non-trivial summands. Thus $\dim\{\mathcal{I}_2^{\mathcal{U}^\star_\pm}(\mathfrak{H}_\pm)\} \geq 4$.

(2) By Lemma 7.7.3, $\dim\{\mathcal{I}_2^{\mathcal{U}^\star_\pm}(\mathfrak{H}_\pm)\} \leq 4$.

(3) Observations (1) and (2) show $\dim\{\mathcal{I}_2^{\mathcal{U}^\star_\pm}(\mathfrak{H}_\pm)\} = 4$. This establishes Remark 7.7.1.

(4) We apply Lemma 2.2.2 and Observation (3) to establish Theorem 1.12.3 by checking that we have an orthogonal direct sum into inequivalent irreducible modules:

$$\oplus_{1\leq i\leq 4} W_{\pm,i}^{\mathfrak{H}} = \mathcal{H}_\pm = \mathfrak{H}_\pm.$$

(5) We apply Observation (4) and Lemma 7.7.5 to see that $\pi_{\pm,1}^{\mathfrak{H}}$, $\pi_{\pm,2}^{\mathfrak{H}}$, and $\pi_{\pm,4}^{\mathfrak{H}}$ are orthogonal projections on $W_{\pm,1}^{\mathfrak{H}}$, $W_{\pm,2}^{\mathfrak{H}}$, and $W_{\pm,4}^{\mathfrak{H}}$, respectively.

(6) Since $\pi_{\pm,3}^{\mathfrak{H}} = \mathrm{Id} - \pi_{\pm,1}^{\mathfrak{H}} - \pi_{\pm,2}^{\mathfrak{H}}$, we may apply Observation (5) to see that $\pi_{\pm,3}^{\mathfrak{H}}$ is orthogonal projection on $U_{\pm,3}^{\mathfrak{H}} = W_{\pm,3}^{\mathfrak{H}} \oplus W_{\pm,4}^{\mathfrak{H}}$.

(7) We have that $\Xi_\pm(\mathfrak{gl}\oplus 0) = \mathcal{H}_\pm = \mathfrak{H}_\pm$. □

Proof of Theorem 1.12.1. Let $(M, g, J_\pm)$ be an almost para-Hermitian manifold $(+)$ or an almost pseudo-Hermitian manifold $(-)$ of dimension $m \geq 6$ (the case of dimension $m = 4$ is analogous). We consider variations $(M, g, J_\pm^\Theta)$. Subtracting $\nabla\Omega_\pm(M, g, J_\pm)(P)$ has no effect on the question of surjectivity. Thus since $\nabla\Omega_\pm$ is linear in the first order jets of the structures involved, we may take g to be flat. The desired result now follows from Observation 7.7.1 (7). □

Proof of Theorem 1.12.2. We have $\Xi_\pm : 0 \oplus \mathfrak{so} \to W_{\pm,3}^{\mathfrak{H}} \oplus W_{\pm,4}^{\mathfrak{H}}$. We must show this map is surjective. We apply Lemma 7.7.4 to see that given $H_\pm \in W_{\pm,4}^{\mathfrak{H}}$, then there exists $f \in C^\infty(V)$ with $f(0) = 0$ so:

$$\nabla\Omega_\pm(V, e^{2f}\langle\cdot,\cdot\rangle, J_\pm) = e^{2f} H_\pm.$$

This completes the proof if $m = 4$ since $W_{\pm,3}^{\mathfrak{H}} = \{0\}$ in that setting. We therefore assume $m \geq 6$. To complete the proof, it suffices to provide a single example taking values in $W_{\pm,3}^{\mathfrak{H}}$. Consider the metrics

$$g_\pm := \varepsilon_\pm + x^1(dx^2 \otimes dx^3 + dx^3 \otimes dx^2 \mp dy^2 \otimes dy^3 \mp dy^3 \otimes dy^2).$$

Clearly $\nabla\Omega_\pm(x, y; z)$ vanishes on a coordinate frame unless

$$x \in \{\partial_{x_i}, \partial_{y_i}\}, \quad y \in \{\partial_{x_j}, \partial_{y_j}\}, \quad z \in \{\partial_{x_k}, \partial_{y_k}\},$$

where $\{i, j, k\}$ is a permutation of $\{1, 2, 3\}$. Thus $\tau_1\nabla\Omega_\pm = 0$. We have:

$$\Omega_\pm = x^1(\pm dx^2 \otimes dy^3 \mp dy^3 \otimes dx^2 \pm dx^3 \otimes dy^2 \mp dy^2 \otimes dx^3).$$

Because $d\Omega_\pm \neq 0$, Lemma 7.7.4 shows that $0 \neq \nabla\Omega_\pm \in W_{\pm,3}^{\mathfrak{H}}$. □

Proof of Theorem 1.12.6. Let (M, g, J_-) be a manifold corresponding to the representation ξ of signature (p, q); this means that $\nabla\Omega_P \in \xi$ for all points P of M. Then $(M, -g, J_-)$ is a manifold corresponding to the representation ξ of signature (q, p). This shows that we may assume that $p \leq q$ in the proof of Theorem 1.12.6. Consequently, as $m \geq 10$, we may assume without loss of generality that $6 \leq q$. This permits us to use Theorem 1.12.5 in the Riemannian setting (see [Gray and Hervella (1980)] for the proof).

The projections $\pi^{\mathfrak{H}}_{-,i}$ for $i = 1, 2, 3$ and the map τ_1 are stable – in other words that they are compatible (or that they commute) with Cartesian product. However, the splitting σ_{J_-} is not stable and this will cause us a small amount of additional technical fuss. For that reason, we suppose first that $W_4 \not\subset \xi$. By Theorem 1.12.5, since $q \geq 6$, we may choose a manifold corresponding to the representation ξ $(M_1, g_1, J_{1,-})$ of Riemannian signature $(0, q)$. Let $(M_2, g_2, J_{2,-})$ be a flat Kähler torus of signature $(p, 0)$. Let

$$M = M_1 \times \mathbb{T}^{(p,0)}, \qquad g := g_1 + g_2, \quad J_- = J_{1,-} \oplus J_{2,-}.$$

Then (M, g, J_-) is an almost pseudo-Hermitian manifold of signature (p, q). We have $\nabla\Omega_g = \nabla\Omega_{\pm,g_1}$ and $\tau_1(\nabla\Omega_g) = \tau_1(\nabla\Omega_{\pm,g_1}) = 0$. Consequently we have that $\pi^{\mathfrak{H}}_{3,-}\nabla\Omega_g$ is projection on $W^{\mathfrak{H}}_{-,3}$; this would not be the case if τ_1 was non-zero. Since $\pi^{\mathfrak{H}}_{-,i}\nabla\Omega_g = \pi^{\mathfrak{H}}_{-,i}\nabla\Omega_{\pm,g_1}$, it now follows that (M, g, J_-) is a manifold corresponding to the representation ξ in this special case.

We complete the proof by considering the remaining case in which we have $\xi = \eta \oplus W^{\mathfrak{H}}_{-,4}$ for some η. Let (M, g, J_-) be an η-manifold of signature (p, q). We set $\tilde{g} := e^{2f}g$ and argue as in the proof of Lemma 7.7.4 to see that:

$$\nabla\Omega_{\pm,\tilde{g}} = e^{2f}\nabla\Omega_g - e^{2f}\sigma_{J_-,g}(df)$$

where we use the original metric to define the splitting $\sigma_{J_-,g}$. This has a non-trivial component in the module $W^{\mathfrak{H}}_{-,4}$ and the components $W^{\mathfrak{H}}_{-,i}$ for $1 \leq i \leq 3$ are not affected. Thus (M, g, J_1) is a manifold corresponding to the representation ξ of signature (p, q). □

Notational Conventions

Basic notation: $\mathbb{N}$ denotes the natural numbers, $\mathbb{Z}$ denotes the integers, $\mathbb{R}$ denotes the real numbers, $\mathbb{C}$ denotes the complex numbers, and $\mathbb{H}$ denotes the quaternions.

Geometric context

(1) (M, ∇) is an affine manifold of dimension m: M is a smooth manifold and ∇ is an affine connection.

   (a) Γ denotes the Christoffel symbols of ∇.
   (b) $\mathcal{R}(x, y) = \nabla_x \nabla_y - \nabla_y \nabla_x - \nabla_{[x,y]}$ is the curvature operator.

(2) (M, g) is a smooth pseudo-Riemannian manifold of dimension m: M is a smooth manifold and g is a pseudo-Riemannian metric of signature (p, q).

   (a) ∇ is the Levi-Civita connection.
   (b) $R(x, y, z, w) = g(\mathcal{R}(x, y)z, w)$ is the $(0, 4)$ curvature tensor.

(3) (M, g, ∇) is a Weyl manifold:

   (a) g is a pseudo-Riemannian metric,
   (b) ∇ is an affine connection,
   (c) g and ∇ are linked by the relation $\nabla g = -2\phi \otimes g$ for some 1-form $\phi \in C^\infty(T^*M)$.

(4) $(M, g, J_\pm)$ is a para-Hermitian manifold $(+)$ of dimension $m = 2n$ or is a pseudo-Hermitian manifold $(-)$ of dimension $m = 2n$.

   (a) Kähler form: $\Omega_\pm(x, y) = g(x, J_\pm y)$.
   (b) (Para)-Nijenhuis tensor:

$$N_\pm(x, y) := [x, y] \mp J_\pm[Jx, y] \mp J_\pm[x, Jy] \pm [J_\pm x, J_\pm y].$$

Algebraic context

(1) $(V, \langle\cdot,\cdot\rangle)$ is an inner product space:

   (a) V is a vector space of dimension m.

      (i) $\{e_i\}$ basis for V,
      (ii) $\{e^i\}$ dual basis of $\{e_i\}$ for the dual vector space V^*,
      (iii) $x^i = e^i(\cdot)$ defines coordinates $(x^1, \ldots, x^m)$ on V.

   (b) $\langle\cdot,\cdot\rangle$ is an inner product of arbitrary signature (p, q).
   (c) J_- is a complex structure and J_+ is a para-complex structure. The Kähler form is given by $\Omega_\pm(x, y) = \langle x, J_\pm y\rangle$.
   (d) A denotes an algebraic curvature tensor.

Spaces of curvature tensors

(1) $\mathfrak{A} = \mathfrak{A}(V)$ is the space of all affine curvature tensors:

$$\mathfrak{A} = \{A \in \otimes^4 V^* : A_{ijkl} + A_{jikl} = 0, A_{ijkl} + A_{jkil} + A_{kijl} = 0\}.$$

(2) $\mathfrak{W} = \mathfrak{W}(V, \langle\cdot,\cdot\rangle)$ is the space of all Weyl curvature tensors:

$$\begin{aligned}\mathfrak{W} = \big\{A \in \mathfrak{A} : &A(x, y, z, w) + A(x, y, w, z)\\ &= \tfrac{2}{m}\{\rho(A)(y, x) - \rho(A)(x, y)\}\langle z, w\rangle\big\}.\end{aligned}$$

(3) $\mathfrak{R} = \mathfrak{R}(V)$ is the space of all Riemannian algebraic curvature tensors:

$$\mathfrak{R} = \{A \in \mathfrak{A} : A_{ijkl} = -A_{jikl} = A_{klij}\}.$$

(4) Special subspaces in (para)-complex geometry:

   (a) $\mathfrak{K}_\pm^{\mathfrak{A}} = \{\mathcal{A} \in \mathfrak{A} : \mathcal{A}(x, y)J_\pm = J_\pm \mathcal{A}(x, y)\}$,
   (b) $\mathfrak{K}_{\pm;+}^{\mathfrak{A}} = \{\mathcal{A} \in \mathfrak{K}_\pm^{\mathfrak{A}} : \mathcal{A}(J_\pm x, J_\pm y) = +\mathcal{A}(x, y)\}$,
   (c) $\mathfrak{K}_{\pm;-}^{\mathfrak{A}} = \{\mathcal{A} \in \mathfrak{K}_\pm^{\mathfrak{A}} : \mathcal{A}(J_\pm x, J_\pm y) = -\mathcal{A}(x, y)\}$,
   (d) $\mathfrak{K}_\pm^{\mathfrak{W}} = \{\mathcal{A} \in \mathfrak{W} : \mathcal{A}(x, y)J_\pm = J_\pm \mathcal{A}(x, y)\} = \mathfrak{K}_\pm^{\mathfrak{A}} \cap \mathfrak{W}$,
   (e) $\mathfrak{R}_+^{\mathcal{U}_\pm} = \{A \in \mathfrak{R} : A(J_\pm x, J_\pm y, J_\pm z, J_\pm w) = A(x, y, z, w)\}$,
   (f) $\mathfrak{R}_-^{\mathcal{U}_\pm} = \{A \in \mathfrak{R} : A(J_\pm x, J_\pm y, J_\pm z, J_\pm w) = -A(x, y, z, w)\}$,
   (g) $\mathfrak{G}_\pm = \{A \in \mathfrak{R} : 0 = A(x, y, z, w) + A(J_\pm x, J_\pm y, J_\pm z, J_\pm w)$
   $\pm A(J_\pm x, J_\pm y, z, w) \pm A(x, y, J_\pm z, J_\pm w) \pm A(J_\pm x, y, J_\pm z, w)$
   $\pm A(x, J_\pm y, z, J_\pm w) \pm A(J_\pm x, y, z, J_\pm w) \pm A(x, J_\pm y, J_\pm z, w)\}$,
   (h) $\mathfrak{K}_\pm^{\mathfrak{R}} = \{A \in \mathfrak{R} : A(x, y, z, w) = \mp A(J_\pm x, J_\pm y, z, w)\}$,

(i) $\mathcal{K}_{\pm}^{\mathfrak{R}} = \{\Theta_{\pm} \in S_{\mp}^{2,\mathcal{U}_{\pm}} \otimes S_{\mp}^{2,\mathcal{U}_{\pm}} : \Theta_{\pm}(x, J_{\pm}y, z, w) + \Theta_{\pm}(y, J_{\pm}z, x, w) + \Theta_{\pm}(z, J_{\pm}x, y, w) = 0\}$,

(j) $\mathfrak{H}_{\pm} = \Lambda_{\pm}^{2,\mathcal{U}_{\pm}} \otimes V^*$.

Operators and tensors defined by the curvature

(1) Affine geometry

(a) $\rho(\mathcal{A})(x, y) = \mathrm{Tr}(z \to \mathcal{A}(z, x)y)$, $\qquad \rho_{13}(A)(x, y) = \varepsilon^{ij} A(e_i, x, e_j, y)$,

(b) $\rho_s(A)(x, y) = \frac{1}{2}\{\rho(A)(x, y) + \rho(A)(y, x)\}$,

(c) $\rho_a(A)(x, y) = \frac{1}{2}\{\rho(A)(x, y) - \rho(A)(y, x)\}$.

(2) Para-Hermitian and pseudo-Hermitian geometry

(a) $\rho(A)(x, y) = \varepsilon^{ij} A(e_i, x, y, e_j)$, $\qquad \tau(A) = \varepsilon^{ij} \rho(A)_{ij}$,

(b) $\rho_J(A)(x, y) = \varepsilon^{ij} A(e_i, x, Jy, Je_j)$, $\qquad \tau_J = \varepsilon^{ij} \rho_J(A)_{ij}$.

(3) Special tensors ($\psi \in \Lambda^2$, $\phi \in S^2$, $S \in \Lambda_{\pm}^{2,\mathcal{U}_{\pm}}$, $H \in \otimes^3 V^*$, $\kappa \in \mathrm{GL}$, $\nu \in V^*$):

(a) $(\sigma_a \psi)(x, y)z = \frac{-1}{1+m}\{2\psi(x, y)z + \psi(x, z)y - \psi(y, z)x\}$,

(b) $(\sigma_s \phi)(x, y)z = \frac{1}{1-m}\{\phi(x, z)y - \phi(y, z)x\}$,

(c) $(\pi_P A)(x, y)z = \mathcal{A}(x, y)z - (\sigma_a \rho_a A)(x, y)z - (\sigma_s \rho_s A)(x, y)z$,

(d) $\{\sigma^{\mathfrak{W}} \psi\}_{ijkl} = 2\psi_{ij}\varepsilon_{kl} + \psi_{ik}\varepsilon_{jl} - \psi_{jk}\varepsilon_{il} - \psi_{il}\varepsilon_{jk} + \psi_{jl}\varepsilon_{ik}$,

(e) $\Psi_{\pm}(S)(x, y, z, w) = 2\langle x, J_{\pm}y\rangle S(z, J_{\pm}w) + 2\langle z, J_{\pm}w\rangle S(x, J_{\pm}y) + \langle x, J_{\pm}z\rangle S(y, J_{\pm}w) + \langle y, J_{\pm}w\rangle S(x, J_{\pm}z) - \langle x, J_{\pm}w\rangle S(y, J_{\pm}z) - \langle y, J_{\pm}z\rangle S(x, J_{\pm}w)$,

(f) $A_\phi(x, y, z, w) = \phi(x, w)\phi(y, z) - \phi(x, z)\phi(y, w)$,

(g) $A_\psi(x, y, z, w) = \psi(x, w)\psi(y, z) - \psi(x, z)\psi(y, w) - 2\psi(x, y)\psi(z, w)$,

(h) $(\tau_1 H)(x) = \varepsilon^{ij} H(x, e_i; e_j)$,

(i) $\sigma_\kappa(\nu)(x, y; z) = \nu(\kappa x)\langle y, z\rangle - \nu(\kappa y)\langle x, z\rangle + \nu(x)\langle \kappa y, z\rangle - \nu(y)\langle \kappa x, z\rangle$.

Lie groups

(1) GL is the general linear group.

(2) $\mathcal{O} = \{T \in \mathrm{GL} : T^*\langle\cdot,\cdot\rangle = \langle\cdot,\cdot\rangle\}$.

(3) $\mathrm{GL}_{\pm} = \{T \in \mathrm{GL} : TJ_{\pm} = J_{\pm}T\}$.

(4) $\mathrm{GL}_{\pm}^{\star} = \{T \in \mathrm{GL} : TJ_{\pm} = J_{\pm}T \text{ or } TJ_{\pm} = -J_{\pm}T\}$.

(5) $\mathcal{U}_{\pm} = \mathcal{O} \cap \mathrm{GL}_{\pm}$, $\qquad \mathcal{U}_{\pm}^{\star} = \mathcal{O} \cap \mathrm{GL}_{\pm}^{\star}$.

Modules for a Lie group

(1) If ξ is a module for the group G, $\mathcal{I}_2^G(\xi)$ is the space of all quadratic invariants.

(2) Modules in an inner product space $(V, \langle\cdot,\cdot\rangle)$:
  (a) $S^2 = \{\theta \in \otimes^2 V^* : \theta(x,y) = \theta(y,x)\}$,
  (b) $S_0^2 = \{\theta \in S^2 : \theta \perp \langle\cdot,\cdot\rangle\}$,
  (c) $\Lambda^2 = \{\theta \in \otimes^2 V^* : \theta(x,y) = -\theta(y,x)\}$,
  (d) $W_6^{\mathcal{O}} = \{A \in \mathfrak{A} \cap \ker(\rho) : A_{ijkl} = -A_{ijlk}\}$,
  (e) $W_7^{\mathcal{O}} = \{A \in \mathfrak{A} \cap \ker(\rho) : A_{ijkl} = A_{ijlk}\}$,
  (f) $W_8^{\mathcal{O}} = \{A \in \otimes^4 V^* \cap \ker(\rho) : A_{ijkl} = -A_{jikl} = -A_{klij}\}$.

(3) Modules in a para-Hermitian or in a pseudo-Hermitian vector space:
  (a) $S_\pm^{2,\mathcal{U}_-} = \{\theta \in S^2 : J_-^*\theta = \pm\theta\}$, $\quad S_\pm^{2,\mathcal{U}_+} = \{\theta \in S^2 : J_+^*\theta = \pm\theta\}$,
  (b) $\Lambda_\pm^{2,\mathcal{U}_-} = \{\theta \in \Lambda^2 : J_-^*\theta = \pm\theta\}$, $\quad \Lambda_\pm^{2,\mathcal{U}_+} = \{\theta \in \Lambda^2 : J_+^*\theta = \pm\theta\}$,
  (c) $S_{0,\mp}^{2,\mathcal{U}_\pm} = \{\theta \in S_\mp^{2,\mathcal{U}_\pm} : \theta \perp \langle\cdot,\cdot\rangle\}$, $\quad \Lambda_{0,\mp}^{2,\mathcal{U}_\pm} = \{\theta \in \Lambda_\mp^{2,\mathcal{U}_\pm} : \theta \perp \Omega_\pm\}$,
  (d) $\mathfrak{Z}_\pm = \{\Theta_\pm \in \otimes^3 V^* \otimes V : \Theta_\pm(x,y,z) = \Theta_\pm(x,z,y)$ and $\Theta_\pm(x, J_\pm y, z) = \Theta_\pm(x,y,J_\pm z) = J_\pm\Theta_\pm(x,y,z)\}$,
  (e) $\mathfrak{Z}_{\pm,h} := \{\Theta_\pm \in \mathfrak{Z}_\pm : \Theta_\pm(J_\pm x, y, z) = J_\pm\Theta_\pm(x,y,z)\}$,
  (f) $\mathfrak{Z}_{\pm,a} = \{\Theta_\pm \in \mathfrak{Z}_\pm : \Theta_\pm(J_\pm x, y, z) = -J_\pm\Theta_\pm(x,y,z)\}$,
  (g) $W_{\pm,7}^{\mathfrak{A}} \approx S_{0,+}^{2,\mathcal{U}_\pm} \subset \mathfrak{K}_{\pm,\mp}^{\mathfrak{A}}$, $\quad W_{\pm,8}^{\mathfrak{A}} \approx \Lambda_{0,\mp}^{2,\mathcal{U}_\pm} \subset \mathfrak{K}_{\pm,\mp}^{\mathfrak{A}}$,
  (h) $W_{\pm,9}^{\mathfrak{A}} = \{A \in \mathfrak{K}_\mp^{\mathfrak{A}} : A(x,y,z,w) = -A(x,y,w,z)\} \cap \ker(\rho)$,
  (i) $W_{\pm,10}^{\mathfrak{A}} = \{A \in \mathfrak{K}_\mp^{\mathfrak{A}} : A(x,y,z,w) = A(x,y,w,z)\} \cap \ker(\rho)$,
  (j) $W_{\pm,11}^{\mathfrak{A}} = \mathfrak{K}_\mp^{\mathfrak{A}} \cap (W_{\pm,9}^{\mathfrak{A}})^\perp \cap (W_{\pm,10}^{\mathfrak{A}})^\perp \cap \ker(\rho_{13}) \cap \ker(\rho)$,
  (k) $W_{\pm,12}^{\mathfrak{A}} = \mathfrak{K}_\pm^{\mathfrak{A}} \cap \ker(\rho)$,
  (l) $W_{\pm,11}^{\mathfrak{W}} = \Omega_\pm \cdot \mathbb{R}$, $\quad W_{\pm,12}^{\mathfrak{W}} = \Lambda_{0,\mp}^{2,\mathcal{U}_\pm}$, $\quad W_{\pm,13}^{\mathfrak{W}} = \Lambda_\pm^{2,\mathcal{U}_\pm}$,
  (m) $W_{\pm,3}^{\mathfrak{R}} = \mathfrak{K}_\pm^{\mathfrak{R}} \cap \ker(\rho)$,
  (n) $W_{\pm,6}^{\mathfrak{R}} = \{\mathfrak{K}_\pm^{\mathfrak{R}}\}^\perp \cap \mathfrak{G}_\pm \cap \ker(\rho \oplus \rho_{J_\pm})$,
  (o) $W_{\pm,7}^{\mathfrak{R}} = \{A \in \mathfrak{R} : A(J_\pm x, y, z, w) = A(x, y, J_\pm z, w)\} = \mathfrak{G}_\pm^\perp$,
  (p) $W_{\pm,10}^{\mathfrak{R}} = \mathfrak{R}_-^{\mathcal{U}_\pm} \cap \ker(\rho \oplus \rho_\pm)$,
  (q) $U_{\pm,3} = \{H_\pm \in \mathfrak{H}_\pm : H_\pm(x,y;z) = \mp H_\pm(x, J_\pm y; J_\pm z)\}$,
  (r) $W_{\pm,1}^{\mathfrak{H}} = \{H \in \mathfrak{H}_\pm : H(x,y;z) + H(x,z;y) = 0\}$,
  (s) $W_{\pm,2}^{\mathfrak{H}} = \{H \in \mathfrak{H}_\pm : H(x,y;z) + H(y,z;x) + H(z,x;y) = 0\}$,
  (t) $W_{\pm,3}^{\mathfrak{H}} = U_{\pm,3} \cap \ker(\tau_1)$, and $W_{\pm,4}^{\mathfrak{H}} = \mathrm{Range}(\sigma_{J_\pm})$.

Bibliography

Abbena E., "An example of an almost Kähler manifold which is not Kählerian", *Boll. Un. Mat. Ital.* **3A** (1984), 383–392.

Alexandrov B. and Ivanov S., "Weyl structures with positive Ricci tensor", *Differential Geom. Appl.* **18** (2003), 343–350.

Apostolov V., Armstrong J., and Drăghici T., "Local models and integrability of certain almost Kähler 4-manifolds", *Math. Ann.* **323** (2002), 633–666.

Apostolov V., Ganchev G., and Ivanov S., "Compact Hermitian surfaces of constant antiholomorphic sectional curvatures", *Proc. Amer. Math. Soc.* **125** (1997), 3705–3714.

Balas A. and Gauduchon P., "Any Hermitian Metric of Constant Non-Positive (Hermitian) Holomorphic Sectional Curvature on a Compact Complex Surface is Kähler", *Math. Z.* **190** (1985), 39–43.

Barberis M., Dotti I., and Fino A., "Hyper-Kähler quotients of solvable Lie groups", *J. Geom. Phys.* **56** (2006), 691–711.

Bejan C., "A classification of the almost para-Hermitian manifolds", *Differential geometry and its applications (Dubrovnik, 1988)*, Univ. Novi Sad, Novi Sad, (1989), 23–27.

Belger M. and Kowalski O., "Riemannian metrics with the prescribed curvature tensor and all its covariant derivatives at one point", *Math. Nachr.* **168** (1994), 209–225.

Besse A. L., "Einstein manifolds", *Ergebnisse der Mathematik und ihrer Grenzgebiete. 3. Folge* **10**, Berlin: Springer-Verlag (1987).

Binder T., "Relative Tchebychev hypersurfaces which are also translation hypersurfaces", *Hokkaido J. Math.* **38** (2009), 1–14.

Biswas I., "Holomorphic principal bundles with an elliptic curve as the structure group", *Int. J. Geom. Methods Mod. Phys.* **5** (2008), 851–862.

Blair D., "Nonexistence of 4-dimensional almost Kaehler manifolds of constant curvature", *Proc. Amer. Math. Soc.* **110** (1990), 1033–1039.

Blaschke W., "Gesammelte Werke", *Affine Differentialgeometrie, Differentialgeometrie der Kreis- und Kugelgruppen* **4**, (ed.: W. Burau et al.), Essen: Thales Verlag (1985).

Blažić N., "Natural curvature operators of bounded spectrum", *Differential Geom. Appl.* **24** (2006), 563–566.

Blažić N., Gilkey P., Nikčević S., and Simon U., "The Spectral Geometry of the Weyl conformal Curvature Tensor", *Banach Center Publications* **69** (2005), 195–203.

Blažić N., Gilkey P., Nikčević S., and Simon U., "Algebraic theory of affine curvature tensors", *Archivum Mathematicum*, Masaryk University (Brno, Czech Republic) ISSN 0044-8753, tomus **42**, supplement: Proceedings of the 26th Winter School of Geometry and Physics 2006 (SRNI) (2006), 147–168.

Blažić N., Gilkey P., Nikčević S., and Stavrov I., "Curvature structure of self-dual 4-manifolds", *Int. J. Geom. Methods Mod. Phys.* **5** (2008), 1191-1204.

Boeckx E., Kowalski O., and Vanhecke L., "Nonhomogeneous relatives of symmetric spaces", *Differential Geom. Appl.* **4** (1994), 45–69.

Bokan N., "On the complete decomposition of curvature tensors of Riemannian manifolds with symmetric connection", *Rend. Circ. Mat. Palermo* **XXIX** (1990), 331–380.

Bokan N., Djorić M., and Simon U., "Geometric structures as determined by the volume of generalized geodesic balls", *Results Math.* **43** (2003), 205–234.

Bokan N., Gilkey P., and Simon U., "Geometry of differential operators on Weyl manifolds", *Phil. Trans. Proc. Royal Soc. London*, Serie A, **453** (1997), 2527–2536.

Bokan N. and Nikčević S., "A characterization of projective and holomorphic projective structures", *Arch. Mat.* **62** (1994), 368–377.

Bokan N., Nomizu K., and Simon U., "Affine hypersurfaces with parallel cubic forms.", *Tôhoku Math. J.* **42** (1990), 101–108.

Bonneau G., "Einstein–Weyl structures and Bianchi metrics", *Classical Quantum Gravity* **15** (1998), 2415–2425.

Bourbaki, N., "Elements of mathematics: Lie groups and Lie algebras", *Elements of Mathematics*, Berlin: Springer-Verlag (2005).

Bredthauer A., "Generalized hyper-Kaehler geometry and supersymmetry", *Nuclear Phys. B* **773** (2007), 172–183.

Brozos-Vázquez M., García-Río E., and Gilkey P., "Relating the curvature tensor and the complex Jacobi operator of an almost Hermitian manifold", *Adv. Geom.* **8** (2008), 353–365.

Brozos-Vázquez M., García-Río E., Gilkey P., and Hervella L., "Geometric realizability of covariant derivative Kähler tensors for almost pseudo-Hermitian and almost para-Hermitian manifolds", to appear *Ann. Mat. Pura Appl.* (2011); DOI: 10.1007/s10231-011-0192-3.

Brozos-Vázquez M., Gilkey P., Kang H., and Nikčević S., "Geometric Realizations of Hermitian curvature models", *J. Math Soc. Japan* **62** (2010), 851–866.

Brozos-Vázquez M., Gilkey P., Kang H., Nikčević S., and Weingart G., "Geometric realizations of curvature models by manifolds with constant scalar curvature", *Differential Geom. Appl.* (2009), 696–701.

Brozos-Vázquez M., Gilkey P., and Merino E., "Geometric realizations of Kaehler and para-Kaehler curvature models", *Int. J. Geom. Methods Mod. Phys* **7** (2010), 505–515.

Brozos-Vázquez M., Gilkey P., and Nikčević S., "Geometric realizations of affine Kähler curvature models", *Results Math.* **59** (2011), 507–521.

Brozos-Vázquez M., Gilkey P., and Nikčević S., "The structure of the space of affine Kähler curvature tensors as a complex module", to appear *Int. J. Geom. Methods Mod. Phys.* **8** (2011b).

Brozos-Vázquez M., Gilkey P., Nikčević S., and Vázquez-Lorenzo R., "Geometric Realizations of para-Hermitian curvature models", *Results Math.* **56** (2009a), 319–333.

Burdík C., Krivonos S., and Scherbakov A., "Hyper-Kaehler geometry via dualization", *Czechslovak J. Physics* **56** (2006), 1099–1103.

Butruille J., "Espace de twisteurs d une variété presque hermitienne de dimension 6", *Ann. Inst. Fourier (Grenoble)* **57** (2007), 1451–1485.

Calabi E., "Hypersurfaces with maximal affinely invariant area" *Amer. J. Math.* **104** (1982), 91–126.

Calderbank D. and Pedersen H., "Selfdual spaces with complex structures, Einstein–Weyl geometry and geodesics", *Annales de l'institut Fourier* **50** (2000), 921–963.

Calderbank D. and Tod P., "Einstein metrics, hypercomplex structures and the Toda field equation", *Differential Geom. Appl.* **14** (2001), 199–208.

Calviño-Louzao E., García-Río E., Gilkey P., and Vázquez-Lorenzo R., "Higher-dimensional Osserman metrics with non-nilpotent Jacobi operators", to appear *Geom. Dedicata.* DOI: 10.1007/s10711-011-9595-y.

Canfes E., "On generalized recurrent Weyl spaces and Wong's conjecture", *Differ. Geom. Dyn. Syst.* **8** (2006), 34–42.

Chevalley C., "Theory of Lie groups", Princeton: Princeton University Press (1946).

Cordero L., Fernández M., and de León M., "Examples of compact non-Kähler almost Kähler manifolds", *Proc. Amer. Math. Soc.* **95** (1985), 280–286.

Cortés V., Lawn M., and Schaefer L., "Affine hyperspheres associated to special para-Kaehler manifolds", *Int. J. Geom. Methods Mod. Phys.* **3** (2006), 995–1009.

Cortés V., Mayer C., Mohaupt T., and Saueressig F., "Special geometry of Euclidean supersymmetry I: vector multiplets", *J. High Energy Phys.* **3** (2004), 028, 73 pp.

Cortés V., Mayer M., Mohaupt T., and Saueressig F., "Special geometry of Euclidean supersymmetry. II. Hypermultiplets and the c-map", *J. High Energy Phys.* **6** (2005), 025, 37 pp; doi: 10.1088/1126-6708/2005/06/025.

Cortés-Ayaso A., Díaz-Ramos J., and García-Río E., "Four-dimensional manifolds with degenerate self-dual Weyl curvature operator", *Ann. Global Anal. Geom.* **34** (2008), 185–193.

Cruceanu V., Fortuny P., and Gadea P., "A survey on paracomplex geometry", *Rocky Mount. J. Math.* **26** (1996), 83–115.

Davidov J., Díaz-Ramos J., García-Río E., Matsushita Y., Muskarov O., and Vázquez-Lorenzo R., "Almost Kaehler Walker 4-manifolds", *J. Geom. Phys.* **57** (2007), 1075–1088.

del Río H. and Simanca S., "The Yamabe problem for almost Hermitian manifolds", *J. Geom. Anal.* **13** (2003), 185–203.

Deprez J., Sekigawa K., and Verstraelen L., "Classifications of Kaehler manifolds satisfying some curvature conditions", *Sci. Rep. Niigata Univ.* Ser. A **24** (1988), 1–12.

De Smedt V., "Decomposition of the curvature tensor of Hyper-Kaehler manifolds", *Letters in Math. Physics* **30** (1994), 105–117.

Díaz-Ramos J., Fiedler B., García-Río E., and Gilkey P., "The structure of algebraic covariant derivative curvature tensors", *Int. J. Geom. Methods Mod. Phys.* **1** (2004), 711–720.

Díaz-Ramos J. and García-Río E., "A note on the structure of algebraic curvature tensors", *Linear Algebra Appl.* **382** (2004), 271–277.

Díaz-Ramos J., García-Río E., and Vázquez-Lorenzo R., "Osserman metrics on Walker 4-manifolds equipped with a para-Hermitian structure", *Mat. Contemp.* **30** (2006), 91–108.

Dunajski M., Mason L., and Tod P., "Einstein–Weyl geometry, the dKP equation and twistor theory", *J. Geom. Phys.* **37** (2001), 63–93.

Dunajski M. and Tod P., "Einstein–Weyl spaces and dispersionless Kadomtsev-Petviashvili equation from Painleve I and II", *Phys. Lett., A* **303** (2002), 253–264.

Dunn C. and Gilkey P., "Curvature homogeneous pseudo-Riemannian manifolds which are not locally homogeneous", *Complex, contact and symmetric manifolds* Progr. Math., **234**, Boston: Birkhäuser (2005), 145–152.

Eisenhart L., "Non-Riemannian Geometry", New York: *AMS Colloquium Publications* (1927).

Eisenhart L., "Riemannian Geometry", London: Oxford University Press (1967).

Evans L., "Partial Differential Equations", *Graduate Texts in Mathematics* **19**, Providence: American Mathematical Society (1998).

Falcitelli M. and Farinola, A, "Curvature properties of almost Hermitian manifolds", *Riv. Mat. Univ. Parma V* (1994), 301–320.

Falcitelli M., Farinola A., and Salamon S., "Almost-Hermitian geometry", *Differential Geom. Appl.* **4** (1994), 259–282.

Ferus D., Karcher H., and Münzner H., "Cliffordalgebren und neue isoparametrische Hyperflächen", *Math. Z.* **177** (1981), 479–502.

Fiedler B., "Determination of the structure of algebraic curvature tensors by means of Young symmetrizers", *Seminaire Lotharingien de Combinatoire* **B48d** (2003). 20 pp. Elect. publ. http://www.mat.univie.ac.at/~slc/.

Fino A., "Almost Kähler 4-dimensional Lie groups with J-invariant Ricci tensor", *Differential Geom. Appl.* **23** (2005), 26–27.

Folland G., "Weyl manifolds", *J. Differential Geometry* **4** (1970), 145–153.

Frobenius, G., "Über das Pfaffsche probleme", *J. für Reine und Agnew. Math.* **82** (1877), 230–315.

Fukami T., "Invariant tensors under the real representation of unitary groups and their applications", *J. Math. Soc. Japan* **10** (1958), 135–144.

Fulton W. and Harris J., "Representation theory. A first course", *Graduate Texts in Mathematics*, Readings in Mathematics, **129**, New York: Springer-Verlag (1991).

Gadea P. and Masque J., "Classification of almost para-Hermitian manifolds", *Rend. Mat. Appl.* **11** (1991), 377–396.

Gadea P. and Oubiña J., "Homogeneous pseudo-Riemannian structures and homogeneous almost para-Hermitian structures", *Houston J. Math.* **18** (1992), 449–465.

Ganchev G. and Ivanov S., "Semi-symmetric W-metric connections and the W-conformal group", *God. Sofij. Univ. Fak. Mat. Inform.* **81** (1994), 181–193.

Ganchev G. and Mihova V., "Kaehler manifolds of quasi-constant holomorphic sectional curvatures", *Cent. Eur. J. Math.* **6** (2008), 43–75.

Ganchev G. and Mihova V., "Warped product Kaehler manifolds and Bochner-Kaehler metrics", *J. Geom. Phys.* **58** (2008a), 803–824.

García-Río E., Gilkey P., Vázquez-Abal M. E., and Vázquez-Lorenzo R., "Four-dimensional Osserman metrics of neutral signature", *Pacific J. Math.* **244** (2010), 21–36.

García-Río E., Kupeli D., and Vázquez-Lorenzo R., "Osserman manifolds in semi-Riemannian geometry", *Lect. Notes Math.* **1777**, Berlin: Springer-Verlag (2002).

Gilkey P., "Curvature and the eigenvalues of the Dolbeault complex for Kaehler manifolds", *Adv. Math.* **11** (1973), 311–325.

Gilkey P., "Geometric properties of natural operators defined by the Riemann curvature tensor", River Edge: World Scientific Publishing Co. Inc. (2001).

Gilkey P., Ivanova R., and Zhang T., "The spectral geometry of the Riemann curvature tensor", *Trends in Mathematics* **5** (2002), 105–114.

Gilkey P. and Nikčević S., "Geometrical representations of equiaffine curvature operators", *Results Math.* **52** (2008), 281–287.

Gilkey P. and Nikčević S., "Kähler and para-Kähler curvature Weyl manifolds", to appear *Math Debrecen* http://arXiv.org/abs/1011.4844 (2011).

Gilkey P. and Nikčević S., "Kähler–Weyl manifolds of dimension 4", http://arxiv.org/abs/1109.4532 (2011).

Gilkey P., Nikčević S., and Simon U., "Geometric theory of equiaffine curvature tensors", *Result. Math.* **56** (2009), 275–317.

Gilkey P., Nikčević S., and Simon U., "Geometric realizations, curvature decompositions, and Weyl manifolds", *J. Geom. Phys.* **61** (2011), 270–275.

Gilkey P., Nikčević S., and Westerman D., "Geometric realizations of generalized algebraic curvature operators", *J. Math. Phys.* **50** (2009), 013515.

Gilkey P., Nikčević S., and Westerman D, "Riemannian geometric realizations for Ricci tensors of generalized algebraic curvature operators", *Differential Geometry – Proceedings of the VIII International Colloquium Santiago de Compostela, Spain.* ed. J. Álvarez-López and E. García-Río. Singapore: World Scientific (2009a), 175–184.

Gilkey P., Park J.H., and Sekigawa K., "Universal curvature relations", *Differential Geom. Appl.* **29** (2011), 770–777.

Gilkey P., Puffini E., and Videv V., "Puffini–Videv Models and Manifolds", http://arXiv.org/abs/math/0605464 (2006).

Gilkey P. and Stavrov I., "Curvature tensors whose Jacobi or Szabó operator is nilpotent on null vectors", *Bulletin London Math Society* **34** (2002), 650–658.

Gray A., "Minimal varieties and almost Hermitian submanifolds", *Michigan Math. J.* **12** (1965), 273–269.

Gray A., "Vector cross products on manifolds", *Trans. Amer. Math. Soc.* **141** (1969), 465–504.

Gray A., "Some examples of almost Hermitian manifolds", *Illinois J. Math.* **10** (1969a), 353–366.

Gray A., "Curvature identities for Hermitian and almost Hermitian manifolds", *Tôhoku Math. J.* **28** (1976), 601–612.

Gray A. and Hervella L., "The sixteen classes of almost Hermitian manifolds and their linear invariants", *Ann. Mat. Pura Appl.* **123** (1980), 35–58.

Hayden H., "Sub-spaces of a space with torsion", *Proc. Lond. Math. Soc. II,* **34** (1932), 27–50.

Higa T., "Weyl manifolds and Einstein–Weyl manifolds", *Comm. Math. Univ. St. Pauli* **42** (1993), 143–160.

Higa T., "Curvature tensors and curvature conditions in Weyl geometry", *Comm. Math. Univ. St. Pauli* **43** (1994), 139–153.

Hitchin N., "Complex manifolds and Einstein's equation", *Springer Lecture Notes* (1982) **970**, 73–99.

Hitchin N., Karlhede A., Lindström U., and Rocek M., "Hyper Kaehler metrics and super symmetry", *Commun. Math. Phys.* **108** (1987), 535–589.

Itoh M., "Affine locally symmetric structures and finiteness theorems for Einstein–Weyl manifolds", *Tokyo J. Math.* **23** (2000), 37–49.

Ivanov S. and Zamkovoy S., "Parahermitian and paraquaternionic manifolds", *Differential Geom. Appl.* **23** (2005), 205–234.

Iwahori N., "Some remarks on tensor invariants of $O(n)$, $U(n)$, $Sp(n)$", *J. Math. Soc. Japan* **10** (1958), 146–160.

Jones P. and Tod P., "Minitwister spaces and Einstein–Weyl spaces", *Classical Quantum Gravity* **2** (1985), 565–577.

Kamada H., "Neutral hyperkähler structures on primary Kodaira surfaces", *Tsukuba J. Math.* **23** (1999), 321–332.

Kath I. and Olbrich M., "New examples of indefinite hyper-Kaehler symmetric spaces", *J. Geom. Phys.* **57** (2007), 1697–1711.

Kim J., "On Einstein Hermitian manifolds", *Monatsh. Math.* **152** (2007), 251–254.

Kirchberg K., "Some integrability conditions for almost Kähler manifolds", *J. Geom. Phys.* **49** (2004), 101–115; see also "Integrability Conditions For Almost Hermitian And Almost Kaehler 4-Manifolds", http://arXiv/math.DG/0605611 (2006).

Kobayashi S. and Nomizu K., "Foundations of differential geometry", *Interscience Tracts in Pure and Applied Mathematics,* **15 Vol. II**, New York: Interscience Publishers John Wiley & Sons, Inc. (1969).

Koto S., "Some theorems on almost Kählerian spaces", *J. Math. Soc. Japan* **12** (1960), 422–433.

Li A., Li H., and Simon U., "Centroaffine Bernstein Problems", *Differential Geom. Appl.* **20** (2004), 331–356.

Li A., Liu H., Schwenk-Schellschmidt A., Simon U., and Wang C., "Cubic form methods and relative Tchebychev hypersurfaces", *Geom. Dedicata* **66** (1997), 203–221.

Li A., Simon U., and Zhao G., "Global affine differential geometry of hypersurfaces", Berlin: de Gruyter (1993).

Manhart F., "Surfaces with affine rotational symmetry and flat affine metric in $\mathbb{R}^3$", *Studia Sci. Math. Hungar.* **40** (2003), 397–406.

Martín-Cabrera F., "Special almost Hermitian geometry", *J. Geom. Phys.* **55** (2005), 450–470.

Martín-Cabrera F. and Swann A., "Almost Hermitian structures and quaternionic geometries", *Differential Geom. Appl.* **21** (2004), 199–214.

Martín-Cabrera F. and Swann A., "Curvature of special almost Hermitian manifolds", *Pacific J. Math.* **228** (2006), 165–184.

Matsuzoe H., "Geometry of semi-Weyl manifolds and Weyl manifolds", *Kyushu J. Math.* **55** (2001), 107–117.

Matzeu P., "Submanifolds of Weyl flat manifolds", *Monatsh. Math.* **136** (2002), 297-311.

Matzeu P. and Nikčević S., "Linear algebra of curvature tensors on Hermitian manifolds", *An. Stiint. Univ. Al. I. Cuza. Iasi Sect. I. a Mat.* **37** (1991), 71–86.

Miritzis J., "Isotropic cosmologies in Weyl geometry", *Classical Quantum Gravity* **21** (2004), 3043–3055.

Mizuhara A. and Shima H., "Invariant projectively flat connections and its applications", *Lobachevskii J. Math.* **4** (1999), 99–107.

Moroianu A. and Ornea L., "Conformally Einstein products and nearly Kaehler manifolds", *Ann. Global Anal. Geom.* **33** (2008), 11–18.

Nagy P., "On nearly-Kähler geometry", *Ann. Global Anal. Geom.* **22** (2002), 167–178.

Narita F. and Satou T., "Conformal transformations of a Weyl manifold", *Kyungpook Math. J.* **44** (2004), 93–99.

Newlander A. and Nirenberg L., "Complex analytic coordinates in almost complex manifolds", *Ann. of Math.* **65** (1957), 391–404.

Nikčević S., "On the decomposition of curvature fields on Hermitian manifolds", *Differential geometry and its applications (Eger, 1989)*, Colloq. Math. Soc. Janos Bolya **56**, Netherlands: North-Holland (1992), 555–568.

Nikčević S., "On the decomposition of curvature tensor", *Proceedings of the Ninth Yugoslav Conference on Geometry (Kragujevac, 1992). Zb. Rad. (Kragujevac)* **16** (1994), 61–68.

Nomizu K., "Lie groups and differential geometry", *The Mathematical Society of Japan* (1956).

Nomizu K., "On the decomposition of generalized curvature tensor fields, Codazzi, Ricci, Bianchi and Weyl revisited", *Differential geometry (In honor of K. Yano)* (1972), Kinokuniya, Tokyo, 335–345.

Nomizu K. and Podestá F., "On affine Kaehler structures", *Bull. Soc. Math. Belg. Sér.* **41** (1989), 275–282.

Nomizu K. and Sasaki T., "Affine differential geometry", Cambridge: Cambridge University Press (1993).

Nomizu K. and Simon U., "Notes on conjugate connections", *Geometry and Topology of Submanifolds* **IV** (ed. F. Dillen et al.), Singapore: World Scientific (1992), 152–172.

Oguro T. and Sekigawa K., "Notes on strictly almost Kähler Einstein manifolds of dimension four", *Yokohama Math. J.* **51** (2004), 19–27.

Oprea T., "On the geometry of Weyl manifolds", *An. Univ. Bucur., Mat.* **54** (2005), 123–126.

Ozdeger A., "On sectional curvatures of a Weyl manifold", *Proc. Japan Acad., Ser.* A **82** (2006), 123–125.

Pedersen H., Poon Y., and Swann A., "The Einstein–Weyl equations in complex and quaternionic geometry", *Differential Geom. Appl.* **3** (1993), 309–321.

Pedersen H. and Swann A., "Riemannian submersions, four manifolds, and Einstein–Weyl geometry", *Proc. Lond. Math. Soc.* (1991) **66**, 381–399.

Pedersen H. and Tod K., "Three-dimensional Einstein–Weyl geometry", *Adv. Math.* **97** (1993), 74–109.

Peter F. and Weyl H. "Die Vollständigkeit der primitiven Darstellungen einer geschlossenen kontinuierlichen Gruppe", *Math. Ann.* **97** (1927), 737–755.

Pinkall U., Schwenk-Schellschmidt A., and Simon U., "Geometric methods for solving Codazzi and Monge-Ampère equations", *Math. Ann.* **298** (1994), 89–100.

Sato T., "On some almost Hermitian manifolds with constant holomorphic sectional curvature", *Kyungpook Math. J.* **29** (1989), 11–25.

Sato T., "Almost Hermitian structures induced from a Kaehler structure which has constant holomorphic sectional curvature", *Proc. Amer. Math. Soc.* **131** (2003), 2903–2909.

Sato T., "Examples of Hermitian manifolds with pointwise constant anti-holomorphic sectional curvature", *J. Geom.* **80** (2004), 196–208.

Schirokow P. A. and Schirokow A. P., "Affine Differentialgeometrie", Leipzig: Teubner (1962).

Schoen R., "Conformal deformation of a Riemannian metric to constant scalar curvature", *J. Differential Geom.* **20** (1984), 479–495.

Scholz E., "Cosmological spacetimes balanced by a Weyl geometric scale covariant scalar field", *Found. Phys.* **39** (2009), 45–72.

Schwenk-Schellschmidt A. and Simon U., "Codazzi-equivalent affine connections", *Results Math.* **56** (2009), 211–229.

Seeley R., "Analytic extension of the trace associated with elliptic boundary problems", *Amer. J. Math.* **91** (1969), 963–983.

Sekigawa K., "On some compact Einstein almost Kähler manifolds", *J. Math. Soc. Japan* **39** (1987), 677–684.

Sekigawa K. and Vanhecke L., "Four-Dimensional Almost Kähler Einstein Manifolds", *Ann. Mat. Pura Appl.* **CLVII** (IV) (1990), 149–160.

Simon U., "Codazzi transformations", *Geometry and Topology of Submanifolds* **VII** (eds. F. Dillen et al.), Singapore: World Scientific (1995), 248–252.

Simon U., "Affine differential geometry", *Handbook of Differential Geometry* **I**, London: Elsevier Science (2000), 905–961.

Simon U., "Affine Hypersurface Theory Revisited: Gauge Invariant Structures", *Russian Mathematics (Izv. vuz)* **48** (2004), 48–73.

Simon U., Schwenk-Schellschmidt A., and Viesel H., "Introduction to the affine differential geometry of hypersurfaces", *Lecture Notes, Science University of Tokyo* (1991).

Singer I., "Infinitesimally homogeneous spaces", *Commun. Pure Appl. Math.* **13** (1960), 685–697.

Singer I. and Thorpe J., "The curvature of 4-dimensional Einstein spaces" *1969 Global Analysis (Papers in Honor of K. Kodaira)*, Tokyo: University Tokyo Press (1969), 355–365.

Strichartz R., "Linear algebra of curvature tensors and their covariant derivatives", *Can. J. Math.* XL (1988), 1105–1143.

Tang Z., "Curvature and integrability of an almost Hermitian structure", *Internat. J. Math.* **17** (2006), 97–105.

Tricerri F. and Vanhecke L., "Curvature tensors on almost Hermitian manifolds", *Trans. Amer. Math. Soc.* **267** (1981), 365–397.

Tricerri F. and Vanhecke L., "Variétés riemanniennes dont le tenseur de courbure est celui d'un espace symétrique riemannien irréductible", *C. R. Acad. Sci. Paris, Sér. I* **302** (1986), 233–235.

Vaisman I., "Generalized Hopf manifolds", *Geom. Dedicata* **13** (1982), 231-255.

Vaisman I., "A survey of generalized Hopf manifolds", *Conference on differential geometry on homogeneous spaces (Turin, 1983)* Rend. Sem. Mat. Univ. Politec. Torino (1983), Special Issue (1984), 205–221.

Vanhecke L., "Some almost Hermitian manifolds with constant holomorphic sectional curvature", *J. Differential Geom.* **12** (1977), 461–471.

Vezzoni L., "On the Hermitian curvature of symplectic manifolds", *Adv. Geom.* **7** (2007), 207–214.

Vrancken L., Li A., and Simon U., "Affine spheres with constant affine sectional curvature", *Math. Z.* **206** (1991), 651–658.

Wang C., "Centroaffine minimal hypersurfaces in R^{n+1}", *Geom. Dedicata* **51** (1994), 63–74.

Watson B., "New examples of strictly almost Kähler manifolds", *Proc. Amer. Math. Soc.* **88** (1983), 541–544.

Weyl H., "Zur Infinitesimalgeometrie: Einordnung der projektiven und der konformen Auffassung", *Gött. Nachr.* (1921), 99–112.

Weyl H., "Space-Time-Matter", New York: Dover Publications (1922).

Weyl H., "The classical groups, their invariants and representations", Princeton: Princeton University Press (1939).

Weyl H., "The classical groups", Princeton: Princeton Univiversity Press (1946) (eighth printing).

Weyl H., "Raum. Zeit. Materie; Vorlesungen über allgemeine Relativitätstheorie", *Heidelberger Taschenbücher* 251, Berlin: Springer-Verlag (1988) (seventh edition).

Yano K., "Differential geometry on complex and almost complex spaces", New York: Pergamon Press (1965).

Index